Lecture Notes in Physics

Springer-Verlag Berlin Heidelberg GmbH

The Editorial Policy for Proceedings

The series Lecture Notes in Physics reports new developments in physical research and teaching – quickly, informally, and at a high level. The proceedings to be considered for publication in this series should be limited to only a few areas of research, and these should be closely related to each other. The contributions should be of a high standard and should avoid lengthy redraftings of papers already published or about to be published elsewhere. As a whole, the proceedings should aim for a balanced presentation of the theme of the conference including a description of the techniques used and enough motivation for a broad readership. It should not be assumed that the published proceedings must reflect the conference in its entirety. (A listing or abstracts of papers presented at the meeting but not included in the proceedings could be added as an appendix.)

When applying for publication in the series Lecture Notes in Physics the volume's editor(s) should submit sufficient material to enable the series editors and their referees to make a fairly accurate evaluation (e.g. a complete list of speakers and titles of papers to be presented and abstracts). If, based on this information, the proceedings are (tentatively) accepted, the volume's editor(s), whose name(s) will appear on the title pages, should select the papers suitable for publication and have them refereed (as for a journal) when appropriate. As a rule discussions will not be accepted. The series editors and Springer-Verlag will normally not interfere with the detailed editing except in fairly obvious cases or on technical matters.

Final acceptance is expressed by the series editor in charge, in consultation with Springer-Verlag only after receiving the complete manuscript. It might help to send a copy of the authors' manuscripts in advance to the editor in charge to discuss possible revisions with him. As a general rule, the series editor will confirm his tentative acceptance if the final manuscript corresponds to the original concept discussed, if the quality of the contribution meets the requirements of the series, and if the final size of the manuscript does not greatly exceed the number of pages originally agreed upon. The manuscript should be forwarded to Springer-Verlag shortly after the meeting. In cases of extreme delay (more than six months after the conference) the series editors will check once more the timeliness of the papers. Therefore, the volume's editor(s) should establish strict deadlines, or collect the articles during the conference and have them revised on the spot. If a delay is unavoidable, one should encourage the authors to update their contributions if appropriate. The editors of proceedings are strongly advised to inform contributors about these points at an early stage.

The final manuscript should contain a table of contents and an informative introduction accessible also to readers not particularly familiar with the topic of the conference. The contributions should be in English. The volume's editor(s) should check the contributions for the correct use of language. At Springer-Verlag only the prefaces will be checked by a copy-editor for language and style. Grave linguistic or technical shortcomings may lead to the rejection of contributions by the series editors. A conference report should not exceed a total of 500 pages. Keeping the size within this bound should be achieved by a stricter selection of articles and not by imposing an upper limit to the length of the individual papers. Editors receive jointly 30 complimentary copies of their book. They are entitled to purchase further copies of their book at a reduced rate. As a rule no reprints of individual contributions can be supplied. No royalty is paid on Lecture Notes in Physics volumes. Commitment to publish is made by letter of interest rather than by signing a formal contract. Springer-Verlag secures the copyright for each volume.

The Production Process

The books are hardbound, and the publisher will select quality paper appropriate to the needs of the author(s). Publication time is about ten weeks. More than twenty years of experience guarantee authors the best possible service. To reach the goal of rapid publication at a low price the technique of photographic reproduction from a camera-ready manuscript was chosen. This process shifts the main responsibility for the technical quality considerably from the publisher to the authors. We therefore urge all authors and editors of proceedings to observe very carefully the essentials for the preparation of camera-ready manuscripts, which we will supply on request. This applies especially to the quality of figures and halftones submitted for publication. In addition, it might be useful to look at some of the volumes already published. As a special service, we offer free of charge LaTeX and TeX macro packages to format the text according to Springer-Verlag's quality requirements. We strongly recommend that you make use of this offer, since the result will be a book of considerably improved technical quality. To avoid mistakes and time-consuming correspondence during the production period the conference editors should request special instructions from the publisher well before the beginning of the conference. Manuscripts not meeting the technical standard of the series will have to be returned for improvement.

For further information please contact Springer-Verlag, Physics Editorial Department II, Tiergartenstrasse 17, D-69121 Heidelberg, Germany

Xing-Wang Pan   Da Hsuan Feng
Michel Vallières (Eds.)

Contemporary Nuclear Shell Models

Proceedings of an International Workshop
Held in Philadelphia, PA, USA, 29–30 April 1996

 Springer

Editors

Xing-Wang Pan
Da Hsuan Feng
Michel Vallières
Department of Physics and Atmospheric Science
College of Arts and Sciences, Drexel University
32nd and Chestnut Streets
Philadelphia, PA 19104, USA

Cataloging-in-Publication Data applied for.
Die Deutsche Bibliothek - CIP-Einheitsaufnahme

Contemporary nuclear Shell models : proceedings of an
international workshop, held in Philadelphia, PA, USA, 29 - 30
1996 / Xing-Wang Pan ... (ed.).
(Lecture notes in physics ; 482)
ISBN 978-3-662-14128-1   ISBN 978-3-540-68068-0 (eBook)
DOI 10.1007/978-3-540-68068-0

NE: Pan, Xing-Wang [Hrsg.]; GT
ISSN 0075-8450
ISBN 978-3-662-14128-1

Typesetting: Camera-ready by authors/editors
Cover design: *design & production* GmbH, Heidelberg
SPIN: 10550675   55/3144-543210 - Printed on acid-free paper

Preface

The nuclear many-body shell model is a formidable problem. However, recent years have witnessed both the development of novel approaches to the solution of this problem and the exploitation by traditional large-scale shell model calculations of unprecedented computing power and algorithms. As a consequence, the nuclear many-body problem is increasingly being addressed in terms of realistic nuclear shell model calculations. With ever-increasing interactions between nuclear structure physics and other fields such as astrophysics, and the fact that new and more accurate information about the atomic nucleus is coming to light from recently established experimental facilities, the nuclear shell model remains not only a fundamental approach for the nuclear many-body system, but its development is entering a new growth phase. It seems timely to have had an intensive workshop to present status reports in this endeavor.

The International Workshop on *Contemporary Nuclear Shell Models*, held in Philadelphia during April 29-30, 1996, brought together experts from various countries to review contemporary methodologies and recent applications of the nuclear shell model. The primary aim of this workshop was to provide improved understanding of some fundamental issues associated with the nuclear shell model. This volume contains the highlights of this workshop.

For these proceedings, the reviewers were requested to present a summary from the origins to the forefront areas in their assigned topics. We attempt to be detail-oriented and incorporate useful references to make the material accessible to a broad audience in nuclear physics and related fields.

We take this opportunity to thank the community for their enthusiastic responses in supporting the workshop. It is a pleasure to thank the invited speakers for their informative talks and all attendees for creating a truly warm atmosphere which fostered much discussion during the workshop. Finally, we acknowledge the help of Springer-Verlag staff in publishing these proceedings.

Philadelphia
November 1996

X.W. Pan
D.H. Feng
M. Vallières

Table of Contents

Nucleon-Nucleon Interactions
R.B. Wiringa 1

Microscopic Theories of Effective Interaction with an Application to Halo Nuclei
T.T.S. Kuo 25

Large No-Core Basis-Space Shell Model Calculations for Light Nuclei
B.R. Barrett, D.C. Zheng, P. Navrátil, J.P. Vary, W.C. Haxton, and C.L. Song 47

Order and Disorder in the Nuclear Shell Model
B.A. Brown, V. Zelevinsky, M. Horoi, and N. Frazier 68

Large-Scale Shell Model Calculations: the Physics in and the Physics out
A.P. Zuker 93

Realistic Shell Model Calculations for Sn Isotopes
A. Covello, F. Andreozzi, L. Coraggio, A. Gargano, and A. Porrino 122

Shell Model Monte Carlo Methods
S.E. Koonin 132

Shell Model for Large Systems and Quantum Monte Carlo Diagonalization Method
T. Otsuka, M. Honma, and T. Mizusaki 133

Shell Model Applications in Nuclear Astrophysics
K. Langanke 155

Large-Scale Continuum Shell Model Calculations for Photonuclear Reactions with Δ Isobars and Exchange Currents
S.R. Cotanch and T.B. Bright 172

Kerman–Klein Method for Nuclear Structure:
Accomplishments and Opportunities
A. Klein and P. Protopapas 201

Solving the Nuclear Shell Model with an Algebraic Method
D.H. Feng, X.-W. Pan, and M. Guidry 231

Challenges to Microscopic Theories of Nuclear Structure
R.F. Casten 254

Projected Shell Model
K. Hara 265

Relativistic Mean Field Theory and Applications in Finite
Nuclei
P. Ring, A.V. Afanasjev, and J. Meng 274

Some Thoughts on the Nuclear Shell Model
M.W. Kirson 289

Conventional Shell Model: Some Issues
M. Vallières, X.-W. Pan, D.H. Feng, and A. Novoselsky 292

List of Participants

B.R. Barrett   bbarrett@ccit.arizona.edu
   University of Arizona, Tucson, AZ 85721 , USA,
B.A. Brown   brown@nscl.msu.edu
   Michigan State University, East Lansing, MI 48824, USA,
R. Casten   rick@riviera.physics.yale.edu
   Yale University, New Haven, CT 06520, USA,
H.C. Chen   htchen@phys730.cycu.edu.tw
   Chung Yuan Christian University, Chung-Li, Taiwan,
S.R. Cotanch   steve@pyrssc.physics.ncsu.edu
   North Carolina State University, Raleigh, NC 27695, USA,
A. Covello   covello@napoli.infn.it
   Università di Napoli "Federico II" and Istituto Nazionale di Fisica Nucleare,
   Mostra d'Oltremare, Pad. 20, 80125 Napoli,Italy,
V. Daniels   daniels@einstein.physics.drexel.edu
   Drexel University, Philadelphia, PA 19104, USA,
J. Dobaczewski   Jacek.Dobaczewski@fuw.edu.pl
   Warsaw University, Hoża 69, 00-681, Warsaw, Poland,
D.H. Feng   Feng@duvm.ocs.drexel.edu
   Drexel University, Philadelphia, PA 19104, USA,
K. Hara   khara@physik.TU-Muenchen.DE
   Technische Universität München, D-85747 Garching, Germany,
R. Haracz   newton@physics.drexel.edu
   Drexel University, Philadelphia, PA 19104, USA,
J.E. Harris   jharris@auc.edu
   Atlanta, GA 30318,
J. Heisenberg   jhh@pauli.sr.unh.edu
   University of New Hampshire, Durham, NH 03824, USA,
M. Honma   m-honma@u-aizu.ac.jp
   University of Aizu, Fukushima 965, Japan,
M. Kirson   fnkirson@weizmann.weizmann.ac.il
   Weizmann Institute of Science, 76100 Rehovot, Israel,
A. Klein   aklein@walet.physics.upenn.edu
   University of Pennsylvania, Philadelphia, PA 19104, USA,
E.S. Koonin   steve_koonin@starbase1.caltech.edu
   California Institute of Technology, Pasadena, CA 91125, USA,

T.T.S. Kuo   kuo@nuclear.physics.sunysb.edu
State University of New York at Stony Brook, Stony Brook, NY 11794, USA,

K. Langanke   karli@almach.caltech.edu
California Institute of Technology, Pasadena, CA 91125, USA,

B. Mihaila   Bogdan.Mihaila@unh.edu
University of New Hampshire, Durham, NH 03824, USA,

T. Otsuka   otsuka@tkyvax.phys.s.u-tokyo.ac.jp
University of Tokyo, Hongo, Bunkyo-ku, Tokyo 113, Japan,

X.W. Pan   pan@einstein.physics.drexel.edu
Drexel University, Philadelphia, PA 19104, USA,

S. Pittel   pittel@bartol.udel.edu
University of Delaware, Newark, DE 19716, USA,

P. Protopapas   pavlos@walet.physics.upenn.edu
University of Pennsylvania, Philadelphia, PA 19104, USA,

P. Ring   peter.ring@physik.TU-Muenchen.DE
Technische Universität München, D-85747 Garching, Germany,

P. Stevenson   paul.stevenson@ox.ac.uk
Oxford University, OX1 3PU Oxford, UK,

W. Sun   fuhy@sun.ihep.ac.cn
Institute of Applied Physics and Computational mathematics, Beijing 100088, China,

Y. Sun   yangsun@utknp3.phys.utk.edu
Joint Institute for Heavy Ion Research, Oak Ridge National Laboratory, Oak Ridge, TN 37831, USA,

S.L. Tabor   tabor@fsulcd.physics.fsu.edu
Florida State University, Tallahassee, FL 32306, USA,

B. Traynor   nucleus@ix.netcom.com
Synergetics International, 1831 Lefthand Circle, Longmont, CO 80501, USA,

M.H. Urin   urin@theor.mephi.msk.su
Moscow Engineering Physics Institute, 115409 Moscow, Russia,

M. Vallières   valliere@einstein.drexel.edu
Drexel University, Philadelphia, PA 19104, USA,

J.P. Vary   jvary@iastate.edu
Iowa State University, Ames, Iowa 50011, USA,

H. Wildenthal   wildenbh@utdallas.edu
The University of Texas at Dallas, Richardson, TX 75083, USA,

R.B. Wiringa   wiringa@theory.phy.anl.gov
Argonne National Laboratory, Argonne, IL 60439, USA,

A.P. Zuker   zuker@crnhp4.in2p3.fr
Université Louis Pasteur, F-67037 Strasbourg Cedex 2, France

International Workshop on Contemporary Nuclear Shell Models
Philadelphia, April 29-30, 1996

Workshop Program

Monday, April 29

Welcoming remarks (Eli Fromm, provost, Drexel University)

Session A1   *Chairperson: A. Klein, University of Pennsylvania*

8:30-9:30   **T.T.S. KUO**, State University of New York at Stony Brook
                  Nucleon-Nucleon Effective Interactions

9:30-10:30   **B.A. BROWN**, Michigan State University
                  Order and Disorder in Shell-Model Wave Functions

10:30-10:45   Coffee Break

Session A2   *Chairperson: A. Klein*

10:45-11:30   **A.P. ZUKER**, Université de Louis Pasteur
                  Large Scale Shell Model Calculations:
                  the physics in and the physics out

11:30-12:15   **B.R. BARRETT**, University of Arizona
                  Large no-core basis-space shell model calculations
                  for light nuclei

Session A3   *Chairpersons: A. Klein and R.F. Casten*
12:30-2:00   Round-table Discussion & box lunch

Session A4   *Chairperson: R.F. Casten, Yale University*

2:00-3:00   **S. E. KOONIN**, California Institute of Technology
                  Monte Carlo Methods for Nuclear Structure

3:00-4:00   **J. DOBACZEWSKI**, University of Warsaw
                  Hartree-Fock Approach

4:00-4:15   Coffee Break

Session A5   *Chairperson: R.F. Casten*

4:15-5:15   **T. OTSUKA and M. HONMA**, University of Tokyo
                  Shell Model for Large Systems and
                  Quantum Monte Carlo Diagonalization Method

Tuesday, April 30

Session B1	*Chairperson: J. Vary, Iowa State University*
8:30-9:30	**R.B. WIRINGA**, Argonne National Laboratory Nucleon-Nucleon Interactions
9:30-10:30	**M. VALLIÈRES**, Drexel University Conventional Shell Model: A Survey
10:30-10:45	Coffee Break

Session B2	*Chairperson: J. Vary*
10:45-11:30	**S. R. COTANCH**, North Carolina State University Large-Scale Continuum Shell Model Calculations for Photonuclear Reactions
11:30-12:15	**A. COVELLO**, University of Napoli Shell-Model Calculations for Sn Isotopes Using Realistic Effective Interactions

Session B3 12:30-2:00	*Chairpersons: J. Vary and M. Kirson* Round-table Discussion & box lunch

Session B4	*Workshop Chairperson: X.W. Pan, Drexel University*
2:00-3:00	**K. LANGANKE**, California Institute of Technology Shell Model Applications in Astrophysics
3:00-4:00	**D.H. FENG**, Drexel University Algebraic Methods in the Shell Model
4:00-4:15	Coffee Break

Session B5	*X.W. Pan*
4:15-5:00	**P. RING**, Technische Universität München Applications of Relativistic Mean Field Theory to Finite Nuclei
5:00-6:00	**H. WILDENTHAL**, The University of Texas at Dallas Workshop Summary

Nucleon-Nucleon Interactions

R. B. Wiringa

Physics Division, Argonne National Laboratory, Argonne, Illinois 60439

Abstract. Nucleon-nucleon interactions are at the heart of nuclear physics, bridging the gap between QCD and the effective interactions appropriate for the shell model. We discuss the current status of NN data sets, partial-wave analyses, and some of the issues that go into the construction of potential models. Our remarks are illustrated by reference to the Argonne v_{18} potential, one of a number of new potentials that fit elastic nucleon-nucleon data up to 350 MeV with a χ^2 per datum near 1. We also discuss the related issues of three-nucleon potentials, two-nucleon charge and current operators, and relativistic effects. We give some examples of calculations that can be made using these realistic descriptions of NN interactions. We conclude with some remarks on how our empirical knowledge of NN interactions may help constrain models at the quark level, and hence models of nucleon structure.

1 Introduction

We have accumulated a great deal of empirical knowledge about the nucleon-nucleon (NN) interaction over the last sixty years. Substantial amounts of experimental data have been added in just the last decade. Partial-wave analyses have improved in sophistication to the point where the NN elastic scattering data below $E_{lab} = 350$ MeV can be fit with a $\chi^2/N_{data} = 1$. There are also now several potential models that fit this data with comparable accuracy.

Realistic NN potentials are very complicated in structure, making nuclear many-body problems the most challenging in all of theoretical physics. Nevertheless, potential representations of the NN interaction can be used to directly study a wide variety of nuclear structure problems, including the ground and excited states of few-body nuclei, light closed-shell nuclei, nuclear matter and neutron stars (Wiringa 1993). In particular, essentially exact calculations of nuclei up to A=8 are now possible, using a variety of Faddeev, correlated hyperspherical harmonic, and quantum Monte Carlo methods. Reactions, including response to electron scattering and various electroweak processes of interest in astrophysics, can also be predicted microscopically at this level.

However, there are many phenomena in nuclear physics that are still far too complicated for present many-body theory to handle with bare interactions. Large or mid-shell nuclei, highly excited states, and nuclei at high spin are examples. These problems are best addressed in terms of effective interactions, which should in principle be derivable from the bare NN interaction. Knowledge gained in the exact solution of smaller nuclei should also be useful in this regime.

In turn, the NN interaction is itself an effective interaction representing more complicated QCD effects between color singlet states. At long range, it may be described in terms of one-pion exchange, while at intermediate distances the exchange of multiple pions, with the possible excitation of nucleon resonances like the $\Delta(1232)$ becomes important. At short range, the exchange of heavier vector mesons, like the ρ and ω, may start to dominate. However, nucleons, mesons, and nucleon resonances have their own internal quark structure and this must come into play at some scale. Since we know empirically a great deal about NN interactions, we may be able to turn this knowledge to good use in modelling quark-quark interactions.

2   Data and Partial-Wave Analyses

Our first consideration in constructing realistic models for NN interactions is the status of NN data. There are more than 5,000 NN elastic scattering data up to $E_{lab} = 350$ MeV that have been accumulated since 1955. These include total cross sections, differential cross sections, polarizations, asymmetries, and transfer parameters of assorted types. About 40% is pp scattering data and the remainder is np data.

In addition there are bound-state properties of the deuteron, including the binding energy, asymptotic normalizations, radius, and magnetic and quadrupole moments. There are also a variety of reactions involving the deuteron, including the elastic structure functions and tensor polarization measured in electron scattering, photo- and electro-disintegration data, and $p(n, \gamma)d$ radiative capture data.

It is convenient to reduce this mass of data to a more manageable set of approximately 200 phase shifts by performing a partial wave analysis (PWA). That is, the scattering amplitude is expanded in partial waves with phase shifts (and mixing parameters in coupled channels) deduced for each channel, typically up to $L = 5$ or 6. A partial wave analysis can be used to check the data for internal consistency; indeed, it is advisable to "weed out" bad data.

The most comprehensive PWA of recent years is that carried out by the Nijmegen group for pp (Bergervoet et al. 1990) and np (Stoks et al. 1993) elastic data in the range $E_{lab} = 0 - 350$ MeV. As an example of data weeding, their procedure is illustrated in Table 1. They collected all the NN data published in regular journal articles between 1955 and 1992 and performed a global minimization within their chosen parametrization of the phase shifts. They rejected data that was more than three standard deviations off from their fit and then readjusted their parameters to the smaller data set. This resulted in the deletion of 15% of the pp data and 28% of the np data. They then added as data to be fit the overall normalizations of many of the data sets, and floated a small number of these where the quoted experimental error added more than nine to the χ^2. The final set has 1787 pp data and 2514 np data.

This weeding of the data may strike some as rather arbitrary (especially if you are an experimentalist whose data has been rejected!) but it is mechanically unbiased and there is no question that a number of data sets in the literature are

Table 1. Data selection in the Nijmegen partial-wave analysis

	pp	np
observables	1947	3298
rejected in 3σ test	−291	−932
remainder (groups)	1656 (215)	2366 (211)
normalizations (floated)	+131 (22)	+148 (16)
total data to fit	1787	2514

flawed. In fact there is little difference between the Nijmegen data base and that maintained by the Virginia Polytechnic Institute and State University group in their SAID program (Arndt *et al.* 1992).

Partial-wave analyses are typically made in either of two modes: 1) a single-energy analysis, where the data is grouped into a number of bins 25–50 MeV wide and treated as if it were at a single discrete energy, or 2) a multi-energy analysis, in which all the data is treated simultaneously, with some assumptions of underlying smoothness. The Nijmegen group has performed both single- and multi-energy analyses in their studies, achieving the impressive $\chi^2/N_{data} = 0.99$ for the multi-energy analysis. The latest update of the VPI&SU group also has both single- and multi-energy PWA's for NN data up to 1.6 GeV. Other recent PWA's include a single-energy analysis of np data between 142 and 800 MeV (Bugg and Bryan 1992), and a single-energy analysis of NN data below 160 MeV (Henneck 1993).

Single-energy analyses can show a large scatter as a function of energy. In this sense they tend to point out problem areas in the NN data where our knowledge is less complete and additional experiments might be valuable. The smooth energy dependence of the phase shifts produced in multi-energy analyses is more physically plausible, and not surprisingly, much easier to fit with a potential model. One of the more poorly-determined phases is the ϵ_1 mixing parameter, which is closely correlated to the $^1P_1$ phase shift. In Fig. 1 we plot these parameters against each other for $E_{lab} = 50$ MeV, where there has been some recent controversy (Machleidt and Slaus 1994, Hammans *et al.* 1994) over the significance of the single-energy PWA. We use this as an illustration of the differences between different single- and multi-energy PWA's and what potential models can fit.

In Fig. 1 the single-energy analyses of the Nijmegen and PSI groups are in reasonable agreement with each other, while the SAID analysis is noticeably different due to the retention of some old data that the other groups have discarded. While the single-energy analyses all have large error bars, the Nijmegen multi-energy analysis has a much smaller error as a consequence of considering simultaneously the data over a much larger range of energy. The older potential models that are shown lie in a band that includes the multi-energy point, but are in clear disagreement with the high values of the single-energy analyses. The new high-accuracy potentials which are not shown here all cluster around the

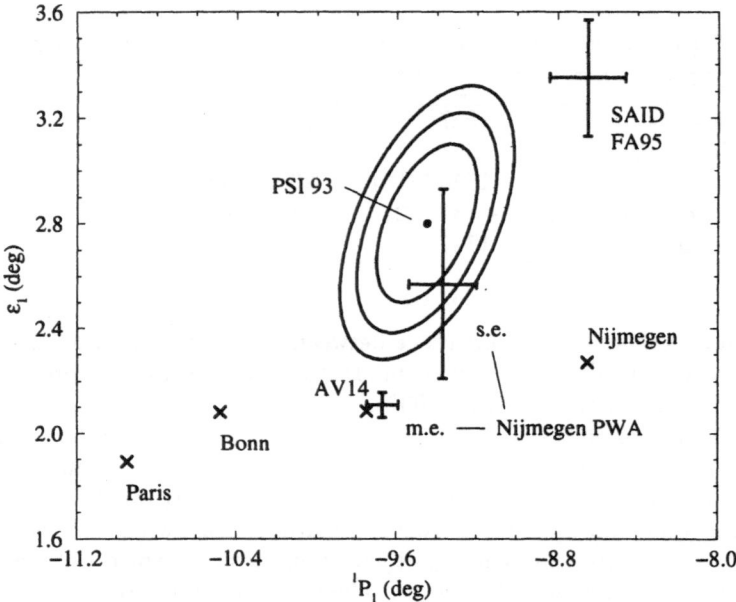

Fig. 1. Partial-wave analyses (dot with ellipses and crosses with error bars) and potential model fits (×'s) at $E_{lab} = 50$ MeV

multi-energy point.

An important ingredient in partial wave analyses and NN potentials is the strength of the long-range one-pion exchange coupling. For many years, the accepted $NN\pi$ coupling constant was $f^2 = 0.0790$ (10) as determined from πN scattering data (Bugg, Carter, and Carter 1973). This value was used in many of the potential models of the '70s and '80s. The recent Nijmegen analyses have called this value into question by trying to determine the pion coupling constants and masses solely from the NN scattering data. They suceeded in obtaining a π^0 mass of 135.6 (13) MeV and a $\pi^{\pm}$ mass of 139.4 (10) MeV, in excellent agreement with accepted values. In the process of fitting these masses, they obtained somewhat weaker $NN\pi$ coupling constants: $f_p^2 = f_{pp\pi^0}^2 = 0.0745$ (6), $f_0^2 = f_{nn\pi^0} f_{pp\pi^0}$ $= 0.0745$ (9), and $f_c^2 = f_{np\pi^-} f_{pn\pi^+} = 0.0748$ (3). This conclusion has caused considerable discussion in the literature. The VPI&SU group has subsequently reexamined πN data (Arndt, Workman, and Pavan 1994) and also obtained a lower value of $f^2 = 0.076$ (1), in agreement with the Nijmegen result. A recent study of NN elastic data from 210 to 800 MeV (Bugg and Machleidt 1995) obtained slightly larger couplings, corresponding to $f^2 = 0.077$ (2), right in between, but with a large enough error bar to touch, both the old and new values. The newest NN potential models, discussed below, which are able to obtain very good fits to the NN scattering data, all use the new lower value of $f^2 = 0.075$.

3   Potential Modelling

In constructing a potential model for NN interactions there are a number of choices that have to be made. The first decision is what the intended purpose of the model is to be. It might be meant for use in many-body problems, in which case the accuracy with which it reproduces the two-body data and the ease with which it can be used in calculations are the primary considerations. Alternatively, the goal may be to elucidate the role of subnucleonic degrees of freedom in the NN interaction, so the inclusion of known mesons and nucleon resonances may become an important feature. Ideally both aspects would be included, but in practice there are always compromises.

Broadly speaking, potentials can be categorized as either phenomenological or meson-theoretic models, with the former empahsizing the fit to data and use in larger systems, while the latter stress the subnucleon aspects. All models have a one-pion-exchange (OPE) character at long range, which is absolutely essential to fit high partial-wave data; thus the labels phenomenological or meson-theoretic refer to the treatment of shorter-range interactions. Examples of phenomenological potentials are the Reid soft-core (Reid 1968), Urbana v_{14} (Lagaris and Pandharipande 1981) and Argonne v_{14} (Wiringa, Smith, and Ainsworth 1984) potentials. Meson-theoretic potentials include the Nijmegen (Nagels, Rijken, and de Swart 1978), Paris (Lacombe et al. 1980), and Bonn (Machleidt, Holinde, and Elster 1987) models. The recent, highly-accurate potentials include several models from the Nijmegen group (Stoks et al. 1994), the Argonne v_{18} potential (Wiringa, Stoks, and Schiavilla 1995), and CD Bonn (Machleidt, Sammarruca, and Song 1996).

For phenomenological potentials, an important choice is the general structure to be used, which may be either a partial-wave or an operator construct. Operator constructs are based on the demonstration (Okubo and Marshak 1958) that the most general form satisfying translational, rotational, Galillean, and time-reversal invariance, and parity conservation can be written as

$$V_{NN} = V_c + V_\sigma \sigma_i \cdot \sigma_j + V_t S_{ij} + V_{ls}(\mathbf{L} \cdot \mathbf{S})_{ij} + V_q Q_{ij} + V_{\sigma p} \sigma_i \cdot \mathbf{p}\, \sigma_j \cdot \mathbf{p} \ , \quad (1)$$

where the tensor and quadratic spin-orbit operators are defined by

$$S_{ij} = 3\,\sigma_i \cdot \hat{\mathbf{r}}\, \sigma_j \cdot \hat{\mathbf{r}} - \sigma_i \cdot \sigma_j \ , \quad (2)$$

$$Q_{ij} = \sigma_i \cdot \mathbf{L}\, \sigma_j \cdot \mathbf{L} + \sigma_j \cdot \mathbf{L}\, \sigma_i \cdot \mathbf{L} \ . \quad (3)$$

The individual V_x here can be functions of $\mathbf{r}$, $\mathbf{p}$, or $\mathbf{L}$, and can have a very general isospin dependence

$$
\begin{aligned}
V_x =\ & V_x^0(\mathbf{r}, \mathbf{p}, \mathbf{L}) + V_x^\tau(\mathbf{r}, \mathbf{p}, \mathbf{L})\, \tau_i \cdot \tau_j \\
& + V_x^T(\mathbf{r}, \mathbf{p}, \mathbf{L}) T_{ij} \\
& + V_x^{\tau z}(\mathbf{r}, \mathbf{p}, \mathbf{L})(\tau_{zi} + \tau_{zj}) \\
& + V_x^M(\mathbf{r}, \mathbf{p}, \mathbf{L})(\tau_{zi} - \tau_{zj})(\sigma_i - \sigma_j) \cdot \mathbf{L} \ ,
\end{aligned}
\quad (4)
$$

with the isotensor operator defined by

$$T_{ij} = 3\tau_{zi}\tau_{zj} - \tau_i \cdot \tau_j \ . \tag{5}$$

The first line in (4) is charge-independent (CI), so that pp, np, nn interactions in the same state of total isospin T are equal. The V_x^T term is charge-dependent (CD) and differentiates np from pp and nn scattering, while the $V_x^{\tau_z}$ term is charge-symmetry-breaking (CSB) and splits pp from nn interactions. These latter two are often referred to as Class II and Class III forces, respectively. The last term is an example of a Class IV CSB force, which mixes isospin in np interactions. The approximate isospin symmetry of nuclear forces makes the V_x^0 and V_x^τ terms dominant, but there is clear evidence for small V_x^T and $V_x^{\tau_z}$ contributions to the nuclear force, while electromagnetic forces give contributions of all kinds.

A partial-wave potential like Reid has the potential specified separately in each $^{2S+1}L_J$ channel, but different channels are not necessarily connected by a consistent operator structure. For example, in the Reid soft-core potential, the $^3P_2 - {}^3F_2$ coupled channels are specified by a $V_c + V_t S_{ij} + V_{ls}(\mathbf{L}\cdot\mathbf{S})_{ij}$ combination, but these same V_c, V_t, and V_{ls} do not produce the potential in $^3P_0$ and $^3P_1$ channels. This kind of channel-by-channel formulation implies some nonlocality in the potential which is not easily characterized.

In many-body applications, like Faddeev calculations of the three- and four-body nuclei, partial-wave expansions for the wave function are common and this kind of potential is straightforward to use. However, for a number of variational and quantum Monte Carlo methods in configuration space that do not use partial-wave expansions, potentials based on a consistent operator structure are far easier to use. Of course, operator-based potentials can easily be projected into partial waves as needed, and all channels are specified, whereas partial-wave potentials are usually truncated at some J. Both kinds of phenomenological potential typically have about 40 parameters to get a reasonable fit to the data.

In meson-theoretic potentials, one has several options for treating the shorter-range parts of the interaction. The simplest is a one-boson-exchange (OBE) format such as the original Nijmegen potential and some of the Bonn models. The Nijmegen potential is constructed using the exchange of π, η, η', ρ, ω, ϕ, ϵ, and S^* mesons, plus Pomeron, f, f', and A_2 contributions. Its parametrization includes the various meson couplings (the masses are taken from particle data), the Pomeron mass, and a cutoff factor for a total of about 12 parameters to be fit, which is typical of meson-theoretic models.

A common problem among meson-theoretic potentials is how to represent the observed scalar attraction at intermediate ranges. The Pomeron provides this attraction in the Nijmegen model. A more common treatment in OBE models is to introduce a fictitious σ meson, a scalar of mass ≈ 500 MeV, as is done in some of the Bonn models. The microscopic source of the intermediate-range attraction is believed to be due to two-pion-exchange (TPE) with the possible excitation of the intermediate nucleons to $\Delta(1232)$ resonance states. Other multiple-meson exchanges and nucleon resonances can also contribute. The full

Table 2. Potential χ^2/N_{data} comparison for E_{lab} = 5–350 MeV; numbers in parentheses indicate data that was not fit at time of construction

Potential	pp	np	Comments
Reid soft-core (1968)	2.5	(10.7)	Partial-wave ($J \leq 2$) local r-space
Nijmegen (1978)	2.0	(.)	OBE nonlocal r- & k-space
Paris (1980)	2.2	(3.8)	OBE+TPE $\Rightarrow$ Operator nonlocal r- & k-space
Argonne v_{14} (1984)	(6.9)	2.1	Operator local r-space
Bonn (1987)	(13.0)	3.0	OBE+TPE E-dependent nonlocal k-space
Nijmegen (1993)	1.78	1.93	OBE CD nonlocal r- & k-space
Reid 93 (1994)	1.00	1.04	Partial-wave ($J \leq 6$) local r-space
Nijm I (1994)	1.00	1.04	Partial-wave CD nonlocal r- & k-space
Nijm II (1994)	1.00	1.04	Partial-wave CD local r- & k-space
Argonne v_{18} (1995)	1.10	1.08	Operator CD-CSB local r-space
CD Bonn (1996)	1.03	1.03	OBE $\Rightarrow$ Partial-wave CD-CSB nonlocal k-space

Bonn model treats these terms in a perturbative manner, drawing the leading-order Feynman diagrams and integrating over the intermediate states. An alternative followed by the Paris group is to use information on $\pi\pi$ and πN scattering and dispersion techniques. Both methods inevitably lead to a certain amount of energy-dependence and nonlocality not present in the OBE models. To make a tractable potential for use in many-body calculations, the energy-dependence may be replaced by a momentum-dependence that is fit to the original model, as was done in the parametrization of the Paris potential.

A final decision in constructing a potential model, whether phenomenological or meson-theoretic, is what data to fit. Through the 1980's the charge-dependence of the nuclear force was neglected and different groups chose to fit either pp or np data in $T = 1$ channels. The pp data is easier to obtain experimentally and has smaller error bars. However there has generally been more np data available and not having to treat the long-range Coulomb interaction makes life a little easier for the potential builder. With the growing size of the NN database and the increasing accuracy of many-body calculations, it has become more important to fit both kinds of data, which necessarily introduces CD into the nuclear potential. The Nijmegen group emphasized this point (Stoks and de Swart 1993, 1995) by comparing a number of potentials to their database and PWA. We show a selection of their results in Table 2, updated to include the new high-accuracy potentials of the last few years.

In looking at the table we see that through the 1980's, the best potentials achieved a $\chi^2/N_{data} \approx 2$ for whichever data set, either pp or np, that the builders chose to fit. However, potentials fit to pp data did not reproduce np data well and *vice versa*. Even when CD is added to a OBE model, as in the case of the updated Nijmegen (1993) model, the χ^2 could not be improved significantly beyond 2 within the limited set (≈ 12) parameters available. The highly accurate potentials of the last few years incorporate a significantly larger number of parameters, not so different from the number required in the Nijmegen PWA itself.

A final consideration is what amount of deuteron data to fit. The binding energy, E_d, is fit in all models, but there are other static properties, such as the asymptotic normalizations A_S and η_d, and the electromagnetic radius, r_d, magnetic moment, μ_d, and quadrupole moment, Q_d, which could be fit. The electromagnetic terms suffer from having two-body charge and current operator contributions from meson-exchange and relativistic effects that are not uniquely specified by the potential. Ideally one would construct such operators simultaneously with the potential model, but this has not been done yet as part of the data fitting procedure.

4   The Argonne v_{18} Potential

As an example of some of the considerations that go into constructing an NN potential, I will discuss in more detail the structure of the new Argonne v_{18} model and some of the "philosophy" behind it. Our primary goal was to produce an accurate potential fit to the latest data that could be easily used in configuration-space variational and quantum Monte Carlo calculations of finite nuclei and nucleon matter. We want to use the two-body potential as part of a many-body Hamiltonian:

$$H = \sum_i K_i + \sum_{i<j} v_{ij} + \sum_{i<j<k} V_{ijk} \; . \tag{6}$$

To fit both pp and np data it was clear that we would have to introduce some CD into the force. We are also interested in looking at CSB in finite nuclei (the Nolen-Schiffer anomaly) so we decided to additionally fit the limited nn data – the singlet scattering length and effective range. To pull the strong-interaction CSB component out of the data, it is essential to first account for all the electromagnetic contributions, including not just Coulomb, but Darwin-Foldy, vacuum polarization, and magnetic moment terms. This had been done in the Nijmegen PWA, where it was essential for fitting low-energy data accurately, but it had not been done in NN potential models before.

After the electromagnetic interaction, the primary source of CD is the effect of the $\pi^{\pm} - \pi^0$ mass difference on the OPE potential. We decided to treat these terms carefully, and to adopt the weaker f^2 OPE coupling constant indicated by the Nijmegen PWA. However, we were not particularly interested in further micrscopic details of the NN force, so we did not attempt to impose a one-boson-exchange or other meson-theoretic model for the short-range structure. We did opt for an operator structure that could immediately be used in our many-body applications where we already have a substantial computational investment. For simplicity, we adopted the general operator structure of the earlier Urbana and Argonne v_{14} models, but found we had to increase the number of functions and parameters to get a good fit to the data, in addition to adding extra CD and CSB operators. One requirement we imposed was that the potential be finite with zero slope at the origin, to represent the finite size of the nucleons. This necessitated form factors on all terms including the electromagnetic ones. We also required that the tensor potential go to zero at the origin.

A final consideration is that we also want to develop the three-body force and two-body charge and current operators in a consistent fashion. This means that the long-range two-pion-exchange three-nucleon force, $V_{ijk}^{2\pi}$, is constructed with the same radial functions and cutoffs as the v_{ij}^{π}. We also investigated the corresponding CD of $V_{ijk}^{2\pi}$, although it turns out to be quite weak. The charge and current operators are constructed in a manner consistent with the v_{ij} following a prescription (Schiavilla, Pandharipande, and Riska 1989,1990) that preserves the continuity equation and their contributions to the deuteron static moments and structure functions are calculated. However, we did not proceed far enough to incorporate these calculations in the fitting procedure, so only the deuteron binding energy was used as a constraint.

The final potential is written as a sum of electromagnetic, OPE, and remaining short-range terms:

$$V_{NN} = v^{\gamma} + v^{\pi} + v^{R} . \tag{7}$$

The electromagnetic terms include one- and two-photon exchange, Darwin-Foldy, vacuum polarization, and magnetic moment contributions:

$$v^{\gamma}(pp) = v_{C1}(pp) + v_{C2} + v_{DF} + v_{VP} + v_{MM}(pp) , \tag{8}$$
$$v^{\gamma}(np) = v_{C1}(np) + v_{MM}(np) , \tag{9}$$
$$v^{\gamma}(nn) = v_{MM}(nn) , \tag{10}$$

where all terms have form factors for finite size effects, derived from the standard nucleon dipole forms:

$$G_E^p = \frac{G_M^p}{\mu_p} = \frac{G_M^n}{\mu_n} = (1 + q^2/b^2)^{-2} , \tag{11}$$
$$G_E^n = \beta_n q^2 (1 + q^2/b^2)^{-3} . \tag{12}$$

The value of b in the form factors is taken from the proton charge radius, and β_n in G_E^n is the experimentally measured slope at $q = 0$. The np Coulomb term $v_{C1}(np)$ arises from the overlap of the nucleon form factors, and while small, it is comparable to the v_{C2}, v_{DF}, and v_{VP} terms. Details of the electromagnetic terms can be found in the paper (Wiringa, Stoks, and Schiavilla 1995).

The OPE terms are given by:

$$v^{\pi}(pp) = f_{pp}^2 v_{\pi}(m_{\pi^0}) , \tag{13}$$
$$v^{\pi}(np) = f_{pp} f_{nn} v_{\pi}(m_{\pi^0}) + (-)^{T+1} 2 f_c^2 v_{\pi}(m_{\pi^{\pm}}) , \tag{14}$$
$$v^{\pi}(nn) = f_{nn}^2 v_{\pi}(m_{\pi^0}) , \tag{15}$$

where T is the isospin and

$$v_{\pi}(m) = \left(\frac{m}{m_s}\right)^2 \tfrac{1}{3} mc^2 \left[Y_{\mu}(r)\sigma_i \cdot \sigma_j + T_{\mu}(r)S_{ij}\right] . \tag{16}$$

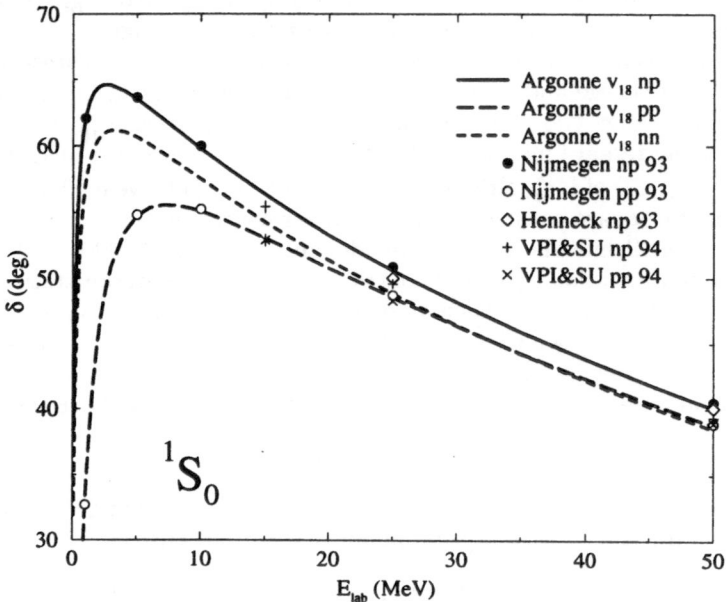

Fig. 2. Phase shifts for np, nn, and pp scattering in $^1S_0$ channel from Argonne v_{18} potential compared to various PWA's

Here $Y_\mu(r)$ and $T_\mu(r)$ are the usual Yukawa and tensor functions with the exponential cutoff of the Urbana and Argonne v_{14} models

$$Y_\mu(r) = \frac{e^{-\mu r}}{\mu r}\left(1 - e^{-cr^2}\right), \tag{17}$$

$$T_\mu(r) = \left(1 + \frac{3}{\mu r} + \frac{3}{(\mu r)^2}\right)\frac{e^{-\mu r}}{\mu r}\left(1 - e^{-cr^2}\right)^2, \tag{18}$$

where $\mu = mc/\hbar$. The scaling mass m_s makes the coupling constant dimensionless, and is taken to be the charged-pion mass, $m_{\pi\pm}$. As mentioned above, the Nijmegen PWA finds very little difference between the different coupling constants, so we take $f_{pp} = -f_{nn} = f_c \equiv f$, and adopt their value $f^2 = 0.075$ (although we did some searching on this value to confirm it was optimal for this model).

An additional term that could have been added to this OPE expression is the short-range δ-function (spread out with a form factor) that is generated when a $(\sigma_i\cdot\mathbf{k})(\sigma_j\cdot\mathbf{k})$ term is Fourier transformed. Such a term would get mixed up with the short-range phenomenology of the v^R term, and was neglected here.

The remaining part of the potential is simply written as a phenomenological

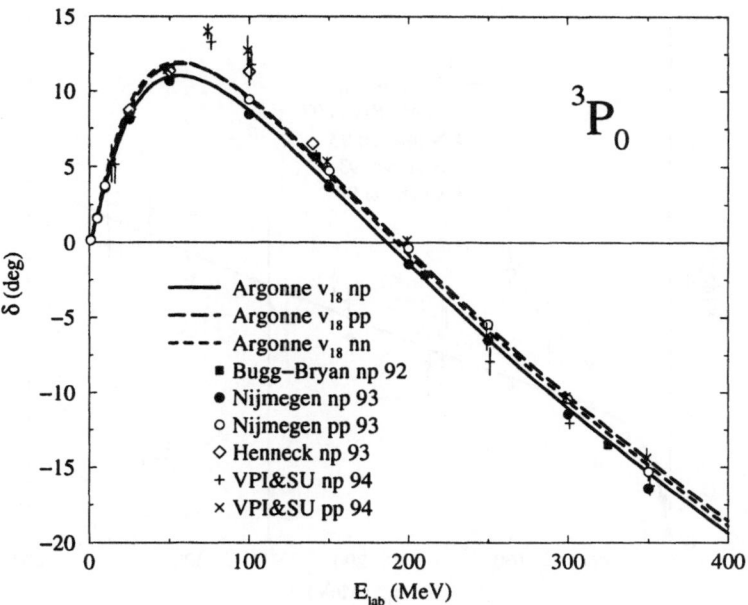

Fig. 3. Phase shifts for np, nn, and pp scattering in $^3P_0$ channel from Argonne v_{18} potential compared to various PWA's

operator structure:

$$v_{ST}^R(NN) = v_{ST,NN}^c(r) + v_{ST,NN}^t(r)S_{12} + v_{ST,NN}^{ls}(r)\mathbf{L \cdot S}$$
$$+ v_{ST,NN}^{l2}(r)L^2 + v_{ST,NN}^{ls2}(r)(\mathbf{L \cdot S})^2 \ . \tag{19}$$

Each of these terms is given the general form

$$v_{ST,NN}^i(r) = I_{ST,NN}^i T_\mu^2(r) + \left[P_{ST,NN}^i + \mu r\, Q_{ST,NN}^i + (\mu r)^2 R_{ST,NN}^i \right] W(r) \ , \tag{20}$$

with μ the average pion mass and $T_\mu(r)$ given by (18). Thus the $T_\mu^2(r)$ term has the range of a two-pion-exchange force. The $W(r)$ is a Woods-Saxon function which provides the short-range core:

$$W(r) = \left[1 + e^{(r-r_0)/a} \right]^{-1} \ . \tag{21}$$

The four sets of constants $I_{ST,NN}^i$, $P_{ST,NN}^i$, $Q_{ST,NN}^i$, and $R_{ST,NN}^i$ are parameters to be fit to data, subject to the regularization condition mentioned above:

$$v_{ST,NN}^t(r = 0) = 0 \ , \tag{22}$$

$$\frac{\partial v_{ST,NN}^{i \neq t}}{\partial r}\bigg|_{r=0} = 0 \ , \tag{23}$$

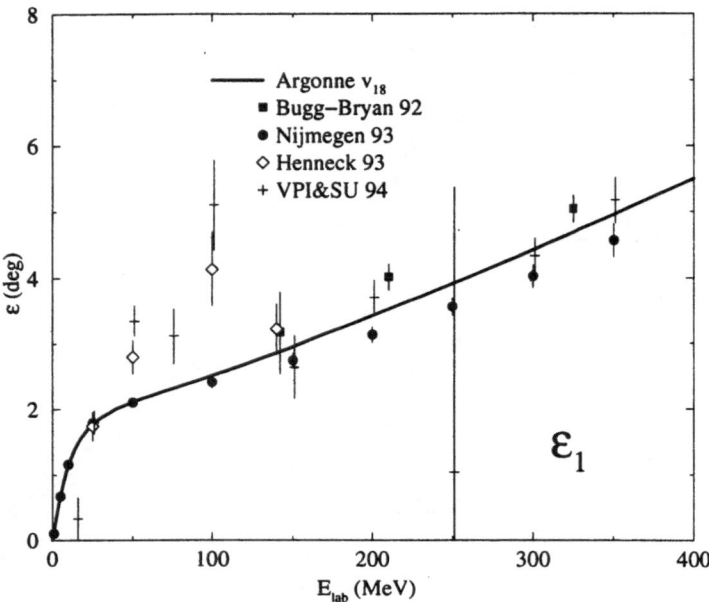

Fig. 4. Phase shifts for np scattering in ϵ_1 channel from Argonne v_{18} potential compared to various PWA's

which effectively constrains one of the parameters.

We explored a number of different parametrizations by fitting to the phase shifts of the Nijmegen PWA. When we settled on the final form, we fixed the f^2, the cutoff parameter, c, in the OPE $Y_\mu(r)$ and $T_\mu(r)$ functions, and the r_0 and a parameters in $W(r)$, and then adjusted some 40 $I^i_{ST,NN}$, $P^i_{ST,NN}$, and $R^i_{ST,NN}$ parameters by making a direct fit to the Nijmegen NN database, rather than the phase shifts. The 1787 pp and 2514 np elastic scattering data from 0–350 MeV were fit with a $\chi^2/N_{data} = 1.09$. In addition we fit the deuteron binding energy $E_d = 2.224575$ MeV, and, by a slight alteration of the $^1S_0$ pp potential, the nn scattering length and effective range. As a check, we also compared our elastic scattering to the SAID database, and got the somewhat larger, but still very respectable $\chi^2/N_{data} = 1.34$.

Some of the more interesting phase shifts are shown in Figs. 2–5. The greatest amount of CD and CSB occurs in the $^1S_0$ channel, as seen in Fig. 2, where we plot the pp, np, and nn phases of the potential and compare to the recent PWA analyses. The bulk of the difference between the pp and nn phases is due to the electromagnetic interaction, whereas the difference between the nn and np phases is primarily a measure of the strong CD, with roughly equal amounts coming from the OPE and the short-range part of the interaction. The next greatest

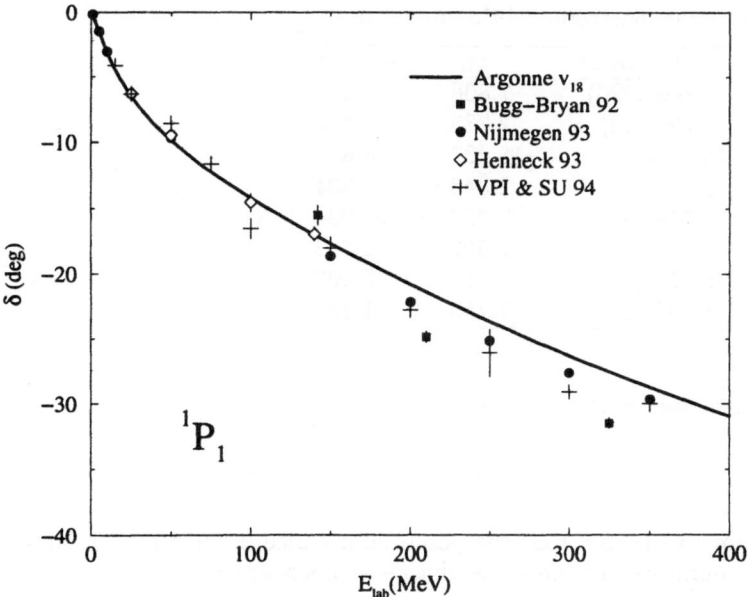

Fig. 5. Phase shifts for np scattering in $^1P_1$ channel from Argonne v_{18} potential compared to various PWA's

amount of CD comes in the $^3P_0$ channel, as seen in Fig. 3; here the data are not good enough to identify any CD beyond that coming from OPE. The ϵ_1 mixing parameter, which is a sensitive indicator of the strength of the tensor force, is given in Fig. 4. Here the various single-energy PWA's show a wide variation, but the Nijmegen multi-energy PWA has a quite smooth behavior, and all the new high-accuracy potentials follow the latter fairly closely. Finally in Fig. 5, we show the $^1P_1$ phases, which are highly correlated with ϵ_1 as discussed earlier.

Scattering lengths and effective ranges produced by Argonne v_{18} are given in Table 3. The experimental values are reproduced fairly well. An interesting point to note is the contribution of the v^γ terms; the importance of the Coulomb terms to $^1a_{pp}$ are well known, but there is also a significant contribution to $^1a_{np}$, coming mostly from the magnetic moment interaction, with some part from the $v_{C1}(np)$ term. Deuteron properties are displayed in Table 4, including meson-exchange contributions (MEC) to the electromagnetic moment operators. The electromagnetic interaction contributes $+18$ keV to E_d, 14 keV from $v_{MM}(np)$ and 4 keV from $v_{C1}(np)$.

The final step in constructing the model is the projection from specific ST, NN channels to an operator format. The strong interaction part can be

Table 3. Scattering lengths and effective ranges in fm for Argonne v_{18}, with and without the electromagnetic potential

	experiment	v_{18}	w/o v^γ
$^1a_{pp}$	−7.8063 (26)	−7.806	−17.164
$^1r_{pp}$	2.794 (14)	2.788	2.865
$^1a_{nn}$	−18.5 (5)	−18.487	−18.818
$^1r_{nn}$	2.8 (1)	2.840	2.834
$^1a_{np}$	−23.749 (8)	−23.732	−23.084
$^1r_{np}$	2.81 (5)	2.697	2.703
$^3a_{np}$	5.424 (3)	5.419	5.402
$^3r_{np}$	1.760 (5)	1.753	1.752

written as a sum of 18 operators:

$$v_{ij} = \sum_{p=1,18} v_p(r_{ij})O_{ij}^p , \tag{24}$$

and it is for this reason that the potential is called a "v_{18}" model. Here the first fourteen operators are the same charge-independent ones used in the Argonne v_{14} potential and are given by

$$O_{ij}^{p=1,14} = 1, \tau_i\cdot\tau_j, \sigma_i\cdot\sigma_j, (\sigma_i\cdot\sigma_j)(\tau_i\cdot\tau_j), S_{ij}, S_{ij}(\tau_i\cdot\tau_j), \mathbf{L}\cdot\mathbf{S}, \mathbf{L}\cdot\mathbf{S}(\tau_i\cdot\tau_j), L^2,$$
$$L^2(\tau_i\cdot\tau_j), L^2(\sigma_i\cdot\sigma_j), L^2(\sigma_i\cdot\sigma_j)(\tau_i\cdot\tau_j), (\mathbf{L}\cdot\mathbf{S})^2, (\mathbf{L}\cdot\mathbf{S})^2(\tau_i\cdot\tau_j) . \tag{25}$$

The four additional operators break charge independence and are given by

$$O_{ij}^{p=15,18} = T_{ij}, (\sigma_i\cdot\sigma_j)T_{ij}, S_{ij}T_{ij}, (\tau_{zi} + \tau_{zj}) . \tag{26}$$

The first three of these operators provide charge-dependence, while the latter provides charge-symmetry-breaking. The projections can be made from the appropriate CI, CD, and CSB combinations:

$$v_{S1} = v_{S1}^{CI} + v_{S1}^{CD}T_{ij} + v_{S1}^{CSB}(\tau_{zi} + \tau_{zj}) \tag{27}$$

where

$$v_{S1}^{CI} = \frac{1}{3}(v_{S1,pp} + v_{S1,nn} + v_{S1,np}) , \tag{28}$$

$$v_{S1}^{CD} = \frac{1}{6}[\frac{1}{2}(v_{S1,pp} + v_{S1,nn}) - v_{S1,np}] , \tag{29}$$

$$v_{S1}^{CSB} = \frac{1}{2}(v_{S1,pp} - v_{S1,nn}) . \tag{30}$$

Overall, the resulting Argonne v_{18} gives a pretty good fit to both pp and np data. The $\chi^2/N_{data} = 1.09$ is not quite as good as the several new Nijmegen or CD Bonn models, which are different in each partial wave, but the latter

Table 4. Deuteron properties for Argonne v_{18}, including meson-exchange contributions to static moments

	experiment	v_{18}	+ MEC
E_d (MeV)	-2.224575 (9)	-2.224575	...
A_S (fm$^{1/2}$	0.8781 (44)	0.8850	...
η_d	0.0256 (4)	0.0250	...
r_d (fm)	1.953 (3)	1.967	...
μ_d (μ_0)	0.857406 (1)	0.847	0.871
Q_d (fm^2)	0.2859 (3)	0.270	0.275
P_D (%)	...	5.76	...

cannot be expressed in the closed operator form that our quantum Monte Carlo many-body methods (discussed below) require. There remain several interesting questions that cannot be resolved with the present *NN* data. The CD of the interaction in S-waves is well established, but poorly constrained in P-waves; consequently the relative strength of T_{ij} and $(\sigma_i \cdot \sigma_j)T_{ij}$ terms is not pinned down. By fitting the *nn* scattering length we have also introduced CSB into the strong interaction, but again, this is only an S-wave quantity, and there is no P-wave data. We chose to use only a $(\tau_{zi} + \tau_{zj})$ CSB term, but one may reasonably expect $(\sigma_i \cdot \sigma_j)(\tau_{zi} + \tau_{zj})$ and $S_{ij}(\tau_{zi} + \tau_{zj})$ terms in addition on the basis of meson-theoretic effects such as ρ-ω mixing. (In fact the v_γ contributes such terms if projected into CI, CD, and CSB components.) However, any careful treatment of CD and CSB effects should only be made after the well-known electromagnetic contributions are included.

Perhaps the least satisfactory feature of this model is the low value for the deuteron quadrupole moment, Q_d. The MEC provide some correction in the right direction, but are too small. This problem is also present in the new Nijmegen potentials and CD Bonn, and may be attributable to the small value of f^2 that all these models use. The error in μ_d is less disturbing because there are two-body current contributions to the electromagnetic operators that are not constrained by v_{ij} through the continuity equation, and these terms can probably be adjusted to fit the data. In future work, it would be nice to develop the two-body potential and charge and current operators simultaneously and fit both the *NN* scattering data and the considerable amount of deuteron data together.

5   Many-Body Calculations

A major reason for constructing new highly-accurate *NN* potentials is the great progress that has been made in many-body calculations in recent years. In the last ten years several groups have developed Faddeev calculations for the trinucleon binding energy with an accuracy of 10 keV or better (Chen *et al.* 1985, Stadler, Glöckle, and Sauer 1991). These methods have since been extended to Faddeev-Yakubovsky studies of the alpha-particle ground state (Glöckle and Kamada 1993). Another recent development is the correlated hyperspherical harmonic (CHH) method (Kievsky, Viviani, and Rosati 1993; Viviani, Kievsky, and Rosati 1995) which also gives three- and four-nucleon ground state energies with very high precision. The Faddeev and CHH methods have also made significant progress in describing problems such as nd and pd elastic scattering (Hüber et al. 1995).

Another line of attack has been the development of variational Monte Carlo (VMC) methods with ever-more sophisticated trial functions (Wiringa 1991) and the Green's function Monte Carlo method (GFMC) which can stochastically evolve these trial functions to the exact ground or lowest excited state of a given quantum number (Carlson 1988). The advantage of these quantum Monte Carlo methods is that extension to larger nuclei is relatively straightforward, and has been made successfully for $A = 5$–7 nuclei (Pudliner et al. 1995, 1996). A cluster-expansion VMC has also been developed for larger nuclei such as ^{16}O (Pieper, Wiringa, and Pandharipande 1992) and it is hoped that a cluster-expansion adaption of GFMC will be forthcoming.

All these calculations, which have been made possible by the tremendous advances in computer technology in the last decade, give us an unprecedented chance to understand nuclear structure at the level of the NN interaction. The accuracy of these methods demands accurate NN potentials. It also allows us to examine small effects such as CD and CSB in many-body systems at a far more meaningful level than heretofore possible.

One challenge is immediately obvious when we calculate the binding energy of ^3H: the realistic NN potentials do not reproduce the observed value of 8.48 MeV when used in the nonrelativistic Schrödinger equation. The local potentials Argonne v_{18}, Nijm II, and Reid 93 all give a binding energy of 7.62 ± 0.01 MeV (Friar et al. 1993). The nonlocal Nijm I gives a little more binding at 7.72 MeV, while CD Bonn, with a more extensive nonlocality, gives a reported 8.00 MeV (Machleidt, Sammarruca, and Song 1996).

In principle one might easily expect the NN potential, as an effective force between nucleons with a composite quark substructure, to be nonlocal. Unfortunately, NN elastic scattering data does not tell us anything about how much nonlocality the NN potential should have. In addition, relativistic effects can be expected to contribute at the level of a few hundred keV to the trinucleon binding, but at present there is no consensus on the sign of the relativistic contribution. Using a relativistic three-dimensional Blanckenbecler-Sugar formalism in the trinucleon, the ^3H binding increases to 8.19 MeV for CD Bonn, a result consistent with earlier work (Rupp and Tjon 1992). However, studies based on relativistic quantum mechanics have consistently found repulsive contributions of order 0.3 MeV in the triton (Glöckle, Lee, and Coester 1986, Forest et al. 1995). Much work remains to be done on these issues.

Because of the composite nature of the nucleon and the corresponding nucleon excitation spectrum we also expect there to be nontrivial three-nucleon (3N) forces. Models for the 3N potential have been discussed for nearly forty years. It was realized early on that they could contribute to the effective spin-orbit splitting in nuclei, and the first models made use of the idea of excitation of an intermediate $\Delta(1232)$ state (Fujita and Miyazawa 1957). A more sophisticated treatment using knowledge of πN scattering and PCAC was made in the Tucson-Melbourne model (Coon et al. 1979).

For the current VMC and GFMC calculations we use one in a series of Urbana $3N$ potentials called model IX, which is constructed with long-range two-pion-exchange (TPE) and a shorter-range phenomenological piece:

$$V_{ijk} = V_{ijk}^{2\pi} + V_{ijk}^{R} , \qquad (31)$$

where

$$V_{ijk}^{2\pi} = A_{2\pi} \sum_{cyclic} \{X_{ij}, X_{jk}\}\{\tau_i \cdot \tau_j, \tau_j \cdot \tau_k\} + \tfrac{1}{4}[X_{ij}, X_{jk}][\tau_i \cdot \tau_j, \tau_j \cdot \tau_k] , \quad (32)$$

$$V_{ijk}^{R} = U_0 \sum_{cyclic} T_\mu^2(r_{ij}) T_\mu^2(r_{jk}) . \qquad (33)$$

and

$$X_{ij} = [Y_\mu(r)\sigma_i \cdot \sigma_j + T_\mu(r)S_{ij}] , \qquad (34)$$

The $Y_\mu(r)$ and $T_\mu(r)$ are the same functions given in (17)-(18). The overall strengths $A_{2\pi}$ and U_0 are adjusted to reproduce the ^{3}H binding energy and to give a reasonable saturation density in variational calculations of nuclear matter. In general, the $V_{ijk}^{2\pi}$ is attractive at all densities, and gives the extra binding needed in light nuclei, while the V_{ijk}^{R} is made repulsive and has a more rapid density dependence, helping to lower the saturation density in nuclear matter which is too large with NN potentials only. The shorter-range piece could in principle have extensive spin- and isospin-dependence, but we have limited constraints from data at present and do not want to introduce any more parameters than necessary.

The binding energy of $A = 3, 4$ nuclei using the Argonne v_{18} with or without Urbana IX is shown in Table 5. The second line gives the energy (with MC statistical error) calculated for Argonne v_{18} alone; the difference with experiment of 0.9 MeV in ^{3}H and 4.2 MeV in ^{4}He must be made up by some combination of nonlocality, relativistic effects, and many-nucleon forces. In the present work, we assume all the difference is due to a V_{ijk}, with the hope that the short-range part may be mocking up some of the nonlocal and relativistic effects. The third (fourth) line gives the energy obtained if we use only the np (pp) potential for $T = 1$ states. This variation of 0.35 MeV in ^{3}H and 1.0 MeV in ^{4}He is a reflection of the amount of CD in the NN force, and is a sizeable fraction of the amount we attribute to V_{ijk}, hence the necessity of getting the CD correct.

The total energy for the full Hamiltonian and the breakdown into kinetic, NN and $3N$ potential terms is given on subsequent lines. It is interesting to note that a very large fraction of the total v_{ij} comes from OPE and the dominant part of V_{ijk} is from TPE. The total V_{ijk} contribution is a significant fraction of the

Table 5. GFMC energy of $A = 3$ & 4 nuclei for Argonne v_{18} + Urbana IX Hamiltonian; numbers in parentheses are Monte Carlo statistical errors

	^{3}H	^{3}He–^{3}H	^{4}He
Experiment	−8.482	0.764	−28.296
Argonne v_{18}	−7.61 (3)	0.739	−24.09 (10)
$(np\ T = 1)$	−7.79	...	−24.69
$(pp\ T = 1)$	−7.44	...	−23.69
+ Urbana IX	−8.47 (2)	0.757	−28.32 (12)
K	51.(1)	0.014	118.(2)
v_{ij}	−58.(1)	0.743	−142.(2)
(v^γ)	0.04	0.677	0.88
(v^π)	−44.1 (5)	0.	−107.(1)
(v^R)	−14.(1)	0.066	−36.
V_{ijk}	−1.20 (3)	0.	−6.6 (3)
$(V^{2\pi})$	−2.25 (6)	0.	−12.6 (3)
(V^R)	1.05 (3)	0.	6.0 (2)

net binding, but it is small compared to the total v_{ij}, so we may hope that any four-nucleon force, which should exist in principle, will be negligible. The second column of Table 5 gives the ^{3}He–^{3}H energy difference, which is a measure of the CSB in the NN force. This difference is fairly well explained within the present force model, especially when considering that the experimental uncertainty in $^1a_{nn}$ is large enough to cover a 20% correction to the strong-interaction CSB term. Basically, we can say the ^{3}He–^{3}H energy difference is completely consistent with the difference in low-energy pp and nn scattering lengths.

We conclude this section with two results from VMC and GFMC studies of light p-shell nuclei. We have calculated the ground and low-lying excited states for A=5–7 nuclei with the same Argonne v_{18} + Urbana IX Hamiltonian (Pudliner et al. 1995, 1996). The results of these calculations are summarized in the spectrum shown in Fig. 6. The VMC and GFMC calculations have an upper-bound property (within their statistical error bars, which are shown by shaded boxes) for the lowest state of given J^π; T quantum numbers. The VMC calculations for A=3,4 nuclei are 2–3% above the exact results given by Faddeev, CHH, or GFMC, but as A increases, they seem to get progressively worse. However, they do get the correct ordering and a reasonable magnitude for the splitting of states for A=5–7. The GFMC calculations, which agree very well with Faddeev-Yakubovsky and CHH methods for $A \leq 4$, give a significant improvement over the VMC calculations in A=5–7. It appears that the $A = 6$ nuclei are underbound by about 1 MeV, and the $A = 7$ nuclei by about 2 MeV with the present Hamiltonian. This may be attributable to our neglect of nonlocal and/or relativistic effects, but most likely to the overly simple form we have used for V_{ijk}^R. We are now investigating whether simple generalizations of this term can improve the spectrum without introducing too many new force parameters. In any

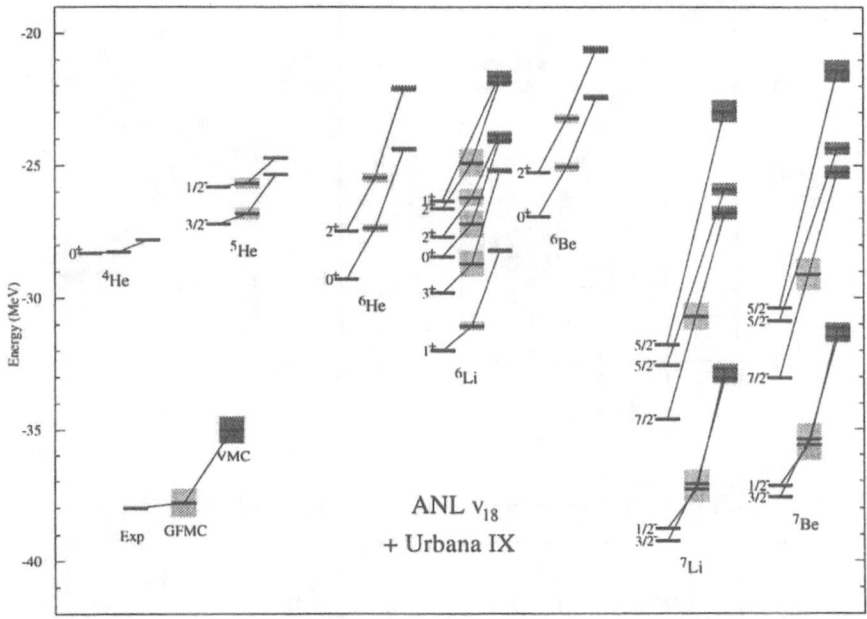

Fig. 6. Spectrum for $A \leq 7$ nuclei calculated using variational and Green's function Monte Carlo methods

event, one can truly see the beginnings of shell structure in these nuclei, arising from realistic bare two- and three-nucleon interactions.

Our second result is a VMC calculation of elastic and transition electromagnetic form factors in ^{6}Li shown in Fig. 7 (Schiavilla and Wiringa 1996). These have been made in both impulse approximation (IA) and with MEC. The comparison with data for the elastic longitudinal form factor, $F_L(q^2)$, is excellent, as is the E2 transition to the 3^+ first excited state. The MEC corrections are small, but stand out noticeably in the first minimum where they significantly improve the fit to data. The M1 transition to the 0^+ isobaric analog state is also reproduced reasonably well. The elastic transverse form factor, $F_T(q^2)$, is good up to the first zero, but then is too large in the region of the second maximum. We plan to repeat these studies with the GFMC wave function to see if the present deficiencies can be removed. Nevertheless, these calculations, which have been done without effective charges as commonly needed in shell-model studies, give a remarkably good explanation of the electron scattering data.

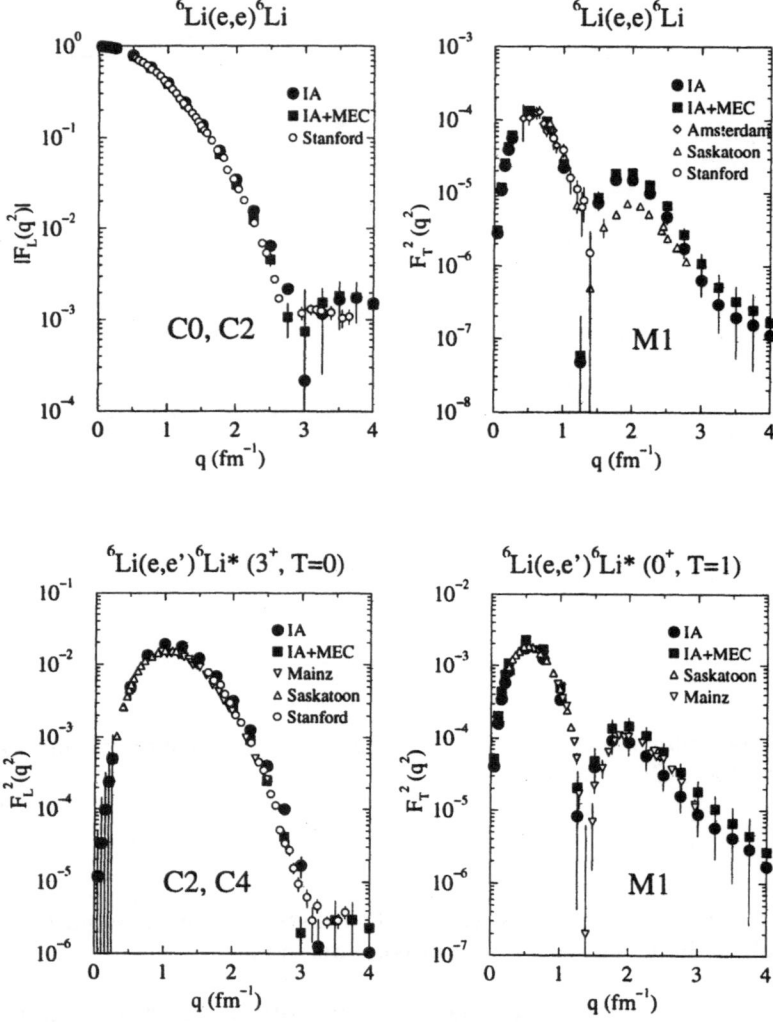

Fig. 7. Calculated elastic and transition form factors in ^{6}Li

6   Influence of NN Interactions on qq Models

Our empirical knowledge of NN interactions is substantial, and our theoretical understanding of the long-range part is long-established. One of the key features we know is that there is a strong tensor force, arising from pion exchange at long range and perhaps modified at short distances by heavier meson-exchange. This leads to a large angular dependence in $S, T = 1, 0$ states and induces a toroidal (dumbell) shape in the deuteron for $M_d = 0$ ($M_d = \pm 1$) at a femtometer

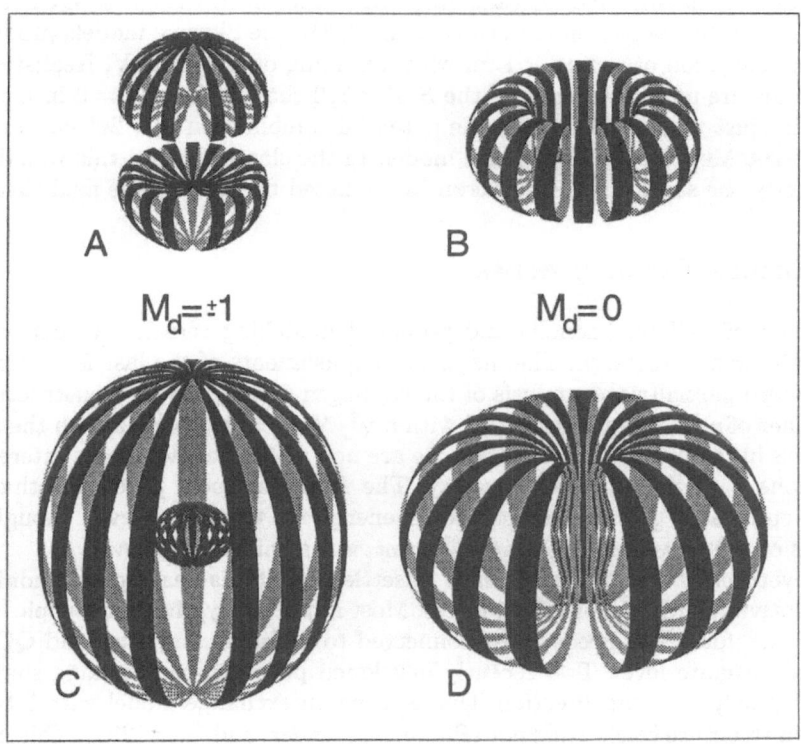

Fig. 8. Surfaces of constant density $\rho = 0.24$ fm^{-3} (A & B) or $\rho = 0.08$ fm^{-3} (C & D) for polarized deuteron

scale (Forest *et al.* 1996). A graphic illustration is the set of constant-density surfaces shown in Fig. 8 for a polarized deuteron. These angular-dependent pair correlations persist in larger nuclei, and are independent of the particular *NN* potential chosen.

If one wanted to construct a constituent quark model for the deuteron, using six valence quarks (and GFMC methods may make such a calculation as feasible as ^{6}Li in the near future) then the quark-quark (qq) interaction V_{qq} must reproduce this feature of the deuteron. Plausible models would have to incorporate a tensor force *between quarks*. While there is a tensor component in one-gluon exchange it is not nearly strong enough to provide the correlations observed in the deuteron. The answer may be to introduce a direct $qq\pi$ coupling, as has been suggested on the basis of systematics in hadronic spectra (Glozman and Riska 1996), leading to a OPE between constituent quarks.

It is also interesting to note that these toroidal structures are predicted by classical Skyrme effective Lagrangians, which have been related to QCD in the limit of infinite colors (Verbaarschot *et al.* 1987). The Skyrme models produce a toroidal deuteron of radius ≈ 1 fm with a binding of ≈ 150 MeV. Realistic NN potentials are most attractive in the $S, T = 1, 0$ channel for $M_S = 0$ in a toroid of radius just under 1 fm, where the potential combination $v^c + 2v^t$ has a depth of 100–200 MeV, depending on the model. In the classical limit, this would lead to exactly the same kind of deuteron as predicted by the Skyrme field theory.

7   Future Developments

The field of NN interactions and potential modelling remains an active and dynamic area of research. The major accomplishments of the last few years are the superb partial-wave analysis of the Nijmegen group, and the construction of a number of new NN potentials, all with a $\chi^2/N_{data} \approx 1$. Coupled with the rapid progress in many-body calculations, we are now able to answer some interesting questions with unparalleled accuracy. The fact that local potentials that are phase-equivalent give the same binding energy for the triton, even though the details of the potentials are quite different, is a significant discovery.

Nevertheless, there remain many unsettled questions dealing with nonlocality, relativity, and many-nucleon forces. Most importantly, the microscopic origin of the NN force still needs to be connected to hadronic structure and QCD at a more intimate level. Two recently developed potentials are making some interesting steps in that direction. One is a meson-exchange model with intrinsic nucleon structure called Ruhrpot (Plümper, Flender, and Gari 1994). This model combines long-range meson exchange with direct NN interactions at short distance and pays particular attention to calculating meson-baryon vertices (form factors) in a self-consistent manner. The χ^2 fit to data is not quite as good as the new high-accuracy models discussed above, but is still quite respectable.

The second model is a potential built on an effective chiral Langrangean (Ordóñez, Ray, and van Kolck 1994) that makes use of an expansion in chiral perturbation theory. One interesting feature of this treatment is a power counting argument that leads to the statements

$$V_{NN} > V_{3N} > V_{4N} \ldots , \tag{35}$$

$$V_{CD} > V_{CSB-III} > V_{CSB-IV} , \tag{36}$$

where we differentiate between class III and IV CSB as given in (4). Both of these features are empirically known facts, but it is nice to see them come out of the same theoretical source. An important development for both of these models is that they are being extended to make consistent predictions for $3N$ forces (Eden and Gari 1996, van Kolck 1994).

For applications to nuclear structure and effective interactions, there also remain many challenging problmes. I have not attempted at all to discuss how one goes about making an effective interaction for use in shell model studies from the realistic NN potentials. Assuming that the shell model will continue to be treated as a nonrelativisitic Schrödinger equation, one of the big challenges will be how to incorporate $3N$ forces, which seem to be a necessary ingredient in the microscopic calculations of light nuclei. In the meantime, we can expect that these direct calculations using realistic potentials will continue to progress to larger and larger nuclei.

8   Acknowledgments

I wish to thank my many colleagues who have contributed to the work discussed here, in particular V. G. J. Stoks and R. Schiavilla for their work on the Argonne v_{18} potential and electromagnetic operators, and also A. Arriaga, J. Carlson, J. L. Forest, J. L. Friar, B. S. Pudliner, V. R. Pandharipande, and S. C. Pieper. Computations have been made possible by generous grants of computer time from the Mathematics and Computer Science Division of Argonne National Laboratory, and the Cornell Theory Center. This work is supported by the U. S. Department of Energy, Nuclear Physics Division, under contract W-31-109-ENG-38.

References

Arndt, R. A. , Roper, L. D. , Workman, R. L. , McNaughton, M. W. (1992): Phys. Rev. D **45**, 3995–4001;
  World Wide Web page at http://clsaid.phys.vt.edu/~CAPS/said_branch.html
Arndt, R. A., Workman, R. L., Pavan, M. M. (1994): Phys. Rev. C **49**, 2729–2734;
Bergervoet, J. R., van Campen, P. C., Klomp, R. A. M., de Kok, J.-L., Rijken, T. A., Stoks, V. G. J., de Swart, J. J. (1990): Phys. Rev. C **41**, 1435–1452
Bugg, D. V., Bryan, R. A. (1992): Nucl. Phys. **A540**, 449–460
Bugg, D. V., Carter, A. A., Carter, J. R. (1973): Phys. Lett. **44B**, 278–280
Bugg, D. V., Machleidt, R. (1995): Phys. Rev. C **52**, 1203–1211
Carlson, J. (1988): Phys. Rev. C **38**, 1879–1885
Chen, C. R., Payne, G. L., Friar, J. L., Gibson, B. F. (1985): Phys. Rev. C **31**, 2266–2273
Coon, S. A., Scadron, M. D., McNamee, P. C., Barrett, B. R., Blatt, D. W. E., McKellar, B. H. J. (1979): Nucl. Phys. **A317**, 242–278
Eden, J. A., Gari, M. F. (1996): Phys. Rev. C **53** 1510–1518
Forest, J. L., Pandharipande, V. R., Carlson, J., Schiavilla, R. (1995): Phys. Rev. C **52**, 576–577
Forest, J. L., Pandharipande, V. R., Pieper, S. C., Wiringa, R. B., Schiavilla, R., Arriaga, A. (1996): Phys. Rev. C **54**, 646–667
Friar, J. L., Payne, G. L., Stoks, V. G. J., de Swart, J. J. (1993): Phys. Lett. **311B**, 4–8
Fujita, J., Miyazawa, H. (1957): Prog. Theor. Phys. **17**, 360–365
Glöckle, W., Lee, T.-S. H., Coester, F. (1986): Phys. Rev. C **33**, 709–716
Glöckle, W., Kamada, H. (1993): Phys. Rev. Lett. **71**, 971–974
Glozman, L. Ya., Riska, D. O. (1996): Phys. Rep. **268** 263–303
Hammans, M., et al. (1994): Phys. Rev. Lett. **72**, 2665

Hüber, D., Glöckle, W., Golak, J., Witala, H., Kamada, H., Kievsky, A., Rosati, S., Viviani, M. (1995): Phys. Rev. C **51**, 1100–1107

Henneck, R. (1993): Phys. Rev. C **47**, 1859–1875

Kievsky, A., Viviani, M., Rosati, S. (1993): Nucl. Phys. **A551**, 241–254

Lacombe, M., Loiseau, B., Richard, J. M., Vinh Mau, R., Côté, J., Pirès, P., de Tourreil, R. (1980): Phys. Rev. C **21**, 861–873

Lagaris, I. E., Pandharipande, V. R. (1981): Nucl. Phys. **A359**, 331–348

Machleidt, R., Holinde, K., Elster, Ch. (1987): Phys. Rep. **149**, 1–89

Machleidt, R., Slaus, I. (1994): Phys. Rev. Lett. **72**, 2664

Machleidt, R., Sammarruca, F., Song, Y. (1996): Phys. Rev. C **53**, R1483–R1487

Nagels, M. M., Rijken, T. A., de Swart, J. J. (1978): Phys. Rev. D **17**, 768–776

Okubo, S., Marshak, R. E. (1958): Ann. Phys. (NY) **4**, 166–179

Ordóñez, C., Ray, L., van Kolck, U. (1994): Phys. Rev. Lett. **72**, 1982–1985

Pieper, S. C., Wiringa, R. B., Pandharipande, V. R. (1992): Phys. Rev. C **46**, 1741–1756

Plümper, D., Flender, J., and Gari, M. F. (1994): Phys. Rev. C **49**, 2370–2378

Pudliner, B. S., Pandharipande, V. R., Carlson, J., Wiringa, R. B. (1995): Phys. Rev. Lett. **74**, 4396–4399

Pudliner, B. S., Pandharipande, V. R., Carlson, J., Wiringa, R. B., Pieper, S. C. (1996): in preparation

Reid, R. V. (1968): Ann. Phys. (NY) **50**, 411–448

Rupp, G., Tjon, J. A. (1992): Phys. Rev. C **45**, 2133–2142

Schiavilla, R., Pandharipande, V. R., Riska, D. O. (1989): Phys. Rev. C **40**, 2294–2309

Schiavilla, R., Pandharipande, V. R., Riska, D. O. (1990): Phys. Rev. C **41**, 309–317

Schiavilla, R., Wiringa, R. B., (1996): in preparation

Stadler, A., Glöckle, Sauer, P. U. (1991): Phys. Rev. C **44** 2319–2327

Stoks, V. G. J., de Swart, J. J. (1993): Phys. Rev. C **47**, 761–767

Stoks, V. G. J., Klomp, R. A. M., Rentmeester, M. C. M., de Swart, J. J. (1993): Phys. Rev. C **48**, 792–815

Stoks, V. G. J., Klomp, R. A. M., Terheggen, C. P. F., de Swart, J. J. (1994): Phys. Rev. C **49**, 2950–2962

Stoks, V. G. J., de Swart, J. J. (1995): Phys. Rev. C **52**, 1698–1701

van Kolck, U. (1994): Phys. Rev. C **49**, 2932–2941

Verbaarschot, J. J. M., Walhout, T. S., Wambach, J., Wyld, H. W. (1987): Nucl. Phys. **A468** 520–538

Viviani, M., Kievsky, A., Rosati, S. (1995): Few-Body Syst. **18**, 25–39

Wiringa, R. B., Smith, R. A., Ainsworth, T. L. (1984): Phys. Rev. C **29**, 1207–1221

Wiringa, R. B. (1991): Phys. Rev. C **43**, 1585–1598

Wiringa, R. B. (1993): Rev. Mod. Phys. **65**, 231–242

Wiringa, R. B., Stoks, V. G. J., Schiavilla, R. (1995): Phys. Rev. C **51**, 38–51

Microscopic theories of effective interaction with an application to halo nuclei

T.T.S. Kuo

Department of Physics, State University of New York,
Stony Brook, New York 11794, U.S.A.

Abstract. The effective interaction used in shell model calculations plays a central role in nuclear structure calculations. We review here several microscopic methods for deriving such effective interactions, starting from the free NN potential. For a chosen model space, there are formal methods for obtaining a model-space effective hamiltonian H_{eff} which can reproduce certain physical properties of the original hamiltonian. These methods are briefly discussed. Among them, the $\hat{Q}$-box folded diagram method initially developed by Kuo-Lee-Ratcliff is relatively more convenient for numerical calculations. To apply this method to nuclear structure calculations, a first step is perform a partial summation of certain $\hat{Q}$-box diagrams so as to express the $\hat{Q}$-box in terms of G-matrix interactions. Accurate calculation of the G-matrix for finite nuclei is now feasible. For a given $\hat{Q}$-box the folded-diagram series for the effective interaction can be summed up to all orders using iterative methods, such as the Lee-Suzuki method and the Krenciglowa-Kuo method. For the $\hat{Q}$-box , however, it seems that one has to adopt some low-order, in the G-matrix, approximation. A highly desirable situation seems to be provided by halo nuclei where the valence nucleons are weakly attached to those of the inner core. In this case the effect of core-polarization is largely weakened, and the $\hat{Q}$-box may be accurately calculated by including only few low-order G-matrix diagrams.

1 Introduction

Let me first thank the organizers, especially my old friend Da Hsuan Feng, for inviting me to this workshop. It is indeed my great honor and pleasure to attend this workshop, and present a report here. The organizers have assigned me to review the various theoretical methods for deriving the effective interaction (V_{eff}) starting from the free NN interaction. This is a heavy assignment, and I am not sure that I can do an adequate job. Nevertheless, I shall try my best. As you all know, currently there is much interest in the study of halo nuclei (Austin and Bertsch 1995, Zhukov *et al.* 1993), using radioactive-beam accelerators. For instance, there were four papers (Nazarewicz *et al.* 1996, Hamamoto *et al.* 1996, Page *et al.* 1996) in a recent issue of Physical Review C dealing with halo nuclei. Thus, it may be of more interest to discuss the various effective interaction theories together with an application to halo nuclei.

Nucleus is a complicated and difficult many-body problem. Although much progress has been made towards our understanding of the nucleon-nucleon (NN) interaction, as reviewed by Wiringa in the preceding article, our knowledge about the free NN interaction V is still not as definite as we like, unlike the atomic situation where the inter-electron interaction is accurately known to be the Coulomb interaction. Taking V as given by a modern potential model such as the Paris (Lacomb *et al.* 1980) or Bonn (Machleidt 1989, Machleidt *et al.* 1987) potential, we may write the Schroedinger equation for the nuclear many-body problem as

$$H\Psi_n(1, 2, ..A) = E_n(1, 2, ..A)\Psi_n(1, 2, ...A) \qquad (1)$$

where the nuclear Hamiltonian is taken as H=T+V, with T denoting the kinetic energy. To define a convenient single-particle basis, we introduce an auxiliary single particle (sp) potential U and rewrite H as $H = H_0 + H_1$; $H_0 = T + U$, $H_1 = V - U$. In the above A denotes the number of nucleons contained in the nucleus. At a more fundamental level, one may treat the nucleus as composed of a system of quarks governed by QCD. We are still far from being able in describing the nucleus in this way.

At the present time, the best available microscopic theory, in my opinion, is still the above nuclear many-body problem consisting of nucleons which interact with each other via a realistic NN potential. In nuclear structure calculations, our purpose is mainly to study the low-energy properties of nuclei, in the sense that the average kinetic energy of each nucleon, which is typically about 50 MeV, is much less than the rest mass of each nucleon. For this situation, it should be adequate to use the above non-relativistic Schroedinger equation to describe the nucleus.

The above nuclear many-body equation is, however, a very difficult problem to solve. In nuclear structure calculations, one almost always deals with a model-space Schroedinger equation of the form

$$P(H_0 + V_{eff})P\Psi_m(A) = E_m(A)P\Psi_m(A); \quad m = 1, d \qquad (2)$$

where P denotes a chosen model space, such as the *sd* shell, and d is the dimension of the model space. H_0 is the unperturbed Hamiltonian. The projection operator P is defined in terms of the eigenfunctions of H_0. Note that the above equation reproduces only a subset of the solutions of the original Eq. (1). For instance, E_m is just a subset of the eigenvalues E_n of Eq.(1).

A large number of nuclear properties have been very successfully described by the nuclear shell model, where one deals with a model-space effective Schroedinger equation which is exactly of the form shown above. To set up this model-space equation, one has to know first the effective interaction V_{eff}. There have been a number of empirical methods in determining the above effective interaction. Often a simple form for V_{eff} is assumed and it is allowed to have several adjustable parameters. These parameters are then adjusted to fit certain observed nuclear properties such as the nuclear binding energies. A number of highly successful effective interactions have been constructed in this way, such as the well known surface-delta interaction and the Skyrme interactions. Instead of determining an analytic form for the effective interaction, one may as well just determine certain matrix elements of it. For example, for the entire sd shell it is sufficient to know the 63 shell-model matrix elements of V_{eff}; it is not necessary to know the underlying form of the effective interaction. As is well known, Wildenthal and his collaborators (Wildenthal 1984) have determined these 63 sd-shell matrix elements and they have been highly successful in describing the sd-shell nuclei. It is clear that there do "exist" such effective interactions.

What is the connection between the shell-model V_{eff} and the free NN interaction? Many physicists have studied this problem, as was investigated by Kuo and Brown (Kuo and Brown 1966) some time ago. It would certainly be highly desirable to have a first-principle theory which can derive the shell-model V_{eff} starting from the free NN interaction. In the following, let me briefly describe the basic tools for this derivation.

2 Folded diagrams

The nuclear many-body problem, as given by Eq.(1), in fact contains too much information than we need to know. (Experimentally we can measure only a small number of selected properties of the nucleus.) By intuition, life should become simpler if we can reduce the full-space many-body problem of Eq.(1) to a smaller model-space problem of the form of Eq.(2). In addition, the familiar shell-model secular eqution is such a model-space equation. This further suggests that the reduction of the full-space equation to a smaller space (model space) effective equation may be a useful first step towards the solution the the nuclear many-body problem. This reduction can be formally achieved by way of the so-called folded-diagram theories.

Let us first recall the well-known Feshbach theory (Feshbach 1958). In matrix form Eq.(1) is written as

$$\begin{pmatrix} PHP & PHQ \\ QHP & QHQ \end{pmatrix} \begin{pmatrix} P\Psi \\ Q\Psi \end{pmatrix} = E \begin{pmatrix} P\Psi \\ Q\Psi \end{pmatrix} \tag{3}$$

where P is the model-space projection operator

$$P = \sum_{n=1,d} |\Phi_n\rangle\langle\Phi_n|. \tag{4}$$

In the above, d is the dimension of the model space and usually it is chosen to be a small number. Φ is the unperturbed eigenfunctions of H_0. Q is the complement of P, namely Q=1-P. We can readily eliminate $Q\Psi$, obtaining a P-space equation

$$H_{eff}(E_n)P\Psi_n = E_n P\Psi_n \tag{5}$$

with

$$H_{eff}(E_n) = PHP + V_{eff}(E_n), \tag{6}$$

and

$$V_{eff}(E_n) = PHQ\frac{1}{E_n - QHQ}QHP. \tag{7}$$

The above is an interesting result. We see that now we only need to deal with an equation which is entirely contained in the P space. However, there is a significant drawback in that the effective interaction is dependent on the eigenvalue E_n. In other words, we need to use different V_{eff} for different eigenstates. This feature is by no means convenient. A main purpose of the folded-diagram method is to remove this eigenvalue dependence. In addition, the effective interaction given by the folded-diagram method is manifestly valence-linked, while that of the preceding equation contains unlinked diagrams.

Folded diagrams have a long history, and many physicists (Morita 1963, Oberlechner et al. 1970, Johnson and Baranger 1972, Kuo et al. 1971, Brandow 1967, Lindgren 1974, Kuo 1981, Kuo and Osnes 1990, Hjorth-Jensen et al. 1995, Krenciglowa and Kuo 1974, Lee and Suzuki 1980, Suzuki and Lee 1980, Suzuki et al. 1994, Kuo et al. 1995) have studied this topic. Basically we may classify them into two categories: the time dependent approach and the time independent approach. The time dependent approach was first investigated by Morita (Morita 1963), followed almost concurrently by Obberlechner et al. (Oberlechner et al. 1970), Johnson and Baranger (Johnson and Baranger 1972) and Kuo, Lee and Ratcliff (Kuo et al. 1971). On the other hand, time independent approaches have been advanced by Brandow (Brandow 1967), Lindgren (Lindgren 1974) and Suzuki, Lee and Okamoto (Lee and Suzuki 1980, Suzuki and Lee 1980, Suzuki et al. 1994). For microscopic many-body calculations, the time formulation is more desirable, for two reasons. First, the time-dependent folded-diagram formulation is based on a factorization of the time-evolution operator U(t,t'), from which we can describe various physical processes using Feynman diagrams. For example, the NN potential is based on meson-exchange Feynman diagrams. And it is more convenient in the time-dependent formulation to study the effect of folded diagrams on the meson-exchange diagrams for the NN potential. In general, the physical meaning of the folded diagrams can be more clearly visualized, using the language of Feynman-Goldstone diagrams. The second reason is that using the time formulation it is more transparent to show the cancellation of disconnected diagrams (Kuo and Osnes 1990).

Among the various folded-diagram methods, it seems that only the KLR (Kuo-Lee-Ratcliff) method (Kuo et al. 1971) has been extensively investigated and applied. In this method, the model-space effective interaction is given rigorously by the folded-diagram expansion

$$V_{eff} = \hat{Q} - \hat{Q}' \int \hat{Q} + \hat{Q}' \int \hat{Q} \int \hat{Q} - \hat{Q}' \int \hat{Q} \int \hat{Q} \int \hat{Q} \cdots, \qquad (8)$$

where each $\hat{Q}$ represents a $\hat{Q}$-box composed of irreducible diagrams and the sign $\int$ represents a generalized folding. The $\hat{Q}'$-box is the same as the $\hat{Q}$-box except that the former begins with diagrams second order in V while the latter contains also diagrams first order in V. The effective interaction given above is energy-independent and is valence linked.

The KLR method employs the time evolution operator U(t,t') to construct the true eigenstate of H, denoted by Ψ_n, starting from an unperturbed eigenstate of H_0 which is denoted by Φ_n. Schematically we write

$$U(t, t')|\Phi_n\rangle \Rightarrow |\Psi_n\rangle \tag{9}$$

where one employs a complex-time (Kuo and Osnes 1990, Kuo *et al.* 1971) limiting process with $t'(1 - i\delta)$, $t' \to -\infty$, $\delta \to 0^+$. In the interaction representation, we have

$$U(t, t') = 1 + \sum_{n=1}^{\infty} (-i)^n \int_{t'}^{t} dt_1 \int_{t'}^{t_1} dt_2 \cdots \int_{t'}^{t_{n-1}} dt_n H_1(t_1) H_1(t_2) \cdots H_1(t_n).$$
$$\tag{10}$$

Consider the wave function generated by the operation $U(0, -\infty)|\Phi_{\alpha\beta}\rangle$ where $|\Phi_{\alpha\beta}\rangle$ is a two-particle state defined by $a_\alpha^+ a_\beta^+|0\rangle$. Here the a^+'s are the sp creation operators and $|0\rangle$ is the vacumn. As indicated in Fig. 1, diagram A is a term generated in this operation, where particles α, β have one interaction at time t_2 (the lower vertex), go into intermediate states γ, δ, have one more interaction at t_1 (the upper vertex), and finally end up in state Φ_{ij}. Here i, j are passive sp states, namely those outside the model space P, and $\alpha, \beta, \gamma, \delta$ are sp states within P.

The concept of folding can be illustrated by the following factorization operation. The time integral contained in diagram (A) has the limits $0 > t_1 > t_2 > -\infty$, i.e. $\int_{-\infty}^{0} dt_1 \int_{-\infty}^{t_1} dt_2$. For (B), the integrations over t_1 and t_2 are independent, both from $-\infty$ to 0. Note that the integrands of (A), (B) and (C) are identical. Clearly (A) is not equal to (B). The folded diagram in this case is defined as the "error" due to the factorization of (A) into (B), namely diagram (C) is the folded diagram given by

$$(A) = (B) - (C). \tag{11}$$

In fact (A), (B), (C) have the values

$$(A) = |\Phi_{ij}\rangle \frac{V_{ij,\gamma\delta} V_{\gamma\delta,\alpha\beta}}{(\epsilon_\alpha + \epsilon_\beta - \epsilon_i - \epsilon_j)(\epsilon_\alpha + \epsilon_\beta - \epsilon_\gamma - \epsilon_\delta)}, \tag{12}$$

$$(B) = |\Phi_{ij}\rangle \frac{V_{ij,\gamma\delta} V_{\gamma\delta,\alpha\beta}}{(\epsilon_\gamma + \epsilon_\delta - \epsilon_i - \epsilon_j)(\epsilon_\alpha + \epsilon_\beta - \epsilon_\gamma - \epsilon_\delta)}, \tag{13}$$

$$(C) = |\Phi_{ij}\rangle \frac{V_{ij,\gamma\delta} V_{\gamma\delta,\alpha\beta}}{(\epsilon_\gamma + \epsilon_\delta - \epsilon_i - \epsilon_j)(\epsilon_\alpha + \epsilon_\beta - \epsilon_i - \epsilon_j)}. \tag{14}$$

When we have a degenerate model space, then $(\epsilon_\alpha + \epsilon_\beta) = (\epsilon_\gamma + \epsilon_\delta)$ and (A) and (B) both become divergent. It is of interest that the folded diagram (C) is still well defined in this case.

Using the above types of folded-diagram factorizations, one can obtain an energy-independent effective interaction given by Eq.(8). It is a formally exact result. It is a folded-diagram expansion for the model-space effective interaction, which is energy-independent and is valence-linked. This expression will not be of any use, if it can not be calculated in a convenient way. We shall demonstrate

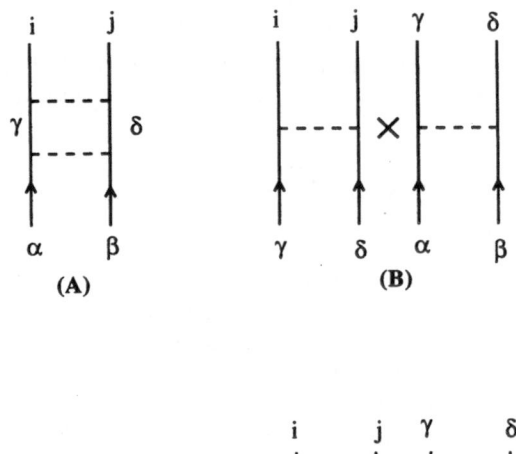

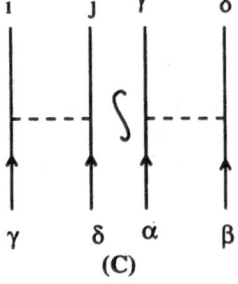

Fig. 1. A folded-diagram factorization.

that this series can in fact be summed up to all orders using iteration methods. Let us first explain the various terms in the equation. Each integral sign in the above denotes a "fold" (Kuo *et al.* 1971, Kuo and Osnes 1990). For instance the last term in the equation is a three-fold term; it has three integral signs. $\hat{Q}$ represents a so-called $\hat{Q}$-box, which may be schematically written as

$$\hat{Q}(\omega) = [PVP + PVQ\frac{1}{\omega - QHQ}QVP]_{linked}. \tag{15}$$

Note that we use Q, without hat, to denote the Q-space projection operator. In fact $\hat{Q}$-box is a irreducible vertex function where the intermediate states between any two vertices must belong to the Q space. It contains linked diagrams only, as indicated by the subscript linked. The $\hat{Q}'$-box of Eq.(8) is defined as ($\hat{Q}$ -PVP). Let us rewrite the expansion of Eq.(8) as

$$V_{eff} = F_0 + F_1 + F_2 + F_3 + F_4 + \cdots \tag{16}$$

where F_n denotes all the diagrams with n folds.

If the model space is degenerate, i.e. $PH_0P=W_0$ which is a constant, then the various F terms can be conveniently written in derivative form as

$$F_0 = \hat{Q}$$
$$F_1 = \hat{Q}_1\hat{Q}$$
$$F_2 = \hat{Q}_2\hat{Q}\hat{Q} + \hat{Q}_1\hat{Q}$$
$$F_3 = \hat{Q}_3\hat{Q}\hat{Q}\hat{Q} + \hat{Q}_2\hat{Q}_1\hat{Q}\hat{Q} + \hat{Q}_2\hat{Q}\hat{Q}_1\hat{Q} + \hat{Q}_1\hat{Q}_2\hat{Q}\hat{Q} + \hat{Q}_1\hat{Q}_1\hat{Q}_1\hat{Q}$$
$$\vdots$$

$$(17)$$

where

$$\hat{Q}_n = \frac{1}{n!}\frac{d^n\hat{Q}}{d\omega^n}\Big|_{\omega=W_0} \tag{18}$$

Note that to calculate the various F's we only need to know the $\hat{Q}$-box and its derivatives, all evaluated at W_0. The number of terms in F_n grows very rapidly with n. For example, there are 14 terms in F_4. Thus, it is not practical to calculate V_{eff} using the above expansion, when one wants to include terms with a large number of folds.

A number of authors (Krenciglowa and Kuo 1974, Lee and Suzuki 1980, Suzuki and Lee 1980) have developed iteration methods for the summation of the above folded-diagram series. However, these methods were specifically designed for the case of a degenerate model space. In general, PH_0P may not be degenerate. For example the well-known sd-shell of the nuclear shell model is not degenerate. (It has been a common practice in treating the sd shell as degenerate in deriving its effective interactions.) Only recently non-degenerate iteration methods (Suzuki et al. 1994, Kuo et al. 1995) for the folded-diagram series have been developed. Let us discuss in the following several non-degenerate methods for summing up the folded diagrams.

3   Lee-Suzuki iteration method

The Lee-Suzuki iteration methods (Lee and Suzuki 1980, Suzuki and Lee 1980) were originally designed for a degenerate model space. We shall demonstrate here that by a simple redefinition of the unperturbed Hamiltonian, their methods can also be used to treat the non-degenerate situation. It is useful to review their basic fomulation, which is based on a similarity transformation. One starts from a many-body Shrödinger equation $H\Psi = E\Psi$ and makes a similarity transformation, leading to

$$\mathcal{H} = X^{-1}HX, \tag{19}$$

$$\begin{pmatrix} P\mathcal{H}P & P\mathcal{H}Q \\ Q\mathcal{H}P & Q\mathcal{H}Q \end{pmatrix} \begin{pmatrix} PX^{-1}\Psi \\ QX^{-1}\Psi \end{pmatrix} = E \begin{pmatrix} PX^{-1}\Psi \\ QX^{-1}\Psi \end{pmatrix}. \tag{20}$$

If we choose X which fullfills the decoupling condition

$$Q\mathcal{H}P = 0, \tag{21}$$

then we have

$$PHP|PX^{-1}\Psi> = E|PX^{-1}\Psi>. \tag{22}$$

This means the model-space effective Hamiltonian H_{eff} is just PHP. Thus to derive H_{eff}, a main step is the derivation of X, which satisfies the above decoupling condition. Suzuki and Lee considered $X = \exp\omega$ where ω, a wave operator, is chosen to have a special property, namely $\omega = Q\omega P$. This implies $X = 1 + \omega$ and $X^{-1} = 1 - \omega$ as by definition $\omega^n = 0$ for $n > 1$. With this choice, the decoupling condition of Eq.(21) becomes

$$QHP - \omega HP + QH\omega - \omega H\omega = 0 \tag{23}$$

and

$$H_{eff} = PH_0P + PVP + PV\omega = PH_0P + V_{eff}. \tag{24}$$

To obtain V_{eff} we need to solve for ω from Eq.(23), which is a non-linear equation. Up to now, we have not imposed any restriction on H_0; it can be either degenerate or non-degenerate. In the following we consider the case of a non-degenerate PH_0P. Let W_0 be a constant. We add and subtract $W_0\omega$ to the decoupling equation, obtaining

$$QHP = (W_0 - QH)\,\omega + \omega\,(H + H\omega - W_0)\,P. \tag{25}$$

Note that W_0 is a constant which therefore commutes with ω. As we discuss later, W_0 is at our disposal and we can choose, in principle, any value for W_0. We can rewrite the above equation as

$$\omega = \frac{1}{W_0 - QHQ}QVP - \frac{1}{W_0 - QHQ}\omega\,(H + V\omega - W_0)\,P. \tag{26}$$

The above is of a form suitable for an iterative solution for ω. Of course there is more than one way to solve it. Following Suzuki and Lee (Lee and Suzuki 1980, Suzuki and Lee 1980) let us use the following iteration scheme

$$\omega_n = \frac{1}{W_0 - QHQ}QVP - \frac{1}{W_0 - QHQ}\omega_{n-1}\left(H_0 - W_0 + V_{eff}^{(n)}\right). \tag{27}$$

Clearly if we have ω_{n-1}, we can calculate ω_n and $V_{eff}^{(n)}$ and continue to iterate until convergence.

Now we multiply, from the left, both sides of the above equation by PV, and add $(H_0 - W_0)$ to each side. This gives

$$\left[H_0 - W_0 + V_{eff}^{(n)}\right] = \left[1 + PV\frac{1}{W_0 - QHQ}\omega_{n-1}P\right]^{-1}\left[H_0 - W_0 + \hat{Q}(W_0)\right], \tag{28}$$

$$\hat{Q}(\omega) = PVP + PVQ\frac{1}{\omega - QHQ}QVP. \tag{29}$$

For simplicity, let us define

$$R_n \equiv H_0 - W_0 + V_{eff}^{(n)}, \tag{30}$$

$$\frac{Q}{e} \equiv Q \frac{1}{W_0 - QHQ} Q. \tag{31}$$

Starting from $\omega_0 = 1$ and $\omega_1 = \frac{Q}{e} V P$, we have from Eqs.(27) and (28)

$$R_2 = \frac{1}{1 - \frac{d\hat{Q}}{d\omega} |_{\omega=W_0}} \times \left[H_0 - W_0 + \hat{Q}(W_0) \right], \tag{32}$$

$$PV \frac{Q}{e} \omega_2 = PV \frac{Q}{e^2} VP - PV \frac{Q}{e^2} \frac{Q}{e} VPR_2. \tag{33}$$

Continuing this iteration process, we obtain

$$R_n = \frac{1}{1 - \hat{Q}_1 - \sum_{m=2}^{n-1} \hat{Q}_m \prod_{k=n-m+1}^{n-1} R_k} \left[H_0 - W_0 + \hat{Q}(W_0) \right], \tag{34}$$

$$\hat{Q}_m = \frac{1}{m!} \frac{d^m \hat{Q}}{d\omega^m} |_{\omega=W_0} \tag{35}$$

In the degenerate case, we may choose $W_0 = PH_0P$ and the above just reduces to the degenerate Lee-Suzuki method (Lee and Suzuki 1980, Suzuki and Lee 1980). What we have done in the above is in fact just a redefinition of the unperturbed Hamiltonian. We begin with $H = H_0 + H_1$. Now we use a degenerate unperturbed Hamiltonian given by $PH_0'P=W_0$. Then H_1 must change accordingly, namely H is now given by W_0+H_1' with $H_1' = [H_0-W_0+H_1]$. If the original H_0 is nearly degenerate, then we can choose a W_0 in its vicinity. In this case $H_0 - W_0$ will be generally small and its presence in Eq.(34) will not adversely affect the convergence of the iteration. But if the non-degeneracy of H_0 is large, then no matter what value we use for W_0, some elements of $H_0 - W_0$ will be large and they will affect the iteration. From studies of model problems, it has been found that the iteration often does not converge in this situation.

In short, when the non-degeneracy of the model space is small, in the sense that PH_0P is confined within a small energy range, the non-degenerate iteration methods for the folded diagrams can still be used with a simple modification of H_0 as discussed above. But for the general case where the non-degeneracy of PH_0P is large, some other iteration methods are required.

4 Non-degenerate iteration methods

We consider here a general non-degenerate model space, namely PH_0P is non-degenerate. In addition to the $\hat{Q}$-box of Eq.(15), we now also need to define a generalized $\hat{Q}$-box as

$$\hat{Q}_n(\epsilon_1\epsilon_2\cdots\epsilon_{n+1}) = (-1)^n[PVQ\frac{1}{\epsilon_1 - QHQ}\frac{1}{\epsilon_2 - QHQ}\cdots\frac{1}{\epsilon_{n+1} - QHQ}QVP]_L.$$
(36)

Note that $\hat{Q}_n$ is defined for $n \geq 1$. In the above there is also the restriction that only the valence-linked diagrams are retained, as indicated by the subscript L.

For the non-degenerate case, the various folded terms of Eq.(8) can no longer be expressed as simple derivatives of the $\hat{Q}$-box as in Eq.(17). But using the above generalized $\hat{Q}$-box, the various folded terms of the effective interaction can still be expressed (Suzuki *et al.* 1994) in a systematic way as

$$F_0 = \sum_b \hat{Q}(\epsilon_b) P_b,$$

$$F_1 = \sum_{bm} \hat{Q}_1(\epsilon_b\epsilon_m) P_m \hat{Q}(\epsilon_b) P_b,$$

$$F_2 = \sum_{bmn} \Big[\hat{Q}_2(\epsilon_b\epsilon_m\epsilon_n) P_m \hat{Q}(\epsilon_n) P_n \hat{Q}(\epsilon_b) P_b$$

$$+\hat{Q}_1(\epsilon_b\epsilon_m) P_m \hat{Q}_1(\epsilon_b\epsilon_n) P_n \hat{Q}(\epsilon_b) P_b\Big]$$

$$F_3 = \hat{Q}_3\hat{Q}\hat{Q}\hat{Q} + \hat{Q}_2\hat{Q}_1\hat{Q}\hat{Q} + \hat{Q}_2\hat{Q}\hat{Q}_1\hat{Q} + \hat{Q}_1\hat{Q}_2\hat{Q}\hat{Q} + \hat{Q}_1\hat{Q}_1\hat{Q}_1\hat{Q},$$
(37)

$$\cdot$$
$$\cdot$$
$$\cdot \, ,$$

where for simplicity we have dropped the subscript L which indicates that each F_n term is valence linked. For the same reason, the energy variables and the projection operators have been also omitted in the expression for F_3. For example the term $\hat{Q}_3\hat{Q}\hat{Q}\hat{Q}$ actually means $\sum_{bmns} \hat{Q}_3(\epsilon_b\epsilon_m\epsilon_n\epsilon_s) P_m \hat{Q}(\epsilon_n) P_n \hat{Q}(\epsilon_s) P_s \hat{Q}(\epsilon_b) P_b$. Here the P's are the projection operator for individual model-space states. For example, $P_m = |\Phi_m\rangle\langle\Phi_m|$. The diagrammatic structure of the various terms is rather simple. For example the twice-folded term $\hat{Q}_1(\epsilon_b\epsilon_m) P_m \hat{Q}_1(\epsilon_b\epsilon_n) P_n \hat{Q}(\epsilon_b)$ P_b corresponds to the time-dependent diagram in the upper part of Fig. 2. Similarly the twice-folded term $\hat{Q}_2(\epsilon_b\epsilon_m\epsilon_n) P_m \hat{Q}(\epsilon_n) P_n \hat{Q}(\epsilon_b) P_b$ corresponds to the lower diagram of Fig. 2.

The above folded-diagram expansion appears to be rather difficult for computation. But actually its calculation is only slightly more complicated than the degenerate case. It can be summed up using the generalized Lee-Suzuki (Suzuki *et al.* 1994) method:

$$R_1 = \sum_\alpha \hat{Q}(\epsilon_\alpha)P_\alpha,$$

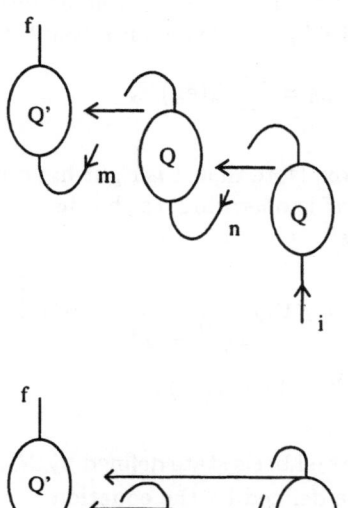

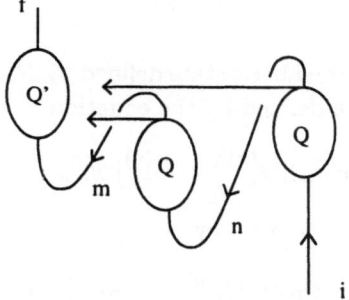

Fig. 2. Q-box folded-diagrams.

$$R_2 = \sum_{\alpha} \left[1 - \sum_{\beta} \hat{Q}_1 \left(\epsilon_{\alpha} \epsilon_{\beta} \right) P_{\beta} \right]^{-1} \hat{Q}(\epsilon_{\alpha}) P_{\alpha},$$

$$R_3 = \sum_{\alpha} \left[1 - \sum_{\beta} \hat{Q}_1 \left(\epsilon_{\alpha} \epsilon_{\beta} \right) P_{\beta} - \sum_{\beta\gamma} \hat{Q}_2 \left(\epsilon_{\alpha} \epsilon_{\beta} \epsilon_{\gamma} \right) P_{\beta} R_2 P_{\gamma} \right]^{-1} \hat{Q}(\epsilon_{\alpha}) P_{\alpha},$$

.

.

.

$$(38)$$

Comparing with the degenerate method, the new ingredient is the generalized $\hat{Q}$-box given by Eq.(36). The generalized $\hat{Q}$-box can be evaluated in terms of the ordinary $\hat{Q}$-box and its derivatives, using a partial fraction method. When PH_0P is degenerate, the generalized $\hat{Q}$-box becomes the energy derivatives of the ordinary $\hat{Q}$-box, and we recover the degenerate iteration method of Eq.(34).

There is another method for summing up the non-degenerate folded-diagram series of Eq.(39), which may be referred to as a generalised Krenciglowa-Kuo iteration method (Kuo *et al.* 1995). We define an initial S-box, S_0 as

$$S_0 = \sum_\alpha \hat{Q}(\varepsilon_\alpha) P_\alpha \tag{39}$$

which is just the ordinary $\hat{Q}$-box. Note that it is right-hand-side on-shell, namely its energy variable is ε_α when it is operating, to the right, on a P-space state α. Then the iteration proceeds as

$$S_{n+1}| \chi_k^{(n)}\rangle = \sum_\alpha \left[PVP + PVQ\frac{1}{E_k^{(n)} - QHQ}QVP \right]_L |\alpha\rangle \langle\alpha| \chi_k^{(n)}\rangle$$
$$= \sum_\alpha \hat{Q}^L(E_k^{(n)}) |\alpha\rangle \langle\alpha| \chi_k^{(n)}\rangle , \tag{40}$$

where $|\alpha\rangle$ denotes the unperturbed basis state defined by H_0. The wave functions $|\chi_k^{(n)}\rangle$ and eigenvalues $E_k^{(n)}$ are defined by the equation

$$(PH_0P + S_n)| \chi_k^{(n)}\rangle = E_k^{(n)}| \chi_k^{(n)}\rangle . \tag{41}$$

The matrix element of S_{n+1} is given by

$$\langle\alpha| S_{n+1} |\beta\rangle = \sum_k \sum_\gamma \langle\alpha| \hat{Q}^L(E_k^{(n)}) |\gamma\rangle \langle\gamma| \chi_k^{(n)}\rangle \langle\tilde{\chi}_k^{(n)}| \beta\rangle \tag{42}$$

where $\langle\tilde{\chi}_k^{(n)}|$ is the biorthogonal state of $|\chi_k^{(n)}\rangle$. If the iteration converges, that is when $S_{n+1}=S_n$, the matrix element of V_{eff} is finally given by

$$\langle\alpha| V_{eff} |\beta\rangle = \sum_k \sum_\gamma \langle\alpha| \hat{Q}^L(E_k) |\gamma\rangle \langle\gamma| \chi_k\rangle \langle\tilde{\chi}_k| \beta\rangle \tag{43}$$

where $E_k = E_k^{(n)} = E_k^{(n+1)}$ and $|\chi_k\rangle = |\chi_k^{(n)}\rangle = |\chi_k^{(n+1)}\rangle$.

The above iteration method is very similar to the original Krenciglowa-Kuo method (Krenciglowa and Kuo 1974); they differ only in the starting point for the iteration. Here the starting point is S_0 of Eq.(39), which is given by the $\hat{Q}$-box evaluated at various P-space energies. In the degenerate case, S_0 is given by the $\hat{Q}$-box evaluated at a common energy W_0.

5   The G-matrix

The folded-diagram methods described earlier provide us a formal framework to derive the effective interaction needed for the shell-model calculation of any chosen model space. In shell model calculations, it is often necessary to employ different model spaces. For example we may use either a four-orbit ($f_{7/2}, f_{5/2}, p_{3/2}, p_{1/2}$) model space or a five-orbit one which includes in addition also the $g_{9/2}$ orbit. The above folded-diagram methods can in principle tell us what are the differences between the V_{eff} for these two choices for the model space.

A first step in calculating the V_{eff} using the folded-diagram methods is the G-matrix. Nucleon-nucleon potentials (V) all have a strong repulsive core and this poses a computational difficulty. Suppose a,b,c,d are standard shell model single-particle orbits, then the matrix elements $< ab|V|cd >$ are typically all very large and repulsive, making perturbation calculations in terms of V meaningless. As is well known, this difficulty can be overcome by way of the Brueckner G-matrix, which was originally designed for nuclear matter calculations.

For finite nuclei, the model-space G matrix is defined by the integral equation (Wong 1967, Tsai and Kuo 1972, Krenglowa *et al.* 1976, Muether and Sauer 1992, Maglione and Ferreira 1994)

$$G(\omega) = V + VQ_2\frac{1}{\omega - Q_2TQ_2}Q_2G(\omega) \tag{44}$$

where Q_2 is the Pauli exclusion operator for the two interacting nucleons, to make sure that the intermediate states of G must not only be above the filled Fermi sea but also be outside the model space within which the model-space secular equation is to be solved. As we shall discuss later, the solution of the above G-matrix equation is rather complicated, mainly because of the presence of the Pauli exclusion operator Q_2.

Recall that we have added and subtracted a single particle potential U to the Hamiltonian, namely $H = T + V = (T + U) + (V - U) = H_0 + H_1$. Hence the G-matrix contains repeated H_1 interactions, between a pair of particles, of the form

$$\cdots + H_1\frac{Q_2}{\omega - H_0}H_1\frac{Q_2}{\omega - H_0}H_1\frac{Q_2}{\omega - H_0}H_1\frac{Q_2}{\omega - H_0}H_1\frac{Q_2}{\omega - H_0}H_1 + \cdots \tag{45}$$

Note that the denominators have H_0 which is (T+U), and the intermediate interactions are H_1 which is (V-U). The -U interactions for the intermediate states can, however, be resummed to all orders, cancelling the U in the denominators. This leads to propagators of the form

$$Q_2\frac{1}{\omega - Q_2TQ_2}Q_2VQ_2\frac{1}{\omega - Q_2TQ_2}Q_2. \tag{46}$$

Clearly with all the intermediate -U interactions resummed to all orders we obtain the G-matrix of Eq(44). If we do not include these -U interactions for the intermediate states, we may have a different G-matrix, denoted as G_B, given by

$$G_B(\omega) = V + V\frac{Q_2}{\omega - H_0}G_B(\omega). \tag{47}$$

Currently both of the above G-matrices, G and G_B, are being used in microscopic calculations for the effective interaction. Several groups (Wong 1967, Krenglowa *et al.* 1976, Muether and Sauer 1992, Maglione and Ferreira 1994,

Jiang *et al.* 1989, Shurpin *et al.* 1983, Hjorth-Jensen and Osnes 1989, Hjorth-Jensen *et al.* 1992) have used the former G-matrix with a QTQ propagator. On the other hand, Barrett and his collaborators (Barrett *et al.* 1971, Zheng *et al.* 1994) have chosen to employ the latter G-matrix G_B, with H_0 taken as an oscillator Hamiltonian (with a shift constant). There are significant differences between the G-matrices of Eq.(44) and Eq.(47). For the former, the intermediate states are orthogonalized plane waves, while for the latter the intermediate states for the G-matrix are entirely oscillator waves. Because of the strong repulsive core contained in the NN interaction, the induced intermediate states in the G-matrix are predominately of rather high momentum. By physical consideration, these states should essentially be plane-wave states, and thus the former G-matrix appears to be physically more appropriate. In addition, for the G_B-matrix the contribution from the -U interactions to the intermediate states remains to be calculated.

A principal difficulty in solving Eq.(44) for the plane-wave G-matrix is about the Pauli exclusion operator Q_2. This operator commutes with H_0 but does not commute with the kinetic-energy operator T, as Q_2 is defined in terms of the eigenfunctions of H_0. In earlier calculations (Wong 1967, Muether and Sauer 1992) one had to make some approximations in treating Q_2, mainly in taking it to be diagonal in the relative and center-of-mass representation. (Note that for the G_B case, the treatment of Q_2 is relatively easier as the propagator here is taken as $\frac{1}{\omega - H_0}$ and Q_2 commutes with H_0.)

Using the Tsai-Kuo method (Tsai and Kuo 1972, Krenglowa *et al.* 1976) an accurate calculation of the plane-wave G-matrix is now feasible. This method gives the exact solution of G as

$$G = G_F + \Delta G. \tag{48}$$

The "free" G-matrix is

$$G_F(\omega) = V + V \frac{1}{\omega - T} G_F(\omega), \tag{49}$$

and the Pauli correction term ΔG is given by

$$\Delta G(\omega) = -G_F(\omega) \frac{1}{e} P_2 \frac{1}{P_2[1/e + (1/e)G_F(\omega)(1/e)]P_2} P_2 \frac{1}{e} G_F(\omega) \tag{50}$$

where $e = \omega - T$ with T being the kinetic energy operator for the two free particles, and ω the so-called starting energy. The projection operator P_2 is defined as $(1-Q_2)$, and Q_2 is specified by a set of numbers (n_1, n_2, n_3) (Krenglowa *et al.* 1976) each representing a shell-model orbital, as to be discussed later. In deriving the above result, one has made use of the matrix identity (Tsai and Kuo 1972)

$$Q \frac{1}{QAQ} Q = \frac{1}{A} - \frac{1}{A} P \frac{1}{P\frac{1}{A}P} \frac{1}{A} \tag{51}$$

where A represents a general matrix, and P and Q are projection operators for two complementary subspaces with P+Q=1.

The calculation of G_F is straightforward, as there is no Pauli exclusion operator. To compute the Pauli correction term, we need the free G-matrix G_F and perform some matrix operations within the model space P_2. We write the projection operator Q_2 as

$$Q_2 = \sum_{all\ ab} Q(ab)|ab\rangle\langle ab|, \tag{52}$$

where $Q(ab) = 0$, if $b \leq n_1, a \leq n_3$ or $b \leq n_2, a \leq n_2$ or $b \leq n_3, a \leq n_1$ and $Q(ab) = 1$ otherwise. The boundary of $Q(ab)$ is specified by the orbital numbers (n_1, n_2, n_3). We denote the shell model orbits by numerals, starting from the bottom of the oscillator well, 1 for orbit $0s_{1/2}$, 2 for $0p_{3/2}$, $\cdots$ 7 for $0f_{7/2}$ and so on. n_1 denotes the highest orbit of the closed core (Fermi sea). n_2 denotes the highest orbit of the chosen model space. Suppose we are calculating ^{18}O with ^{16}O treated as a closed core. Then we have $n_1 = 3$. If we use a model space including all the 6 orbits in the s, p and sd shells. Then for this case $n_2 = 6$. In principle one should take $n_3 = \infty$ (Krenglowa *et al.* 1976), because we include only particle (above Fermi sea) states for the G-matrix intermediate states. For instance, any intermediate states with one nucleon in the $0p_{3/2}$ orbit is disallowed for the G-matrix. In practice, this $n_3 = \infty$ requirement is not feasible, and one can only use a large n_3 determined by an empirical procedure. Namely, we perform calculations with increasing values for n_3 until numerical results become stable.

It has been found that when n_3 is sufficiently large, G is approaching a stable value. We give some representative results in Table 1. (Here the values used for n_1 and n_2 are respectively 3 and 6.) As shown, the values of G appear to be quite stable after $n_3=21$. Thus a cutoff of $n_3=21$ should provide an accurate calculation for the G-matrix. In solving the G-matrix from Eq.(44), we need to make only one approximation, namely the finite-n_3 approximation. This approximation appears to be adequately accurate if a large n_3 is employed.

The above results are supported by the following qualitative argument, that the contributions from large n_3 states are supressed by momentum conservation. The purpose of having large n_3 is to disallow the G-matrix from having intermediate states with one nucleon in the n_3 orbit and the other nucleon below the Fermi sea. Such states usually have rather high momentum. Since the external lines of G are within the model space which has usually relatively low momentum, the contribution from the above type of intermediate states is supressed by momentum conservation when n_3 is very large.

It may be pointed out that the effect of the Pauli operator is usually quite important. For example, the (045451), $\omega = -5$ matrix element of Table 1 is -18.0522 MeV for the G_F case, while it becomes -5.1594 MeV with the Pauli correction included. This indicates that it is essential to treat the Pauli operator for finite nuclei as accurately as one can. In short, practically for any chosen model space one can now calculate the G-matrix for finite nuclei in an accurate way.

Table 1. Typical n_3 dependence of the G-matrix of Eq.(4). Listed are the matrix elements $\langle abJT|G(\omega)|cdJT\rangle$, in MeV. The orbitals (4,5) represent respectively $(0d_{5/2}, 0d_{3/2})$. The first row in each group is the free-space G matrix G_F. The next 4 rows are obtained with various n_3 values. The results shown are obtained with the Paris potential, $\hbar\omega=14$ MeV, and starting energy $\omega=-10$ MeV.

T	a	b	c	d	J	n_3	-5	-10	ω -20	-40	-85
0	4	4	4	4	1		-3.0405	-1.8427	-1.0603	-0.4344	+0.1898
0	4	4	4	4	1	10	-0.3625	-0.2870	-0.1555	+0.0607	+0.4222
0	4	4	4	4	1	15	-0.3590	-0.2843	-0.1535	+0.0621	+0.4232
0	4	4	4	4	1	21	-0.3474	-0.2742	-0.1457	+0.0672	+0.4258
0	4	4	4	4	1	28	-0.3474	-0.2742	-0.1457	+0.0672	+0.4258
0	4	5	4	5	1		-18.0522	-12.0828	-8.6647	-6.4159	-4.6291
0	4	5	4	5	1	10	-5.3884	-5.2321	-4.9698	-4.5456	-3.8412
0	4	5	4	5	1	15	-5.3003	-5.1592	-4.9148	-4.5091	-3.8220
0	4	5	4	5	1	21	-5.1594	-5.0348	-4.8127	-4.4341	-3.7768
0	4	5	4	5	1	28	-5.1594	-5.0348	-4.8127	-4.4341	-3.7768

6   Effective interaction for halo nuclei

There are three main steps in carrying out microscopic calculations of the effective interaction. First we choose a model space P. Then the model-space effective interaction is formally given by the $\hat{Q}$-box folded-diagram series of Eq.(8). To proceed, we need to evaluate the irreducible vertex function $\hat{Q}$-box starting from the underlying NN interaction V. Because of the short range repulsion contained in V, we need to "convert" V into the G-matrix. This conversion corresponds to a partial summation, to all orders, of certain $\hat{Q}$-box diagrams of a common structure so that each irreducible diagram is expressed in terms of G-matrix vertices. For example, diagrams of type (a) of Fig. 3 lead to the G-matrix diagram (a'). And those of type (b) leads to the well known core polarization diagram (b'). Diagrams (a') and (b') are often referred to as G and G_{3p1h} respectively. Thus the second main step in microscopic effective interaction calculations is the evaluation of the G-matrix. As demonstrated earlier, this step can now be performed highly accurately. In other words, starting from a chosen NN potential, we now can calculate accurately each individual $\hat{Q}$-box diagram in terms of the G-matrix vertices.

In most calculations (Jiang et al. 1989, Shurpin et al. 1983, Dean et al. 1996, Radha et al. 1996, Covello et al. 1996), the effective interaction is obtained with a $\hat{Q}$-box including only the one- and two-body diagrams first- and second-order in G. Although the results obtained are generally satisfactory, one is still concerned about the importance of higher-order $\hat{Q}$-box diagrams. Indeed the study of such

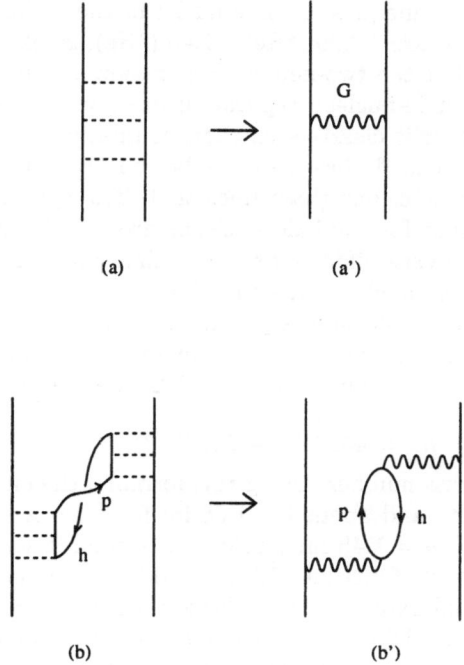

(a)   (a')

(b)   (b')

Fig. 3. Q-box diagrams with bare interaction and G-matrix interaction.

higher order diagrams remains to be a major to-be-resolved problem in the microscopic effective interaction theory. Osnes and his collaborators (Hjorth-Jensen and Osnes 1989, Hjorth-Jensen *et al.* 1992) have calculated the third-order, in G, $\hat{Q}$-box diagrams. Their calculation is already quite complicated and is an extensive project; it would require a major further effort to calculate the $\hat{Q}$-box diagrams beyond third order.

Recently there has been much interest in halo nuclei, and a two-frequency shell model (TFSM) approach has been employed by several authors (Kuo *et al.* (1996), Kuo *et al.* 1996) to study the structure of halo nuclei. Following their approach, let us discuss here the effective interaction for halo nuclei. And it seems likely that one may be able to calculate the effective interaction for halo nuclei more acurately than for ordinary nuclei.

In ordinary nuclei, the valence nucleons are rather close to the core, and hence the effect of core polarization is expected to be large. For sd-shell nuclei, the magnitude of the core polarization diagram G_{3p1h} is quite comparable to the bare G diagram (Jiang *et al.* 1989, Shurpin *et al.* 1983). For example, G is -1.72 MeV while G_{3p1h} is -1.10 MeV for the diagonal T=1, J=0 $d_{5/2}^2$ matrix element as given in Ref. (Jiang *et al.* 1989).

The physical environment in halo nuclei is different. Here the halo nucleons are "far away" from the core. An example is the valence neutrons in ^6He. Let us also look at the binding energies of some nuclei in the vicinity of ^6He. The observed values (Tuli 1995) are: 28.29 (^{4}He), 27.40 (^{5}He), 26.32 (^{5}Li), and 29.26 (^{6}He), in MeV. We see that the two-neutron separation energy for ^{6}He is only about 1 MeV, and the single-nucleon separation energies for ^{5}He and ^{5}Li are both negative (unbound). It is clearly seen that, comparing to ordinary nuclei, ^{6}He is a very "special" breed. It has a tightly-bound ^{4}He core with a pair of very loosely bound outer nucleons (halo nucleons). The spatial extent of the core wave function and that for the halo nucleons are drastically different. For this situation, it would be very difficult for the ordinary shell model to give an adequate description for the nuclear wave function.

In ordinary shell model, it is customary to use a common $\hbar\omega$ value for all the wave functions. To fit the observed rms radii for ordinary (stable) nuclei, one usually employs an empirical formula (Bertsch 1972) to determine the oscillator parameter $\hbar\omega$, namely

$$\hbar\omega = 45A^{-1/3} - 25A^{-2/3} \qquad (53)$$

where A is the nuclear mass number. Using this formula, the empirical value of b is about 1.45 fm for ^4He and about b= 1.55 fm for ^6He. A simple $0s^4$ shell model wave function with $b = 1.45$ fm would give a good description for ^4He. But a simple $0p - shell$ wave function with $b = 1.55fm$ is by far inadequate in describing the large spatial extent of the valence nucleons in ^6He. Thus when using the ordinary shell model for halo nuclei, one needs to include several major shells in order to provide an adequate description for the wave function of the halo nucleons. Such multi-shell calculations are quite complicated.

In the TFSM approach for halo nuclei, one adopts a short cut by employing single particle wave functions of two different oscillator constants $\hbar\omega_{in}$ and $\hbar\omega_{out}$ (or length parameters b_{in} and b_{out}), the former for the inner orbits while the latter for the halo (outer) orbits. This provides an economic and physically reasonable way for describing the halo nuclei. To illustrate, let us consider the halo nucleus ^6He. It is reasonable to consider its "core" as an ordinary alpha particle. Hence we take in the present work the $0s_{1/2}$ orbit as an inner orbit with oscillator length parameter b_{in}=1.45 fm. The outer, or halo, nucleons of ^6He are spatially extended, and for them we use the halo orbits $0p_{3/2}$ and $0p_{1/2}$ with oscillator constant $\hbar\omega_{out}$ or b_{out}, which is treated as a variation parameter. Clearly one needs to have all the sp wave functions $\{\phi^{in}, \phi^{out}\}$ be orthonormal to each other, and one may be concerned as to how can one fulfill this condition, as b_{in} is in general not equal to b_{out}. To fullfill this requirement, we need to use a common b value for all the orbits with the same l and j values. Thus, for our present ^6He case, we use b_{in} not only for the $0s_{1/2}$ orbit but also for all the other $s_{1/2}$ orbits. (Note that the $ns_{1/2}$ orbits, with $n > 0$, are needed for the intermediate states in the evaluation of the G-matrix.)

An advantage of this TFSM approach is that the G-matrix for halo nuclei can also be calculated accurately, using the method described in Section 5. The calculation is more complicated than for ordinary nuclei, because we now have wave functions of two different oscillator parameters. We have calculated the effective interaction for the two halo neutrons of ^6He using several b_{out} values, with b_{in} fixed as 1.45 fm.

An interesting result is that the magnitude of the core polarization diagram G_{3p1h} is rapidly deminishing as b_{out} increases. (Kuo et al. 1996) For example, for b_{out}=1.45, 1.75, 2.00 and 2.25 fm, the diagonal $(p^2_{3/2}, T = 1, J = 0)$ G_{3p1h} has values -.625, -.248, -.128, and -.073 MeV respectively. They were obtained with the Bonn-A potential, and with b_{in} fixed as 1.45 fm.

In ordinary nuclei, the valence nucleons are close to the nuclear core. There is a strong valence-core coupling and hence a large core polarization effect. The halo nucleons in a halo nucleus are located quite far away from the core. As we increase b_{out}, we are increasing the average distance between the halo nucleons and the core and so reducing the coupling between them. For sufficiently large b_{out}, the total core polarization effect must be small and it should be sufficiently accurately given by the second-order (lowest order) core polarization diagram alone. This is a very promising situation, implying that the effective interaction for halo nucleons can be rather accurately calculated by including only the first-order G-matrix diagram and the lowest order core polarization diagram.

¿From mass tables (Tuli 1995), the valence interaction energy for ^6He is

$$E_v^{exp}(^6He) = -[BE(^6He) + BE(^4He) - 2BE(^5He)] = 2.77MeV. \tag{54}$$

We have calculated this valence energy, by diagonalizing a two-neutron ($T = 1, J = 0$) matrix in a $p_{3/2}$-$p_{1/2}$ space, using the folded-diagram effective interaction described earlier. Our result is $E_v^{th} = -2.77$ MeV at $b_{out} = 2.25$ fm and $b_{in} = 1.45fm$ for the Paris potential. As the core polarization effect is strongly suppressed for such a b_{out} value, this result almost entirely comes from the bare G-matrix. Recall that we have fixed $b_{in} = 1.45$ fm. With these values of b_{in} and b_{out} and assuming a pure s^4p^n wave function, our calculated rms radius is $R^{th}(^6He) = 2.51$ fm, in good agreement with the empirical value $R^{exp}(^6He) = 2.57 \pm 0.1$ fm (Zhukov et al. 1993).

7   Summary and outlook

Our aim has been to derive the shell-model effective interaction starting from the basic NN potentials. This is clearly not an easy task, although some progress has been made. We have shown that a general many-body problem specified by Hamiltonian H can be formally reduced to a model-space problem defined by $PH_{eff}P$. The model-space Hamiltonian is given by $(H_0 + V_{eff})$ where the effective interaction V_{eff} is given by the $\hat{Q}$-box folded-diagram series of Eq.(8). This approach has been widely applied to ordinary nuclei. Recently it has also been applied to hypernuclei (Hao et al. 1993, Kuo and Hao 1994) and to halo nuclei (Kuo et al. (1996), Kuo et al. 1996). Previously the calculated folded-diagram effective interaction was non-hermitian. Some progress towards the construction of a realistic hermitian effective interaction has been recently achieved (Suzuki 1982, Suzuki and Okamoto 1983, Kuo et al. 1993, Suzuki et al. 1993).

In actual calculations, we first choose a model space. Here the spirit is that we should choose a "good" model space in the sense that the model space should have a large overlap with the physical wave function. If the overlap is small, then one expects all kinds of problems such as bad convergence and large non-hermiticity (Suzuki *et al.* 1993). The next step is to calculate the corresponding model-space G-matrix. As demonstrated, we can now calculate the G-matrix quite accurately. The G-matrix, as seen from Eq.(44), has an independent energy variable ω, and similarly the $\hat{Q}$-box also is dependent on the energy variable ω. A remarkable role of the folded diagrams is to remove this energy dependence. The input $\hat{Q}$-box is energy-dependent, yet the output effective interaction given by Eq.(8) is rigorously energy independent.

To proceed, we need now to calculate the $\hat{Q}$-box in terms of the G-matrix. Let us discuss this point a little later. Suppose we have already had the $\hat{Q}$-box , then the entire folded-diagram series can be summed up to all orders by iteration methods. Basically there are two types of iteration methods, the Krenciglowa-Kuo (KK) method (Krenciglowa and Kuo 1974) and the Lee-Suzuki (LS) method (Lee and Suzuki 1980, Suzuki and Lee 1980). The main differences of this two methods are summarized below. For the KK method one needs as input the $\hat{Q}$-box evaluated over a range of energy variables. In contrast, for the LS method one needs the $\hat{Q}$-box and its energy derivatives, at only one fixed starting energy. When the $\hat{Q}$-box can be obtained exactly, as in simple matrix models, it can be proved (Suzuki and Lee 1980, Krenciglowa and Kuo 1974) that the KK method and the LS method converge to different states of the original Hamiltonian: the KK method converges to the states whose wave functions have maximum overlaps with the model space, while the LS method converges to states whose eigenvalues are closest to the starting energy used to initialize the iteration.

In actual calculations for nuclear systems, we can only calculate the $\hat{Q}$-box approximately. As we mentioned earlier, for most calulations the $\hat{Q}$-box is calculated only with diagrams 1st- and 2nd-order in G. How to get a more accurate $\hat{Q}$-box remains to be a major open problem. The ultimate check of any theory is to compare its results with experiments. In nuclear physics, the most successful model is the nuclear shell model. It may be pointed out that the model-space approach discussed here provides a natural microscopic foundation for the nuclear shell model. (Kuo and Osnes 1990) For instance, the nucleus ^{18}O is described in the nuclear shell model as 2 valence nucleons outside an inert ^{16}O core. In the present approach, this just corresponds to the choice of a (2p0h) model space. (npmh means n-particle m-hole, namely we have n valence prticles outside the ^{16}O core and m holes inside the core. The entire configurational space for ^{18}O consists of 2p0h, 3p1h, 4p2h,.....components.)

Employing a low-order $\hat{Q}$-box and with the G-matrix interactions derived from modern NN potentials, the above model-space approach has been widely applied to a large number of nuclei, such as the Monte Carlo calculations of the Caltech group (Dean *et al.* 1996, Radha *et al.* 1996), the fp-shell calculations of Nakada, Sebe and Otsuka (Nakada *et al.* 1994), the very-large-space shell model calcualtions of Zucker et al. (see his contribution in this proceedings and Ref.

(Caurier 1996)), the recent Drexel shell model calculations (see contribution of Feng et al. in this proceedings), and the extensive calculations of Covello et al. for the tin region (see his contribution in this proceedings and Covello *et al.* 1996). Generally speaking the results obtained are indeed remarkably good as compared with experiments, for both nuclear energies and nuclear transition rates. Being an optimist, I believe that we must be doing something which is in the right direction, not perfectly but to a large extent correct. In fact one usually just needs to make some small adjustments to the derived matrix elements to obtain the "best-fit" empirical matrix elements.

But still, we are just using a low-order approximation for the $\hat{Q}$-box . How to obtain a better $\hat{Q}$-box is a basic difficulty in microscopic calculations of the effective interaction. A possible help may come from the choice of the auxiliary sp potential U. Recall that we add and then subtract U to redefine the Hamiltonain as $H = H_0 + H_1$ with $H_0 = T + U$ and $H_1 = V - U$. U is at our disposal, and usually U is chosen as the oscillator potential, which is perhaps not realistic for high-lying single particle orbitals. By a suitable choice of U, one may be able to minimize the effect of the higher-order diagrams of the $\hat{Q}$-box, and in this way the $\hat{Q}$-box may be accurately given by including only some low-order diagrams.

The situation for halo nuclei is different and may be promising. Initial G-matrix calculations for halo nuclei, within the framework of the TFSM approach, have led to encouraging results. The long-standing problem of the higher order core polarization diagrams is avoided here, because of the large physical separation between the halo nucleons and the core nucleus. This may enable us to derive, starting from realistic NN interactions, the effective interaction for the valence nucleons in halo nuclei more accurately than for ordinary nuclei.

8   Acknowledgements

This work is supported in part by the USDOE Grant DE-FGO2-88ER40388.

References

Austin S.M. Bertsch G.F. (1995): Scientific American **272**, 90

Zhukov M.V. *et al.* (1993): Phys. Rep. **231**, 151, and refs. therein

Nazarewicz, W. *et al.* (1996): Phys. Rev. **C53**, 740

Hamamoto I. Sagawa H. and Zhang X.L. (1996): Phys. Rev. **C53**, 765; Hamamoto I. Sagawa H. (1996): Phys. Rev. **C53**, 1492

Page R.D. *et al.* (1996): Phys. Rev. **C53**, 660

Lacomb M. *et al.* (1980): Phys. Rev **C21**, 861

Machleidt R. (1989): Adv. Nucl. Phys. **19**, 189

Machleidt R. Holinde K. Elster C. (1987): Phys. Reports **149**, 1

Wildenthal B.H. (1984): Prog. Part. Nucl. Phys. **11**, 5; Brown B.A. Richter W.A. Julies R.E. Wildenthal B.H. (1988): Ann. Phys. (N.Y.) **182**, 191

Kuo T.T.S. and G.E. Brown G.E. (1966): Nucl. Phys. **85**, 40

Feshbach H. (1958): Ann. Rev. Nucl. Sci. **8**, 49

Morita T. (1963): Prog. Theor. Phys. **29**, 351

Oberlechner G. Owono-N'-Guema F. J. Richert J. (1970): Nuovo Cimento **B68**, 23

Johnson M.B. and Baranger M. ()1972): Ann. Phys. (N.Y.) **62**, 172

Kuo T.T.S. Lee S.Y. K.F. Ratcliff K.F. (1971): Nucl. Phys. **A176**, 172

Brandow B.H. (1967): Rev. Mod. Phys. **39**, 771

Lindgren I. (1974): J. Phys. **B7**, 2441

Lee S.Y. and Suzuki K. (1980): Phys. lett. **91B**, 173

Suzuki K. and Lee S.Y. (1980): Prog. Theor. Phys. **64**, 2091

Suzuki K. Okamoto R. Ellis P.J. Kuo T.T.S. (1994): Nucl. Phys. **A567**, 576

Kuo T.T.S. Krmpotic F. Suzuki K. Okamoto R. (1995): Nucl. Phys. **A582**, 205

Kuo T.T.S. (1981): Lecture Notes in Physics (Springer-Verlag) **Vol. 144**, 248

Kuo T.T.S. Osnes E. (1990): Lecture Notes in Physics (Springer-Verlag) **Vol.364**, 1

Hjorth-Jensen M. Kuo T.T.S. Osnes E. (1995): Phys. Rep. **261**, 126

Krenciglowa E.M. Kuo T.T.S. (1974): Nucl. Phys. **A235**, 171

Tsai S.F. and Kuo T.T.S. (1972): Phys. Lett. **39B**, 427

Krenciglowa E.M. Kung C.L. Kuo , T.T.S. Osnes E. (1976): Ann. Phys. (N.Y.) **101**, 154

Wong C.W. (1967): Nucl. Phys. **A104**, 417

Muether H. Sauer P.U. (1992): Computational Nuclear Physics 2(ed. by K. Langanke, J.A. Maruhn and S. Koonin), Springer-Verlag ,30

Maglione E. Ferreira L.S. (1994): Phys. Rev. **C50**, 1240

Jiang M.F. Machleidt R. Stout D.B. Kuo T.T.S. (1989) Phys. Rev. **C40**, R1857; Phys. Rev. **C46**, 910(1992)

Shurpin J. Kuo T.T.S. Strottman D. (1983): Nucl. Phys. **A408**, 310

Hjorth-Jensen M. Osnes E. (1989): Phys. Lett. **B228**, 281

Hjorth-Jensen M. Osnes E. Müther H. Ann. of Phys. **213**, 102

Barrett B.R. Hewitt R.G.L. McCarthy R.J. (1971): Phys. Rev. **C3**, 1137

Zheng D.C. Barrett B.R. Vary J.P. McCarthy R.J. (1994): Phys. Rev. **C49**, 1999

T.T.S.Kuo T.T.S. , H. Muether H. and K. Azimi-Nili K. (1996): Nucl. Phys. **A606**, 15

Kuo T.T.S. Krmpotic F. Tzeng Y. (1996): Preprint

Bertsch G. (1972): Practitioner's Shell Model (North Holland)

Tuli J.K. (editor) (1995): Nuclear Wallet Cards, Nuclear Data Center, Brookhaven National Lab.

Dean D.J. Koonin S.E. Kuo, T.T.S. Langanke K. Radha P.B. (1996): Phys. Lett. B367 (1996) 17.

Radha P.B. Dean D.J. Koonin S.E. Kuo T.T.S. Langanke K. Poves A. Retamosa J. Vogel P. (1996): Phys. Rev. Lett. /bf 76, 2642

Nakada N. Sebe T. Otsuka T. (1994): Nucl. Phys. **A571**, 467

Caurier E. Novaski F. Poves A. Retamosa J. (1996): Phys. Rev. Lett. **77**, 1954

Covello A. Andreozzi F. Coraggio L. Gargano A. Porrino A. (1996): New Perspectives in Nuclear Structure (edited by A. Covello), World Scientific p.147.

Suzuki K. (1982): Prog. Theo. Phys. **68**, 246

Suzuki K. Okamoto R. (1983): Prog. Theo. Phys. **70**, 439

Kuo T.T.S. Ellis P.J. Hao J. Li Z. Suzuki K. Okamoto R. Kumagai K. (1993): Nucl. Phys. **A560**, 621

Suzuki K. Okamoto R. Ellis P.J. Hao J. Li Z. Kuo T.T.S. (1993): Phys. Lett. **B308**, 1

Hao J. Kuo T.T.S. Reuber A. Holinde K. Speth J. Millener J.D. (1993): Phys. Rev. Lett. **71**, 1498

Kuo T.T.S. Hao J. (1994): Prog. Theo. Phys. Suppl. **117**, 351

Large No-Core Basis-Space Shell Model Calculations for Light Nuclei

B. R. Barrett[1], D. C. Zheng[1,2], P. Navrátil[1,3], J. P. Vary[4], W. C. Haxton[5], and C. L. Song[5]

[1] Department of Physics, Building 81, University of Arizona, Tucson, Arizona, USA
[2] W. K. Kellogg Radiation Laboratory, 106–38, California Institute of Technology, Pasadena, CA 91125, USA*
[3] Institute of Nuclear Physics, Academy of Sciences of the Czech Republic, 25068 Řež near Prague, Czech Republic**
[4] Department of Physics and Astronomy, Iowa State University, Ames, Iowa 50011, USA
[5] Institute for Nuclear Theory and Department of Physics, University of Washington, Seattle, WA 98195, USA.

Abstract. One of the major outstanding problems in nuclear physics is to determine the nature of the interaction between two nucleons in the nuclear medium. This problem has been investigated since the early days of the nuclear shell-model with considerable progress phenomenologically. Microscopic approaches suffer from the problem that the perturbation expansion for the interaction in the model space diverges due to intruder states. This difficulty might be circumvented by performing the shell-model calculation in a no-core model space, in which all A nucleons are active and all core-polarization processes are eliminated. The effective interaction for $A = 2$ in a $(0s_{1/2})^2$ model space is simply the Brueckner G-matrix evaluated at a starting energy equal to the $A = 2$ eigenvalue. For $A > 2$, exact results for the eigenenergies can be obtained only if the generalized, A-nucleon G-matrix and associated folded diagrams can be constructed. For sufficiently large model spaces, the perturbation expansion for the effective interaction may be reasonably expressed in terms of only the Brueckner reaction matrix G in the no-core space plus all folded diagrams developed from it. A method for doing this is described, along with techniques for treating some of the neglected many-body effects, such as using multi-valued G-matrices. The results of calculations for light nuclei ($A = 2$ to 7) are presented.

* Present address
** Permanent address

e-mail addresses:
bbarrett@ccit.arizona.edu
zhengdc@krl.caltech.edu
navratil@physics.arizona.edu
jvary@iastate.edu
haxton@phys.washington.edu
song@phys.washington.edu

1   Introduction

One of the most significant, outstanding problems in nuclear-structure physics is the microscopic determination of the effective shell-model interaction between two nucleons inside a nucleus. This problem has a long and rich history, making it impossible to discuss and reference all possible work on this topic. In this regard, the interested reader is referred to the review articles on this subject given in Refs. [1,2].

The central problem has to do with the form of the effective Hamiltonian (or effective interaction) to use in a truncated-space calculation. In our case, this is a nuclear shell-model calculation; however, this problem is common to many fields of physics [3].

One would like to start with the free nucleon-nucleon (NN) interaction and many-body quantum mechanics and solve for the properties of finite nuclei. In principle, this simply involves solving the many-body Schrödinger equation:

$$H|\Psi_\alpha\rangle = E_\alpha|\Psi_\alpha\rangle \tag{1}$$

for the eigenenergies E_α and the eigenstates $|\Psi_\alpha\rangle$ of the many-particle system, where α is some label characterizing the states. But it is impossible to solve this problem in the full Hilbert space S, when the number of particles in the system exceeds a certain limit, because it contains too many degrees of freedom. Consequently, one wishes to truncate the problem to a smaller space S of dimension d, in which it becomes tractable to carry out the calculation. Now let $|\Phi_\beta\rangle$ represent the projections of d of the states $|\Psi_\beta\rangle$ into S. Thus, we define the effective Hamiltonian $\mathcal{H}$ in S to satisfy

$$\mathcal{H}|\Phi_\beta\rangle = E_\beta|\Phi_\beta\rangle, \tag{2}$$

where the eigenvalues $\{E_\beta\}$ are d of the exact eigenvalues $\{E_\alpha\}$ in Eq.(1). Because the $|\Phi_\beta\rangle$ are projections of the $|\Psi_\alpha\rangle$, they are, in general, *not* orthogonal.

The question then arises whether an appropriate $\mathcal{H}$ exists for any given truncation. One can show this to be true by constructing the biorthogonals to $|\Phi_\beta\rangle$, namely, $|\tilde{\Phi}_\gamma\rangle$, which satisfy $\langle\tilde{\Phi}_\gamma|\Phi_\beta\rangle = \delta_{\gamma\beta}$. It then follows that the effective Hamiltonian $\mathcal{H}$ always exists and is of the form

$$\mathcal{H} = \sum_{\beta \in S} |\Phi_\beta\rangle E_\beta \langle\tilde{\Phi}_\beta|, \tag{3}$$

which automatically satisfies Eq.(2). As Kirson [3] has emphasized, the question is *not* whether $\mathcal{H}$ exists, but whether it has a *simple* enough form so as to be *useful*.

The form of $\mathcal{H}$, given by Eq.(3), is not very helpful with regard to understanding the relationship of $\mathcal{H}$ to H [or of the effective interaction $\mathcal{V}$ (to be defined) to the free NN interaction]. There are many different ways in which to formulate this problem: in terms of time-independent perturbation theory [1,2,4,5], time-dependent perturbation theory [1,2,6,7], the coupled-cluster or e^S

method [8], moment methods [9], and variational approaches [10]. In our work, we use the time-independent-perturbation-theory approach in establishing the connection between $\mathcal{H}$ and H. This formalism will be described in Section 2. Readers interested in other techniques may look at the appropriate references listed above. In Section 2 we also discuss the relationship between the Brueckner G-matrix [11] and our formulation of the two-body effective interaction. Our results for $A = 2$ to 7 are presented and discussed in Section 3, and we give our conclusions in Section 4.

2   Formalism

2.1   Time-Independent Perturbation Theory

The basic idea involves the separation of the Hilbert space of the A active nucleons into two parts, using the projection operators P and Q. Here, P projects onto the truncated or shell-model space, defined by the eigenstates of an unperturbed Hamiltonian H_0, and Q defines the excluded space outside the shell-model space. The projection operators P and Q are A-particle operators; they define non-overlapping spaces, so that $PH_0Q = 0$.

In the full Hilbert space, the conventional choice for H is of the form

$$H = \sum_{i=1}^{A} t_i + \sum_{i<j}^{A} v_{ij} = T + V = (T + U) + (V - U) = H_0 + H_I, \qquad (4)$$

where U is some single-particle potential, $H_0 = T + U$, and $H_I = V - U$ is the residual interaction. Only two-body interactions v_{ij} have been assumed among the particles, but the method can be generalized to many-body forces.

Using the Feshbach projection method [12], one can explicitly project H into the P and Q spaces and rewrite the P space equation (omitting the subscript α everywhere) in the form

$$\left[PHP + PHQ \frac{1}{E - QHQ} QHP \right] P|\Psi\rangle = EP|\Psi\rangle, \qquad (5)$$

where $P|\Psi\rangle = |\Phi\rangle$. The term in brackets defines the effective Hamiltonian

$$\mathcal{H} = PH_0P + \mathcal{V}(E), \qquad (6)$$

where

$$\mathcal{V}(E) = PH_IP + PH_IQ \frac{1}{E - QHQ} QH_IP \qquad (7)$$

is the effective interaction. It should be noted that, in general, $\mathcal{V}$ is an A-nucleon operator and the energy E in the denominator corresponds to one of the exact eigenenergies of the A-nucleon system.

Although $\mathcal{V}(E)$ is an A-nucleon interaction, a standard assumption is to approximate it in terms of two-body interactions. Numerical calculations are performed in the basis space of the eigenfunctions Φ_β of Eq.(2), which are usually taken to be Slater determinants of eigenfunctions of the one-body harmonic-oscillator (HO) Hamiltonian

$$H_0 = \sum_{i=1}^{A} h_i = \sum_{i=1}^{A} \left(\frac{\mathbf{p}_i^2}{2M} + \tfrac{1}{2} M \Omega^2 r_i^2 \right) \ . \tag{8}$$

These many-body basis states can be labeled according to the number of oscillator quanta they contain, $N_{\text{sum}} = \sum_{i=1}^{A} N_i$, or equivalently, the unperturbed energies

$$\sum_{i=1}^{A} \left(N_i + \tfrac{3}{2} \right) \hbar\Omega \ , \tag{9}$$

where N_i is the number of oscillator quanta $(2n_i + l_i)$ of the ith single-particle (SP) state. Conventionally, the labeling is relative to the minimum energy configuration, so that states are partitioned into "$0\hbar\Omega$", "$1\hbar\Omega$", "$2\hbar\Omega$", etc. configurations.

Early shell-model configurations were generally restricted to a single shell, such as the 0p or 1s0d shells, and thus involved only $0\hbar\Omega$ valence nuclear configurations. The effective interaction is then introduced to account for excluded configurations, such as very high energy excitations associated with the hard core in the NN interaction.

There are a number of uncertainties associated with the usual perturbation-theory shell model approach; for example:

1. the choice (i.e., the nature) of NN potential,
2. the choice of SP basis (i.e., the choice of U or H_0),
3. the treatment of the spurious center-of-mass (CM) motion,
4. the choice of the starting energy ω in the two-particle reaction matrix G,
5. the size of the model space,
6. many-body forces for systems with more than two particles, and
7. the possible divergence of the perturbation-theory expansion.

By far the most serious problem is the last one, divergence of the perturbation expansion due to intruder configurations into the model space [13]. These intruder states have dominant components from the Q-space but their exact energies, E, occur in the range of eigenvalues of states with dominant P-space components. Perhaps one should avoid doing a perturbation expansion for the effective interaction altogether and look for a different approach to this problem. One possible way is a hierarchical scheme as an approximation to the exact effective interaction, in which first two-body G-matrix input is included, followed by three-body G-matrix input, etc.. Our present efforts are consistent with this hypothesis.

2.2   Multi-Major Shell Calculations

In recent years shell-model calculations involving two or more major shells have been frequently performed. A full multi-$\hbar\Omega$ basis is one that includes all many-body configurations, such that $N_{sum} \leq N_{max}$ for some N_{max}. For example, a calculation of the positive parity states in ^{16}O might include all $(0 + 2)\hbar\Omega$ or $(0+2+4)\hbar\Omega$ many-body configurations, relative to the closed core (fully occupied 0s and 0p shells).

Such full multi-$\hbar\Omega$ bases have other appealing properties. If HO SP states are employed, the model space wave functions can be decomposed so that the relative degrees of freedom are separated from a pure oscillator CM component. Thus the over-completeness of the Slater determinants (which depend on $3A$ coordinates, while intrinsic wave functions depend on $3(A-1)$) can be cured by retaining only those linear combinations, which keep the CM in the 0s state.

A second property has to do with technical difficulties in evaluating the effective interaction. If, in addition to defining the basis states, H_0 of Eq.(8) plays the role of the unperturbed Hamiltonian, then the unperturbed energies of configurations in the excluded space always exceed those in the model space.

2.3   No-Core Shell-Model Calculations

In an attempt to circumvent the divergence problem, we propose to calculate $\mathcal{H}$ or $\mathcal{V}(E)$ in a "no-core" model space, in which *all* A nucleons in a nucleus are active [14,15], for a complete $N\hbar\Omega$ basis space and a large value for N. As pointed out in Section 2.2, by working in a complete $N\hbar\Omega$ space with basis states defined by H_0 of Eq. (8), it is guaranteed that all excluded configurations involve an energy of at least $2\hbar\Omega$ and, hence, should not intrude into the model space. The large value for N guarantees that we have included the major configurations making up the physical low-lying states. At the same time, the no-core basis simplifies the form of the effective interaction, since there are no hole states, and Eq.(7) may be interpreted as a generalized A-nucleon G-matrix equation, which allows us to calculate energies relative to the vacuum. For a one-dimensional model space, the exact solutions for the eigenvalues are given by

$$E = E_0 + \mathcal{V}(E), \tag{10}$$

where E_0 is the eigenenergy of the unperturbed Hamiltonian H_0 in Eq.(8). If $\mathcal{V}(E)$ can be constructed for the A-nucleon system, then Eq.(10) can be solved diagrammatically or iteratively [16,17], to obtain all the eigenenergies of the A-nucleon system whose eigenstates have non-vanishing projections on the one-dimensional model space. For examples of these procedures with solvable models, see Ref. [18].

It is not generally possible to construct the full A-nucleon G-matrix. We approximate it by the two-nucleon nuclear reaction matrix G (Ref. [11]) plus higher-order terms. The two-nucleon G-matrix is simply the infinite sum of two-particle ladder terms.

The perturbation-theory expansion for $\mathcal{V}(E)$ is now rewritten as a perturbation series in $G(\omega)$, where ω replaces E and is called the starting energy. In our application to no-core model spaces, corrections at the two-particle level are all of the folded-diagram type [1,2,6,7]. The remaining corrections generate effective many-body interactions.

Regarding the other uncertainties associated with the perturbation-expansion approach, we have shown in earlier investigations that numerical results for nuclear properties are essentially the same for several choices of the NN potentials [19] and that one can essentially neglect the SP insertions, $-U$, so long as the model space is large enough [20]. However, a straightforward method exists for calculating G exactly including the SP insertions [21].

As stated earlier, since we neglect effective many-body forces, it is necessary to take a sufficiently large model space so that these effective many-body forces have a minimal residual effect. In this connection, we have shown that it helps convergence to take a physically reasonable oscillator parameter in defining H_0 [22,23].

Within a chosen model space we can examine results as a function of a further truncation to a subset of the many-body states in the space. For example, we may keep all $N\hbar\Omega$ excitations up to some value less than the total N value allowed in the model space. Then, we learn of the convergence with respect to the retained excitations by tracking results as N increases towards its maximum within the space. If significant changes in an observable of interest still occur when N is near maximum, then we would try to increase the size of the model space.

For light nuclei [15,24], we have demonstrated how well the numerical results have converged for low-lying states in model spaces of $4\hbar\Omega$ and $6\hbar\Omega$ excitations.

2.4   Multi-valued-G vs. Single-G Calculations

Investigators performing large-basis shell model calculations in multi-$\hbar\Omega$ spaces have consistently chosen effective two-body interactions of the form $\mathcal{V}(ab; cd)$, just as in traditional $0\hbar\Omega$ calculations. To be more complete, the effective interaction [25] in such spaces may carry an additional index $N_{\mathrm{sum}}^{\mathrm{spectators}}$; i.e. $\mathcal{V}(ab; cd; N_{\mathrm{sum}}^{\mathrm{spectators}})$, where $N_{\mathrm{sum}}^{\mathrm{spectators}}$ labels the total oscillator quanta of the "spectator" (i.e., non-interacting) nucleons in the many-body states connected by the matrix element $\mathcal{V}(ab; cd)$. It is given by

$$N_{\mathrm{sum}}^{\mathrm{spectators}} = N_{\mathrm{sum}} - N_c - N_d = N'_{\mathrm{sum}} - N_a - N_b \ , \tag{11}$$

where N_{sum} and N'_{sum} are the numbers of the total oscillator quanta of the initial and final many-body states, respectively. In the case of traditional $0\hbar\Omega$ calculations, all basis states are characterized by the same $N_{\mathrm{sum}}^{\mathrm{spectators}}$, so this additional index is unnecessary. But for multi-$\hbar\Omega$ bases, the $N_{\mathrm{sum}}^{\mathrm{spectators}}$ dependence arises: if model-space configurations exist with different unperturbed energies, the gaps and interactions coupling these configurations to the excluded space will differ. The appropriate energy denominators in the G-matrix ladder sum are not given just by the initial and final two-particle labels, but also depend on the energies of the $A - 2$ "spectator" nucleons.

The omission of the $N_{\text{sum}}^{\text{spectators}}$ dependence in our previously reported large-space shell-model calculations [23,24] amounts to neglect of certain many-body processes of the same unperturbed energy as some retained many-body processes. While the effects of these neglected many-body processes are expected to decrease in importance as the number of shells included in the model space increases, our investigation, which retains them through the $N_{\text{sum}}^{\text{spectators}}$ dependence of the two-body effective interaction, will reveal that these neglected effects are important in present-day calculations.

We shall see in the following calculations that the resulting shifts can be large, amounting to about 5 MeV for some diagonal matrix elements. The approximation in present-day multi-shell calculations to neglect the $N_{\text{sum}}^{\text{spectators}}$ dependence can lead to unattractive consequences. One example is the apparent need for unrealistic SP energies to reduce the splittings between the "$0\hbar\Omega$" and "$2\hbar\Omega$" states, as required by experiment.

Recently model calculations [26] have been performed to compare the multi-valued-G approach with exact calculations as well as with the single-G approach. The results of these calculations clearly show the fully converged (i.e., energy-independent) multi-valued-G approach does markedly improve agreement with the results of the exact calculations.

2.5   The Effective Interaction in a No-Core Model Space

For a no-core model space, the *two-body* effective interaction is simply the G-matrix [11] plus the folded diagrams series arising from it [16,17]. The G-matrix is the sum of the ladder diagram series, which represents the multiple scattering processes of two nucleons in a nuclear medium. We continue to follow our philosophy given in Ref. [27] for the no-core G-matrix in large spaces, which treats two-particle scattering via a realistic NN interaction v_{12} in an "external" field, u, which is provided by the other nucleons in the same nucleus. Thus, we write

$$G(\omega) = v_{12} + v_{12}\frac{Q}{\omega - (h_1 + h_2)}G(\omega) = v_{12} + v_{12}\frac{Q}{\omega - (h_1 + h_2 + v_{12})}v_{12}, \quad (12)$$

where $h = t + u$ is the one-body Hamiltonian and u is the nuclear mean field, i.e., $H_0 = \sum_{i=1}^{A} h_i$, as in Eq.(8). The quantity ω is the starting energy, which represents the initial energy of the two in-medium nucleons. The Pauli operator Q excludes the scattering of the two nucleons into the intermediate states, which are inside the model space. It is, therefore, related to the choice of the model space and will be specified in the next section.

A rigid prescription for the nuclear mean field u is not necessary since the results will be independent of u once all convergence criteria are satisfied. In most practical calculations, u is approximated by a one-body potential of a simple and convenient form. The two most common choices for u are a shifted HO potential and zero:

$$u(r) = -V_0 + \tfrac{1}{2}m\Omega^2 r^2, \tag{13}$$

$$u(r) = 0. \tag{14}$$

The latter choice corresponds to a plane-wave basis. One should note (see Refs. [28,29]) that the two seemingly very different one-body potentials in Eqs. (13,14) actually led to rather similar G-matrix elements, provided one makes a careful choice for the starting energy (related to the choice of u).

We will approximate the nuclear mean field by the HO potential (13) not only because this simplifies the G-matrix calculation [29,30], but, more importantly, for the reason that this makes possible an exact removal of the effects of the spurious CM motion from our many-body wavefunctions, as discussed in Section 2.2. Once $G(\omega)$ is obtained as a function of the starting energy, it is straightforward to evaluate the folded diagrams using the techniques developed by Lee and Suzuki [16] and by Kuo and Krenciglowa [31] and to obtain a starting-energy-independent effective two-body interaction (denoted by $\mathcal{V}^{(2)}$).

In Ref. [27] it is shown that $\mathcal{V}^{(2)}$ can be well approximated by the G-matrix calculated at starting energies ω', which are related to the initial unperturbed energy of the two nucleons in the ladder scattering processes in a simple way:

$$\omega' = \omega + 2V_0 = \epsilon_a + \epsilon_b + \Delta, \tag{15}$$

where the ϵ_i are the HO SP energies given by Eq.(8) (a and b are the SP states that the two nucleons initially occupy). Such a state-dependent choice for ω' will lead to a non-hermitian G-matrix, but the non-hermicity is found to be small. Here we will follow Ref. [27] and adopt the average of $G(\omega')$ and its conjugate as our estimate of $\mathcal{V}^{(2)}$.

In general, Δ is a state-dependent quantity, which yelds an energy-independent value for $\mathcal{V}^{(2)}$, when combined with $\epsilon_a + \epsilon_b$. In our calculations we take the quantity Δ to be a state-independent phenomenological adjustment to the two-body effective interaction that allows us to reproduce binding energies reasonably well. The need for such a correction is due to the emission of higher-order corrections to the effective interaction, those beyond the two-body ladders that involve the multiple scattering of clusters of three or more nucleons.

The model calculations [26], referred to earlier, also included results for the multi-valued-G approach using ω', as given by Eq. (15). These results were almost identical with the energy-independent multi-valued-G results and were also a good approximation to the exact results. Consequently, these model calculations tend to serve as a justification of the use of Eq. (15) in calculating a reasonable approximation to the energy-independent G-matrix or, equivalently, $\mathcal{V}^{(2)}$.

Our shell-model Hamiltonian can now be written as

$$H_{\text{SM}} = \left(\sum_{i=1}^{A} t_i - T_{\text{CM}} \right) + \sum_{i<j}^{A} G_{ij} + V_{\text{Coulomb}} + \lambda(H_{\text{CM}} - \tfrac{3}{2}\hbar\Omega), \tag{16}$$

where T_{CM} is the CM kinetic energy, so the first term in brackets is simply the relative kinetic energy. The proton and neutron masses are taken to be the same. The last term in Eq.(16) is included in order to project out spurious CM motion: inclusion of this term with a large value of λ produces low-lying excitations with the CM in the 0s state. For this procedure to work properly, the model space must be exactly separable, as is the case for complete multi-$\hbar\Omega$ bases. The bare Coulomb interaction $V_{Coulomb}$ is diagonalized only within the model space. For the two-body effective interaction $\mathcal{V}^{(2)}$ within the model space we have used two choices. The first is to take $\mathcal{V}^{(2)} = G(\omega')$ as given by eqs. (12) and (15), which we refer to as the single-G approximation. The second choice sets

$$\mathcal{V}^{(2)} \cong G(\omega, N_{\text{sum}}^{\text{spectators}})$$

$$= v_{12} + v_{12} Q(N_{\text{sum}}^{\text{spectators}}) \frac{1}{\omega - (h_1 + h_2 + v_{12})} Q(N_{\text{sum}}^{\text{spectators}}) v_{12}. \quad (17)$$

Equation (17) is referred to as the multi-G approximation because there exists a separate set of G-matrix elements for each value of $N_{\text{sum}}^{\text{spectators}}$, as given by Eq.(11). There is no $N_{\text{sum}}^{\text{spectators}}$ dependence in Q for single shell calculations but, as discussed above, this $N_{\text{sum}}^{\text{spectators}}$ dependence arises in multi-$\hbar\Omega$ calculations.

The index $N_{\text{sum}}^{\text{spectators}}$ signifies the role of the full model space many-body configuration in controlling the intermediate two-particle states available for scattering. In a shell model calculation whose model space includes all many-body states with $N_{\text{sum}} \leq N_{\text{max}}$, the allowed intermediate states for the two particles, "1" and "2", scattered by v_{12}, are specified by:

$$N_1 + N_2 + N_{\text{sum}}^{\text{spectators}} > N_{\text{max}} , \quad (18)$$

which corresponds to the following Pauli operator:

$$Q(N_{\text{sum}}^{\text{spectators}}) = \begin{cases} 0 & \text{if } N_1 + N_2 \leq N_{\text{max}} - N_{\text{sum}}^{\text{spectators}}, \\ 1 & \text{otherwise.} \end{cases} \quad (19)$$

In Fig. 1, we depict the various spectator-dependent Pauli operators appropriate for a full $6\hbar\Omega$ calculation of ^{6}Li ($N_{\text{max}} = 8$). For a single-G calculation of ^{6}Li in a full $6\hbar\Omega$ model space ($N_{\text{max}} = 8$), Q is simply given by

$$Q = \begin{cases} 0 & \text{if } N_1 + N_2 \leq N_{\text{max}} \\ 1 & \text{otherwise} \end{cases} \quad (20)$$

The fact that we introduce a spectator dependence to the G-matrix raises interesting possibilities for identifying specific Pauli-violating processes. Some two-particle scattering states in the excluded space will place a nucleon in an SP state that may be occupied by a spectator nucleon in a given model space wave function. One might avoid these Pauli violating processes in a full multi-$\hbar\Omega$ calculation by labeling Q with the full set of quantum numbers on which G operates. This, of course, is impractical. However, for the specific case of these light nuclei and for $N_{\text{sum}}^{\text{spectators}} = 0$, we can easily eliminate the Pauli violating

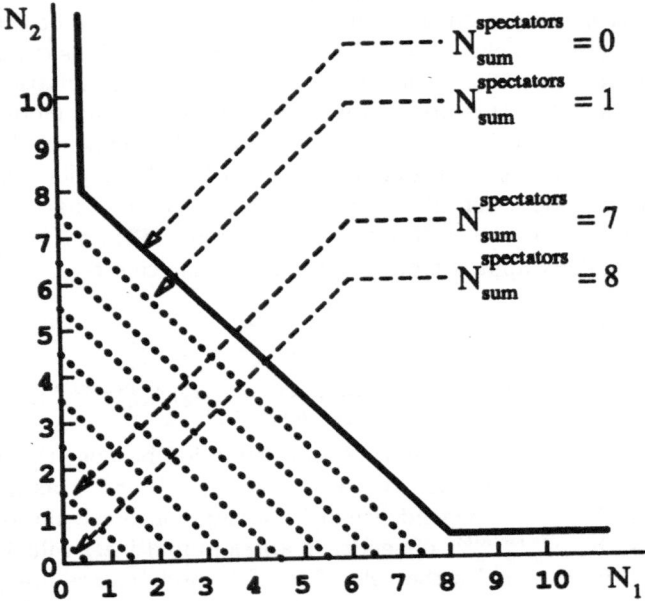

Fig. 1. An illustration of the Q operator appropriate for a full $6\hbar\Omega$ calculation of ^{6}Li. The regions interior to the lines are the $Q = 0$ regions defined in Eq.(19). The lines correspond to the possible values of $N_{\mathrm{sum}}^{\mathrm{spectators}}$, which range from 0 to $N_{\max}$. The contour for $N_{\mathrm{sum}}^{\mathrm{spectators}} = 0$ is given as a solid line. The wings result from the fact that the spectator nucleons are in a unique configuration (closed 0s shell), in this case, forbidding scattering into the 0s shell. The wings make a negligible contribution numerically and can be ignored. The contours for other values of $N_{\mathrm{sum}}^{\mathrm{spectators}}$ are denoted by dashed l ines. Note that a single-valued G-matrix would employ a single contour and th us neglect some of the physics governing $\mathcal{V}$ in a multi-$\hbar\Omega$ spa ce.

processes involving the $0s$ nucleons by including the "wings" as depicted in Fig. 1. However, we have found that the presence or absence of the wings in the case $N_{\mathrm{sum}}^{\mathrm{spectators}} = 0$ results in minor differences in our results due to the large size of the model spaces.

To provide the reader with some measure of the size of the effects associated with $N_{\mathrm{sum}}^{\mathrm{spectators}}$, we give in Table 1 the matrix elements

$$\langle (0s_{1/2}\, 0s_{1/2})|\mathcal{V}^{(2)}|(0s_{1/2}\, 0s_{1/2})\rangle,$$

$$\langle (0s_{1/2}\, 0p_{3/2})|\mathcal{V}^{(2)}|(0s_{1/2}\, 0p_{3/2})\rangle,$$

and

$$\langle (0p_{3/2}\, 0p_{3/2})|\mathcal{V}^{(2)}|(0p_{3/2}\, 0p_{3/2})\rangle,$$

that we evaluated for a full $6\hbar\Omega$ calculation of the positive-parity states in ^{6}Li. In this calculation, $N_{\max} = 8$ and N_{sum} can take on four values (2, 4, 6, 8). The Table shows that the values of these diagonal matrix elements can shift by up to 3.3 MeV, when $N_{\text{sum}}^{\text{spectators}}$ dependence is properly treated.

This "multi-valuedness" is a bookkeeping complication in shell-model studies. However, its inclusion builds in important contributions previously missing from multi-$\hbar\Omega$ calculations. Model-space states of higher unperturbed energy are now more strongly repelled downwards by effects of states in the excluded space, which are included in G for the first time.

Table 1. Some diagonal matrix elements $\langle(ab : JT)|\mathcal{V}^{(2)}|(ab : JT)\rangle$ (in MeV) for four possible values of N_{sum} in a full $6\hbar\Omega$ ($\hbar\Omega = 14$ MeV) calculation of the positive-parity states in ^{6}Li. The effective interaction $\mathcal{V}^{(2)}$ is defined in Eq.(17) with the Pauli operator Q specified in Fig.1.

N_{sum}[a]	2	4	6	8
$(ab : JT) = (0s_{1/2}\,0s_{1/2} : 01)$	-6.689	-6.734	-6.894	-7.371
$(ab : JT) = (0s_{1/2}\,0s_{1/2} : 10)$	-8.272	-9.006	-9.969	-11.554
$(ab : JT) = (0s_{1/2}\,0p_{3/2} : 10)$	-1.144	-1.415	-1.769	-2.344
$(ab : JT) = (0s_{1/2}\,0p_{3/2} : 11)$	-3.768	-3.812	-3.935	-4.273
$(ab : JT) = (0s_{1/2}\,0p_{3/2} : 20)$	-8.272	-9.006	-9.969	-11.554
$(ab : JT) = (0s_{1/2}\,0p_{3/2} : 21)$	-1.006	-1.029	-1.058	-1.090
$(ab : JT) = (0p_{3/2}\,0p_{3/2} : 01)$	-3.227	-3.256	-3.342	-3.588
$(ab : JT) = (0p_{3/2}\,0p_{3/2} : 10)$	-1.272	-1.575	-1.950	-2.522
$(ab : JT) = (0p_{3/2}\,0p_{3/2} : 21)$	-1.364	-1.389	-1.439	-1.545
$(ab : JT) = (0p_{3/2}\,0p_{3/2} : 30)$	-4.179	-4.528	-5.021	-5.821

[a] N_{sum} is related to $N_{\text{sum}}^{\text{spectators}}$ through Eq.(11). Here we have $(ab) = (cd)$, therefore, $N_{\text{sum}}^{\text{spectators}} = N_{\text{sum}} - N_a - N_b$.

3   Results and Discussion

We employ the method of Barrett, Hewitt and McCarthy [29] to calculate the G-matrices. For the bare $N\text{-}N$ interaction v_{12}, we use the latest versions of the potentials from the Nijmegen group, such as the Reid-like potential (Reid 93) or the Nijm II potential [32]. The HO basis parameter $\hbar\Omega$ is usually fixed at 14 MeV, although calculations for a different choice of $\hbar\Omega$ are performed for selected cases for the purpose of comparison.

Over the last few years we have performed a number of different large-basis, no-core, shell-model calculations for $A = 2$ to 7, by diagonalizing the Hamiltonian H_{SM} in Eq.(16) with the use of Many-Fermion-Dynamics Shell-Model code

[33]. During this time we have checked for convergence, as discussed in Section 2.3, and found that numerical results have converged reasonably well for model spaces of size $8\hbar\Omega$, particularly for ^{4}He. In these calculations we also treat all possible particle excitations within the given model space, so as to obtain a fully-converged result for that space.

3.1 Single-G Calculations

In our earlier single-G calculations for $A = 2$ to 6 [23] we used the Reid 93 potential [32] and a no-core model size with $N_{max} = 5$ (i.e., the lowest six HO major shells), with different Q operators in Eq.(12) for $A \le 4$ and $A > 4$:

$$\text{For} \quad A \le 4: \quad Q = 1 \quad \text{for} \quad n_1 \ge 6, \; n_2 \ge 6, \quad \text{or} \quad n_1 + n_2 \ge 8, \tag{21}$$
$$= 0 \quad \text{otherwise};$$
$$\text{For} \quad A > 4: \quad Q = 1 \quad \text{for} \quad n_1 + n_2 \ge 6, \tag{22}$$
$$= 0 \quad \text{otherwise}.$$

In the above equations, $n=2n_r+l$ is the principal quantum number for the HO SP states. It starts from 0 with $n=0$ representing the first major shell ($0s$).

The results of our single-G calculations are given in Table 2. These calculations involve a free parameter Δ which has been fixed at -35 MeV for all the nuclei considered. For $A > 2$, we can obtain smaller binding energies (in better agreement with exact calculations) by decreasing Δ (i.e., making it more negative), since the binding energies decrease monotonically with the decreasing Δ. Our adoption of a Δ value to fit experimental binding energies stems from an assumption that we can account for neglected effective many-body forces (as well as true many-body forces) by phenomenologically adjusting the two-body G-matrix. It would, of course, have been preferable to calculate the needed adjustment. For example, had we been able to evaluate the full three-body ladder sum, this quantity could be averaged over a nuclear "core" to produce density-dependent corrections to the two-body G-matrix. Presumably, much of the need for Δ would then be removed.

3.2 Multi-G Calculations

Recently we have performed large-basis, no-core shell-model calculations for $A = 4$ to 7. We properly evaluate the G-matrix for full multi-$\hbar\Omega$ spaces resulting in a multi-valued two-body effective interaction, as described in Section 2.4. Only the results for ^{4}He and ^{6}Li will be presented here. The full results can be found in Ref. [37].

A. ^{4}He. For the positive-parity states in ^{4}He, we use a 9-major-shell model space which allows us to include all configurations with $N_{sum} = N_1 + N_2 +$

Table 2. The results for ^{2}H, ^{3}H, ^{4}He, ^{5}He and ^{6}Li obtained in large no-core (consisting of 6 HO major shells) shell-model calculations. The experimental data are taken from Refs. [34,35,36]. In the Table, E_B is the binding energy (in MeV); $E_x(J_n^\pi, T)$ the excitation energy (in MeV) of the J_n^π, T state. The ground-state rms *point* radius for protons $\sqrt{\langle r_p^2 \rangle}$ (in fm), electric quadrupole moment Q (in efm^2) and magnetic dipole moment μ (in μ_N) are also listed.

Observable	Calc.	Exp't	Observable	Calc.	Exp't
Deuteron			**Triton**		
E_B	2.103	2.2246	E_B	8.589	8.4819
$\sqrt{\langle r_p^2 \rangle}$	1.653	1.95	$\sqrt{\langle r_p^2 \rangle}$	1.573	1.41–1.62
μ	0.857	0.8573	μ	2.659	2.9790
Q	0.242	0.2859	$E_x(\frac{5}{2}^-_1, \frac{1}{2})$	12.716	unbound
$E_x(0^+_1, 1)$	3.754	unbound	$E_x(\frac{1}{2}^-_1, \frac{1}{2})$	12.868	unbound
^{4}He			**^{5}He**		
E_B	28.757	28.296	E_B	25.960	27.410
$\sqrt{\langle r_p^2 \rangle}$	1.488	1.46	$\sqrt{\langle r_p^2 \rangle}$	1.659	
$E_x(0^+_1, 0)$	0.000	0.00	μ	-1.864	
$E_x(0^+_2, 0)$	26.135	20.21	Q	-0.332	
$E_x(0^-_1, 0)$	22.848	21.01	$E_x(\frac{3}{2}^-_1, \frac{1}{2})$	0.000	0.00
$E_x(2^-_1, 0)$	24.351	21.84	$E_x(\frac{1}{2}^-_1, \frac{1}{2})$	3.112	4 ± 1
$E_x(2^-_1, 1)$	25.739	23.33	$E_x(\frac{1}{2}^+_1, \frac{1}{2})$	7.437	See $^{a)}$
$E_x(1^-_1, 1)$	26.338	23.64	$E_x(\frac{5}{2}^+_1, \frac{1}{2})$	14.206	See $^{a)}$
$E_x(1^-_1, 0)$	27.337	24.25	$E_x(\frac{3}{2}^+_1, \frac{1}{2})$	14.439	See $^{a)}$
$E_x(0^-_1, 1)$	27.418	25.28	$E_x(\frac{3}{2}^+_2, \frac{1}{2})$	20.445	16.75$^{b)}$
$E_x(1^-_2, 1)$	27.905	25.95	$E_x(\frac{3}{2}^-_2, \frac{1}{2})$	21.499	N/A
^{6}Li			$E_x(\frac{1}{2}^+_2, \frac{1}{2})$	23.563	N/A
E_B	30.648	31.996	$E_x(\frac{7}{2}^+_1, \frac{1}{2})$	23.592	N/A
$\sqrt{\langle r_p^2 \rangle}$	2.050	2.38	$E_x(\frac{1}{2}^-_2, \frac{1}{2})$	24.045	N/A
μ	0.851	0.822	$E_x(\frac{3}{2}^+_3, \frac{1}{2})$	24.398	N/A
Q	-0.116	-0.082	$E_x(\frac{1}{2}^+_3, \frac{3}{2})$	25.861	N/A
$E_x(1^+_1, 0)$	0.000	0.000	$E_x(\frac{1}{2}^+_3, \frac{1}{2})$	26.240	N/A
$E_x(3^+_1, 0)$	2.959	2.186	$E_x(\frac{3}{2}^+_4, \frac{1}{2})$	27.359	N/A
$E_x(0^+_1, 1)$	3.607	3.563	$E_x(\frac{7}{2}^-_1, \frac{1}{2})$	27.681	N/A
$E_x(2^+_1, 0)$	5.485	4.31			
$E_x(2^+_1, 1)$	6.505	5.366			
$E_x(1^+_2, 0)$	7.828	5.65			

$^{(a)}$ Low-lying positive-parity states (e.g. a $J^\pi = \frac{1}{2}^+$, $T = \frac{1}{2}$ state at $\sim$5 MeV and $J^\pi = \frac{3}{2}^+$, $T = \frac{1}{2}$ and $J^\pi = \frac{5}{2}^+$, $T = \frac{1}{2}$ states at $\sim$12 MeV) are predicted to exist. See Ref. [36] for more details. $^{(b)}$ We identify the calculated 20.445 MeV state as the experimental 16.75 MeV state, because the calculated state is dominated by the $(0s)^3(0p)^2$ configuration.

$N_3 + N_4 \leq 8$ [i.e., $N_{\text{max}} = 8$]. For the negative-parity states, we use a 8-major-shell space and include all configurations with $N_{\text{sum}} \leq N_{\text{max}} = 7$. The lowest configuration in this nucleus is $(0s)^4$, which has $N_{\text{sum}} = 0$, so we are performing a full $8\hbar\Omega$ ($7\hbar\Omega$) calculation for the positive-parity (negative-parity) states. The calculations involve ($N_{\text{max}} + 1$) G-matrices, corresponding to ($N_{\text{max}} + 1$) possible values of $N_{\text{sum}}^{\text{spectators}}$ (from 0 to N_{max}).

The parameter Δ in the starting energy is chosen to be -55 MeV, which, for ^{4}He, yields a reasonable binding energy of 26.3 MeV compared to the experimental value of 28.8 MeV. It should be pointed out that due to the large size of the model space, the G-matrix elements are a very smooth function of Δ. The binding energy of ^{4}He increases by less than 1 MeV when Δ is increased by 10 MeV from -60 MeV to -50 MeV.

The calculated results are given in Table 3 along with the experimental data, taken from a recent compilation of Tilley et al. [35] and Tanihata et al. [38]. Very good agreement with experiment is obtained for the energy spectrum. In particular, the experimental low-lying negative-parity ("$1\hbar\Omega$") states are reproduced to within 1.2 MeV with a correct level sequence. The first excited 0^+ (predominantly "$2\hbar\Omega$") state is obtained at an excitation energy of 21.8 MeV, only 1.6 MeV higher than experiment.

The major differences in the spectra resulting from the present work and Refs. [23,24] appear in the lowering of the excited states due to increased admixtures of higher lying configurations for the reasons mentioned above. For example, the 0_2^+ state is lowered by 11.9 MeV from its excitation energy in Ref. [24] and by 4.3 MeV from its excitation energy in Ref. [23]. On the other hand, the 0_1^- state is lowered by only 0.8 MeV and 1.3 MeV relative to its excitation energy in Ref. [24] and Ref. [23], respectively.

Since there are a number of differences between the present work and our previous efforts, we will discuss in Section 3.3 the dependence of our results on model space size alone with all other ingredients in the calculations held fixed.

B. ^{6}Li. For this nucleus, we perform a full $6\hbar\Omega$ calculation ($N_{\text{max}} = 8$) for the positive-parity states and a full $5\hbar\Omega$ calculation ($N_{\text{max}} = 7$) for the negative-parity states. The results are shown in Table 4. The six low-lying states known experimentally are nicely reproduced except that the $J^\pi = 2^+$, $T = 1$ state at 5.37 MeV and the $J^\pi = 1^+$ $T = 1$ state at 5.65 MeV are obtained at excitation energies about 1 MeV too high. The other four "$0\hbar\Omega$" states are obtained at excitation energies of 9.94, 10.74, 11.38 and 12.93 MeV. The new results presented here again show some improvement over the previous results [23,24]. In particular, the member of the 0^+ isospin triplet state is obtained at an excitation energy of 3.79 MeV, close to the experimental value of 3.56 MeV. This state is of some interest for the study of the isospin and parity violations [39].

The g.s. magnetic dipole moment is calculated to be 0.840 μ_N, slightly larger than the experimental value of 0.822 μ_N. The g.s. quadrupole moment is calculated to be $-0.067\, e\,\text{fm}^2$, very close to the experimental value of $-0.082\, e\,\text{fm}^2$. These results are obtained by using bare electromagnetic operators. In principle,

Table 3. The results for ^{4}He from a full $8\hbar\Omega$ $[N_{\max} = 8]$ calculation for the positive-parity states and a full $7\hbar\Omega$ $[N_{\max} = 7]$ calculation ($\hbar\Omega = 14$ MeV) for the negative-parity states. In the Table, E_B is the binding energy and $E_x(J_n^\pi, T)$ the excitation energy of the J_n^π, T state. All energies are in MeV. The dominant major-shell configuration for each state is given in the column labeled "Main Conf.". The g.s. rms *point* radius for protons $\sqrt{\langle r_p^2 \rangle}$ is also given. The "experimental" g.s. point particle rms radius is deduced from the charge rms radius $\sqrt{\langle r_c^2 \rangle}$ through $\langle r_p^2 \rangle = \langle r_c^2 \rangle - 0.81^2$ to correct for the finite proton charge radius contribution. We have removed CM effects from the theoretical rms radius, but we have ignored the neutron charge distribution and other higher-order effects.

Observable	Main Conf.	Multi-valued G	Experiment[a]
E_B	—	26.459	28.296
$\sqrt{\langle r_p^2 \rangle}$ (fm)	—	1.492	1.46
$E_x(0_1^+, 0)$	$0\hbar\Omega$	0	0
$E_x(0_2^+, 0)$	$2\hbar\Omega$	21.824	20.21
$E_x(0_1^-, 0)$	$1\hbar\Omega$	21.566	21.01
$E_x(2_1^-, 0)$	$1\hbar\Omega$	23.003	21.84
$E_x(2_1^-, 1)$	$1\hbar\Omega$	24.214	23.33
$E_x(1_1^-, 1)$	$1\hbar\Omega$	24.418	23.64
$E_x(1_1^-, 0)$	$1\hbar\Omega$	25.286	24.25
$E_x(0_1^-, 1)$	$1\hbar\Omega$	25.370	25.28
$E_x(1_2^-, 1)$	$1\hbar\Omega$	25.671	25.95

[a] From Ref. [35], except for the rms radius, which is from Ref. [38].

these electromagnetic operators should also be renormalized in a way consistent with how the effective interaction is derived from the bare NN potential. This is particularly important when the model space is small. While we hope that our model spaces are large enough to permit the use of bare operators, we are aware that this assumption ought to be verified by explicit calculations of effective operators.

3.3   Dependence on the Size of the Model Space

We now examine, as a function of model space size, the differences arising from the use of a multi-valued G-matrix, rather than a conventional single-valued effective interaction. These differences are expected to diminish as the model space is increased due to increasing energy denominators in Eq.(17). In Table 5, the calculated energy and root-mean-square (rms) proton point radius of the ground state and the excitation energy of the first excited state in ^{4}He are given for four different model spaces ($N_{\max} = 2, 4, 6$ and 8) and two choices of $\hbar\Omega$

Table 4. The results for ^{6}Li from a full $6\hbar\Omega$ [$N_{\max} = 8$] calculation for the positive-parity states and a full $5\hbar\Omega$ [$N_{\max} = 7$] calculation ($\hbar\Omega = 14$ MeV) for the negative-parity states. Calculated states with an excitation energy larger than 18 MeV are not shown. See the caption of Table 3 for more explanations.

Observable	Main Conf.	Multi-valued G	Experiment[a]
E_B	—	30.525	31.996
$\sqrt{\langle r_p^2 \rangle}$ (fm)	—	2.11	2.41
$\mu(\mu_N)$	—	0.840	0.822
$Q(e\,\mathrm{fm}^2)$	—	-0.067	-0.082
$E_x(1_1^+,0)$	$0\hbar\Omega$	0	0
$E_x(3_1^+,0)$	$0\hbar\Omega$	2.619	2.186
$E_x(0_1^+,1)$	$0\hbar\Omega$	3.786	3.563
$E_x(2_1^+,0)$	$0\hbar\Omega$	4.713	4.31
$E_x(2_1^+,1)$	$0\hbar\Omega$	6.406	5.366
$E_x(1_2^+,0)$	$0\hbar\Omega$	6.764	5.65
$E_x(2_2^+,1)$	$0\hbar\Omega$	9.942	N/A
$E_x(1_1^+,1)$	$0\hbar\Omega$	10.742	N/A
$E_x(2_1^-,0)$	$1\hbar\Omega$	10.863	N/A
$E_x(1_1^-,0)$	$1\hbar\Omega$	11.082	N/A
$E_x(1_3^+,0)$	$0\hbar\Omega$	11.382	N/A
$E_x(0_2^+,1)$	$0\hbar\Omega$	12.934	N/A
$E_x(0_1^-,0)$	$1\hbar\Omega$	13.147	N/A
$E_x(1_1^-,1)$	$1\hbar\Omega$	13.706	N/A
$E_x(2_1^-,1)$	$1\hbar\Omega$	14.242	N/A
$E_x(1_4^+,0)$	$2\hbar\Omega$	14.716	N/A
$E_x(1_2^-,0)$	$1\hbar\Omega$	15.422	N/A
$E_x(3_2^+,0)$	$2\hbar\Omega$	16.083	15.8
$E_x(2_2^-,0)$	$1\hbar\Omega$	16.950	N/A
$E_x(0_1^-,1)$	$1\hbar\Omega$	17.328	N/A
$E_x(0_3^+,1)$	$2\hbar\Omega$	17.515	N/A

[a] From Ref. [36], except for the rms radius, which is from Ref. [40].

(14 and 20 MeV). As expected, the differences between the excitation energies obtained in the conventional and multi-valued G-matrix calculations diminish as the model spaces increase.

Similarly, the choice of $\hbar\Omega$ becomes less important in the larger model spaces. Note in particular that the calculated g.s. rms radius is about the same (~ 1.49 fm) in the $8\hbar\Omega$, multi-valued G calculations for the two values of $\hbar\Omega$, indicating good convergence for this quantity.

It is clear from Table 5 that the increased size of the model space and the use of an appropriate (multi-valued) G-matrix both contribute to the improved results for the 0_2^+ state in ^{4}He. For example, in a conventional (single-valued) G-matrix calculation with $\hbar\Omega = 14\,\text{MeV}$, the excitation energy of this state decreases by 0.55 MeV from 22.93 MeV to 22.38 MeV when we go from a $6\hbar\Omega$ space to a $8\hbar\Omega$ space; in the $8\hbar\Omega$ space, the use of the multi-valued G-matrix further decreases the result by another 0.56 MeV to 21.82 MeV.

Table 5. The results for the g.s. energy (in MeV), proton rms radius (in fm) and the excitation energy (in MeV) of the first excited state in ^{4}He obtained in the multi-valued G (m-G) and single-G (s-G) calculations in different model spaces with two choices of $\hbar\Omega$ (14 and 20 MeV). The difference between the s-G and m-G results is also given.

$\hbar\Omega$	$N_{\max}$	Approach	$E(0_1^+)$	$\sqrt{\langle r_p^2 \rangle}$	$E_x(0_2^+)$
14	2	s-G	-23.18	1.57	26.38
		m-G	-23.64	1.56	25.17
		diff.	0.46	0.01	1.21
	4	s-G	-25.23	1.57	26.73
		m-G	-25.95	1.56	25.78
		diff.	0.72	0.01	0.95
	6	s-G	-25.62	1.51	22.93
		m-G	-26.44	1.49	22.27
		diff.	0.82	0.02	0.66
	8	s-G	-25.62	1.51	22.38
		m-G	-26.46	1.49	21.82
		diff.	0.84	0.02	0.56
20	2	s-G	-25.62	1.38	33.05
		m-G	-25.94	1.37	30.56
		diff.	0.32	0.01	2.49
	4	s-G	-26.34	1.46	31.84
		m-G	-26.84	1.45	30.23
		diff.	0.50	0.01	1.61
	6	s-G	-25.73	1.46	26.93
		m-G	-26.27	1.46	25.49
		diff.	0.54	0.00	1.44
	8	s-G	-25.21	1.49	24.71
		m-G	-25.82	1.48	23.35
		diff.	0.61	0.01	1.36
Experiment			-28.30	1.46	20.21

As a further comparison, we present in Table 6 some very recent results by Navrátil et al. [21] for ^{4}He using both the single-G and multi-valued-G approaches in an $N_{max} = 8$, $\hbar\Omega = 22$ MeV calculation. These calculations clearly

Table 6. Comparison of the g.s. energy (in MeV) and the excitation energies (in MeV) for ^{4}He from single-G and multi-valued-G calculations, including -U insertions, but not the Coulomb interaction, for $N_{max} = 8$ (a full $8\hbar\Omega$ calculation) and $\hbar\Omega = 22$ MeV.

$E(J^\pi, T)$	Single-G	Multi-G	Expt. [a]
E_B	29.552	30.325	28.296
$0_1^+, 0$	0	0	0
$0_2^+, 0$	26.383	24.458	20.21
$0_1^-, 0$	24.926	23.309	21.01
$2_1^-, 0$	26.639	25.009	21.84
$2_1^-, 1$	28.299	26.603	23.33
$1_1^-, 1$	29.150	27.059	23.64
$1_1^-, 0$	30.721	28.205	24.25
$0_1^-, 1$	30.512	28.319	25.28
$1_2^-, 1$	31.257	28.636	25.95

[a] From Ref. [35]

show the big shift in first excited $J = 0^+$, $T = 0$ state in going from the single-G to the multi-valued-G, bringing it almost below the $J = 0^-$, $T = 0$ level. Similarly, significant shifts are also obtained for the "$1\hbar\Omega$" negative parity excited states. The binding energies are larger in these calculations, because the Kuo-Krenciglowa technique [31] is used to obtain starting-energy independent G-matrices; there is no Δ parameter for making a fit to the binding energy. The -U SP insertions have been included in these calculations to all orders, but there is no Coulomb potential. The 1 to 2 MeV overbinding again shows the importance of omitted, repulsive many-body forces in obtaining the correct binding energy, although the result is rather good considering that only a two-body effective interaction is used in the calculations.

4   Conclusions

We have presented a formalism for constructing the two-body effective interaction to be used in multi-major-shell no-core model spaces, which should be free of the intruder-state problem due to the use of large $N\hbar\Omega$ basis spaces. In particular, a modification has been introduced to take account of the dependence of the two-body G-matrix on the unperturbed energy of the other $A - 2$ nucleons. We have used such a multi-valued G-matrix in large, no-core, shell-model calculations for light nuclei. When compared to conventional calculations, proper treatment of the $N_{\text{sum}}^{\text{spectators}}$ dependence of the G-matrix tends to lower the energies of the "$1\hbar\Omega$" and "$2\hbar\Omega$" excited states more than the "$0\hbar\Omega$" states, bringing energies into better agreement with experiment.

Applying this approach to large, no-core, shell-model calculations, we have achieved a reasonable description of the "low-lying" states (including "$1\hbar\Omega$" and "$2\hbar\Omega$" states) in light nuclei. With model spaces consisting of as many as nine HO major shells, the experimentally known states in ^{4}He, ^{5}He, ^{6}Li, and ^{7}Li have been obtained with improved agreement between theory and experiment.

The magnetic dipole and electric quadrupole moments, calculated using bare operators with meson-exchange-current effects neglected, are also in reasonable agreement with experiment. The Coulomb interaction accounts for the bulk part of the differences in the experimental binding energies of mirror pairs (^{3}H-^{3}He and ^{5}He-^{5}Li).

By using large, no-core model spaces, we have eliminated the need for adjustable SP energies conventionally involved with shell-model calculations using effective interactions. In other words, the SP behavior must arise from the underlying microscopic effective Hamiltonian. In this sense we have performed *ab initio* shell model calculations. However, it should be emphasized that in calculating the G matrices, we have used an empirical prescription for the starting energy, which involves a parameter Δ. This parameter is adjusted to yield a reasonable binding energy. For this reason, our calculated binding energies should not be interpreted as exact results, which can only be obtained through a parameter-free approach. Nevertheless, we observe that once this parameter is adjusted to reproduce the binding energy, other nuclear properties are predicted with satisfying outcomes.

There are important improvements that could be incorporated into future calculations of the type reported here. Our use of very large model spaces was motivated by the hope that bare operators and effective interactions approximated by a two-body G-matrix might be successful in such spaces. Presumably the need for large values of Δ is connected with the omission of the folded diagrams and neglected interactions of three-body and higher clusters in the excluded space. As there are prospects for improving these aspects of the calculations [25], we consider the present effort a first step toward the ultimate goal of accurate shell model calculations based on realistic NN interactions.

If one were able to generate the exact $\mathcal{V}$, energy eigenvalues should not depend on the choice of the model space. Thus perhaps the most important result from present investigations using multi-valued G-matrices is that some improvement was achieved in the rate of convergence of energy eigenvalues, as a function of the complexity of the model space (see, for example, Table 5). We would argue that the degree to which our results can be further improved is an open question: clearly we have the capacity to put substantial new physics into calculations of $\mathcal{V}$ and to generate the corresponding effective operators. Work along these lines is in progress.

5 Acknowledgments

B.R.B., D.C.Z. and P.N. acknowledge partial support by the National Science Foundation, Grant No. PHY-93-21668. J.P.V. acknowledges partial support by the U.S. Department of Energy under Grant No. DE-FG02-87ER-40371, Division of High Energy and Nuclear Physics and the Alexander von Humboldt Foundation. W.C.H. and C.L.S acknowledge partial support of this work by the U.S. Department of Energy.

References

1. B.R. Barrett and M.W. Kirson, *in:* "Advances in Nuclear Physics", Vol. 6, M. Baranger and E. Vogt, ed., (Plenum Press, New York, 1973), 219; T. T. S. Kuo, *Ann Rev. Nucl. Sci* 24 (1974) 101; P. J. Ellis and E. Osnes, *Rev. Mod. Phys.* 49 (1977) 777; "Proc. International Conference Effective Interactions and Operators in Nuclei," ed. B. R. Barrett, Lecture Notes in Physics, Vol. 40 (Springer, Verlag, Berlin, 1975), and references therein.
2. M. Hjorth-Jensen, T. T. S. Kuo and E. Osnes, *Phys. Rep.* 261 (1995) 125.
3. M.W. Kirson, *in:* "Nuclear Shell Models", M. Vallieres and B.H. Wildenthal, ed., (World Scientific, Singapore, 1985), 290
4. C. Bloch and J. Horowitz, *Nucl. Phys.* 8 (1958) 91.
5. B.H. Brandow, *Rev. Mod. Phys.* 39 (1967) 711.
6. M. B. Johnson and M. Baranger, *Ann. Phys. (N.Y.)* 62 (1971) 172.
7. T. T. S. Kuo, S. Y. Lee and K. F. Ratcliff, *Nucl. Phys.* A176 (1971) 65.
8. H. Kümmel, K. H. Lührmann and J. G. Zabolitzky, *Phys. Rep.* C36 (1978) 1.
9. B. D. Chang and J. P. Draayer, *Phys. Rev. C* 20 (1979) 2387; B. D. Chang, *Nucl. Phys.* A304 (1978) 217.
10. J. M. Irvine et al., *Ann. Phys.* (N.Y.) 102 (1976) 129.
11. K. A. Brueckner, *Phys. Rev.* 97 (1955) 1353; ibid 100 (1955) 36.
12. H. Feshbach, *Ann. Phys. (N. Y.)* 19 (1962) 287.
13. T. Schucan and H. A. Weidenmüller, *Ann. of Phys.* 73 (1972) 108; ibid 76 (1973) 483.
14. J. P. Vary, in: "Theory and Applications of Moment Methods in Many-Fermion Systems", B. J. Dalton, S. M. Grimes, J. P. Vary, and S. A. Williams, ed., (Plenum Press, New York, 1980) 423.
15. L. Jaqua, P. Halse, B. R. Barrett and J. P. Vary, *Nucl. Phys.* A571 (1994) 242.
16. S. Y. Lee and K. Suzuki, *Phys. Lett.* 91B (1980) 79; K. Suzuki and S. Y. Lee, *Prog. of Theor. Phys.* 64 (1980) 2091.
17. T. T. S. Kuo, in "Lecture Notes in Physics", Vol. 144, T. T. S. Kuo and S. S. M. Wong, ed., (Springer, Berlin, 1981), 248.
18. D. C. Zheng, J. P. Vary and B. R. Barrett, *Nucl. Phys.* A560 (1993) 211.

19. D. C. Zheng and B. R. Barrett, *Phys. Rev.* C49 (1994) 3342.
20. L. Jaqua, D. C. Zheng, B. R. Barrett and J. P. Vary, *Phys. Rev.* C48 (1993) 1765.
21. P. Navrátil, B. R. Barrett and D. C. Zheng, in preparation.
22. D. C. Zheng, B. R. Barrett, J. P. Vary and H. Müther, *Phys. Rev.* C51 (1995) 2471.
23. D. C. Zheng, J. P. Vary, and B. R. Barrett, *Phys. Rev.* C50 (1994) 2841.
24. D. C. Zheng, B. R. Barrett, L. Jaqua, J. P. Vary and R. J. McCarthy, *Phys. Rev.* C48 (1993) 1083.
25. W. C. Haxton, C. L. Song, J. P. Vary, B. R. Barrett and D. C. Zheng, in preparation.
26. P. Navrátil and B. R. Barrett, *Phys. Lett.* B369 (1996) 193.
27. D. C. Zheng, B. R. Barrett, J. P. Vary, and R. J. McCarthy, *Phys. Rev.* C49 (1994) 1999.
28. T. T. S. Kuo and G. E. Brown, *Nucl. Phys.* 85 (1966) 40.
29. B. R. Barrett, R. G. L. Hewitt and R. J. McCarthy, *Phys. Rev.* C3 (1971) 1137.
30. J. P. Vary and S. N. Yang, *Phys. Rev.* C15 (1977) 1545.
31. T. T. S. Kuo and E. M. Krenciglowa, *Nucl. Phys.* A342 (1980) 454.
32. V. J. G. Stoks, R. A. M. Klomp, C. P. F. Terheggen and J. J. de Swart, *Phys. Rev.* C49 (1994) 2950.
33. J. P. Vary and D. C. Zheng, "The Many-Fermion-Dynamics Shell-Model Code", Iowa State University (1994) (unpublished).
34. D. R. Tilley, H. R. Weller and H. H. Hasan, *Nucl. Phys.* A474 (1987) 1.
35. D. R. Tilley, H. R. Weller and H. H. Hasan, *Nucl. Phys.* A541 (1992) 1.
36. F. Ajzenberg-Selove, *Nucl. Phys.* A490 (1988) 1.
37. D. C. Zheng, B. R. Barrett, J. P. Vary, W. C. Haxton and C.-L. Song, *Phys. Rev.* C52 (1995) 2488.
38. I. Tanihata et al., *Phys. Rev. Lett.* 55 (1985) 2676.
39. R. G. H. Robertson et al., *Phys. Rev.* C29 (1984) 755; R. G. H. Robertson and B. A. Brown, *Phys. Rev.* C28 (1983) 443.
40. R. C. Barrett and D. F. Jackson, "Nuclear Sizes and Structure", (Clarendon, Oxford, 1977), 146.

Order and Disorder in the Nuclear Shell Model

B. Alex Brown[1,2], Vladimir Zelevinsky[1,2,3], Mihai Horoi[1,4,5] and Njema Frazier[1,2]

[1] National Superconducting Cyclotron Laboratory, Michigan State University,
    East Lansing, MI 48824-1321
[2] Department of Physics and Astronomy, Michigan State University,
    East Lansing, MI 48824-1116
[3] Budker Institute of Nuclear Physics, Novosibirsk 630090, Russia
[4] Physics Department, Central Michigan University,
    Mount Pleasant, MI 48859
[5] Institute of Atomic Physics, Bucharest, Romania

Abstract. We discuss how the large-basis shell-model wave functions can be used to understand the transition from order in the low-lying nuclear levels to disorder at high excitation energy. The level spacing at high excitation energy shows the usual Wigner (GOE) distribution. But other properties at both low and high excitation show deviations from the GOE. The strength distribution of basis states amplitudes in the eigenfunctions is not uniform but evolves from a Breit-Wigner shape at weak off-diagonal interaction strengths to a Gaussian shape for the normal off-diagonal interaction strength. We define an information entropy in the shell-model basis for each eigenstate which shows a smooth increase at low excitation energy to a value near the GOE limit at high excitation energy. This information entropy has a temperature associated with it which is nearly identical to the temperature obtained from the level density and from the Fermi distribution function for the single-particle occupancies. The interrelation between quantum chaos, Fermi liquid theory and thermalization is discussed.

1 Introduction

The first ideas about nuclear structure were dominated by the concept of the complex motion of the nucleons in the compound nucleus introduced by Niels Bohr (Bohr 1936). But by the late 1940's much experimental data had accumulated on the properties of low-lying nuclear states which showed many regularities, and this eventually led to the nuclear shell model. The nuclear shell model with its simple motion of a single nucleon in the mean-field of other nucleons is a model of order. At high excitation energy, disorder begins to take over and the compound nucleus picture is more appropriate. There are many interesting problems which arise in understanding the transition from order at low excitation energy to disorder at higher excitation energy. How fast does the transition occur? How can it be measured? How much "disorder" is there in the low-lying states? How much "order" is left at high excitation energy? What is the physics behind this transition?

One can try to make an analogy to the transition from classical periodic (regular) motion to classical chaos. The main problem is to understand the relationship between classical mechanics and quantum mechanics in this respect. This relationship is best understood for very simple systems such as (one-body) stadium billiards (Bohigas and Giannoni 1984, Berry and Robnik 1986, Gräf et al. 1992). In this case the classical system with regular (chaotic) trajectories is found to correspond to the quantum system with Poisson (Wigner) level spacings. The Wigner level spacing first emerged from Wigner's analysis of the compound nucleus in terms of matrices with random matrix elements – the Gaussian Orthogonal Ensemble (GOE). This provides a starting point for making some connections between the understood aspects of one-body quantum chaos and the actual properties of quantum many-body states at high level density. Even though many-body quantum chaos is not yet well defined, the number of publications related to it increases exponentially, see for example (Porter 1965, Bohr and Mottelson 1969, Brody et al. 1981, Elyutin 1988, Bohigas and Weidenmüller 1988, Schuster 1989, Izrailev 1990, Haake 1991) and references therein.

One way to answer these questions is to analyze the results from extended (large-basis) shell-model calculations. Such calculations not only reproduce the experimental properties of the low-lying states, but also have a dense spectrum of states at higher excitation energy. Although there are some aspects of the shell-model matrix which resemble the GOE, there are also many differences, and it is these differences which lead to the transition between order and disorder (from order to chaos). Some results from these shell-model matrices have already been investigated. In particular, it is known that the level spacing changes from Poisson at low-excitation energy to Wigner (GOE) at high excitation energy, and the basis vectors distribution changes from exponential at low-excitation energy to Gaussian (Porter-Thomas) at high excitation energy (Brown and Bertsch 1984, Ormand and Broglia 1992, Auerbach and Brown 1994, Lopac et al. 1990 , Alhassid and Novoselsky 1992). In the present work (Zelevinsky et al. 1995, Horoi et al. 1995, Zelevinsky et al. 1996) we investigate other properties which will help to define many-body quantum chaos as well as to make a connection to thermodynamics. The advantages of using a nuclear physics model are:

- The nucleus is a strongly interacting Fermi system.
- Nuclear spectroscopy shows an onset of chaos in the local level statistics at relatively low excitation energy of 3-5 MeV (Mitchell et al. 1988, Shriner et al. 1990, Raman et al. 1991, Garrett et al. 1991).
- Many experimental results have been interpreted in terms of chaotic dynamics, for example, the Porter-Thomas distribution of widths of neutron and proton resonances (Porter 1965, Bohr and Mottelson 1969, Brody et al. 1981), enhancement of weak interaction and parity nonconservation in fission (Sushkov and Flambaum 1980, Sushkov and Flambaum 1982, Flambaum and Gribakin 1995), spreading width of isobaric analog states (Harney et al. 1986, Zelevinsky and von Brentano 1991), saturation of widths of giant resonances (Gaardhøje 1992, Lauritzen et al. 1995), the narrowing of multiple giant excitations (Bertulani and Zelevinsky 1994, Lewenkopf and Zelevinsky 1994), and fluctuations in rotational cascades (Lauritzen et al.

1986, Matsuo et al. 1993, Døssing et al. 1996).
- Shell-model calculations determine the semiempirical hamiltonians which
  work well in the region of available spectroscopic information (Brown and
  Wildenthal 1988) and can be extrapolated beyond this region.
- It is possible to satisfy the conservation laws and ensure that the eigenstates
  have correct exact quantum numbers.
- The corresponding dimensions, of the order 10^3, are sufficient for obtaining statistically reliable results, and, at the same time, are practical to be
  effectively and rapidly handled (Brown et al. 1988).

2   Shell-model calculations

We base our analysis on the exact diagonalization of the effective semiempirical
hamiltonian H in a large $N \times N$ Hilbert space spanned by a truncated set of
shell-model configurations. Each configuration is characterized by the distribution ("partition") of independent fermions over available spherical single-particle
orbitals. Within a configuration, various ways to occupy the magnetic substates
of the j-levels give rise to the "m-scheme" Slater determinants.

The hamiltonian H keeps rotational and isospin invariance. The necessity
of using the appropriate $J^\pi T$ states was demonstrated in the first studies of
quantum chaos in the shell model (Brown and Bertsch 1984). The basis states
$|k\rangle$ have good quantum numbers of the total angular momentum J, its projection
M, parity π, isospin T and its projection T_3. Therefore they are far from being
simple Slater determinants. This "premixing" is absent in the analysis of high
spin rotational bands in the framework of the cranking model (Matsuo et al.
1993, Døssing et al. 1996).

The effective shell-model hamiltonian H consists of the independent particle
(one-body) part H_0 and the residual interaction H' of the two-body type. The
unperturbed hamiltonian H_0 describes noninteracting fermions in the mean field
of the appropriate spherical core. In the projected basis $|k\rangle$, the residual interac-
tion H' has both diagonal and off-diagonal matrix elements. The diagonal part
already lifts some degeneracy within a partition. Full diagonalization in each
sector with given exact quantum numbers leads to the stationary states $|\alpha\rangle$,

$$H|\alpha\rangle = E_\alpha|\alpha\rangle, \tag{1}$$

which can be represented by superpositions of unperturbed states $|k\rangle$,

$$|\alpha\rangle = \sum_k C_k^\alpha |k\rangle. \tag{2}$$

The orthonormalized amplitudes C_k^α can be taken as real in the case of the
interaction invariant under time reversal. A number N_α of significant compo-
nents $|k\rangle$ characterizes the delocalization of a state $|\alpha\rangle$ in the given basis. The
corresponding amplitudes have an order of magnitude $N_\alpha^{-1/2}$. A completely delocalized function would have N_α close to the space dimension. Direct estimates

(Sushkov and Flambaum 1980, Sushkov and Flambaum 1982) show that the matrix elements of simple operators between such complicated states are suppressed $\sim N_\alpha^{-1/2}$ if the coefficients (2) are chaotic ("N-scaling"), the phase coherence is absent and only the weights $W_k^\alpha \sim N_\alpha^{-1}$ are important for the estimates. With the level density enhanced roughly by a factor N_α, this qualitatively explains enhancement of weak interactions and saturation of the spreading widths.

We will discuss mostly the results for a system of valence particles in one major shell, for example 12 particles in the sd shell (24 states including $0d_{5/2}$, $0d_{3/2}$ and $1s_{1/2}$) when the one-body part of the total hamiltonian is due to the core ^{16}O. This system mimics (for $T_3 = 0$) a subset of states in the ^{28}Si nucleus. We use the Wildenthal hamiltonian (Brown and Wildenthal 1988) obtained by fitting more than 400 binding energies and excitation energies for the sd-shell nuclei; for the recent comparison of experiment and theory in ^{28}Si see (Brenneisen et al. 1995). Similar studies were performed using the "proton-neutron" $(p-n)$ formalism with no explicit isospin where we used, along with the Wildenthal interaction rewritten for the $p-n$ scheme (WPN), the interaction (WPNC) which explicitly violates isospin on the experimentally allowed level (Ormand and Brown 1989). All calculations were carried out with the computer program OXBASH (Brown et al. 1988), which uses the m-scheme basis states together with the projection operators to construct and diagonalize matrices with good J and T in the isospin formalism and good J in the $p-n$ formalism.

Taking into account J and T conservation, there are 63 non-vanishing two-body matrix elements $\langle (j_1 j_2)_{JT} | H' | (j_3 j_4)_{JT} \rangle$ in sd-shell space. Being in general of the order of an MeV, the two-body matrix elements are not random; for example they show pair correlations in the $T = 1$ states with even angular momenta. In our calculations, we mostly use the $(0d1s)^{12}$, $J^\pi T = 2^+0$ and 0^+0 classes of states, with dimensions 3276 and 839, respectively.

The signatures of quantum chaos come exclusively from the particle interaction (H_0 is very simple and does not lead to one-body chaos). Some relevant properties of the hamiltonian can be noticed prior to actual diagonalization. Long ago such analyses were carried out by statistical spectroscopy (French and Ratcliff 1971, Ratcliff 1971).

Diagonal matrix elements are dominated by the one-body part H_0. The two-body diagonal contributions split the degenerate levels within the partitions so that each partition partly overlaps with the other partitions. In our $A = 28$ example, the diagonal part of the hamiltonian is spread from -120 MeV to -60 MeV. The basis states $|k\rangle$ will be mixed by the off-diagonal part H'. Due to the premixing related to J and T projection, the two-body matrix elements between the many-body basis states are reduced by an order of magnitude. as it should be according to the N-scaling (Sushkov and Flambaum 1980, Sushkov and Flambaum 1982): the dimension of a partition is typically 10^2 and the reduction factor is $\sim N^{-1/2}$.

Without the full knowledge of the strength function, one can describe the fragmentation of simple states by its lowest moments (French and Ratcliff 1971, Ratcliff 1971). The centroid $\bar{E}_k$ of the energy distribution of the basis state

$|k\rangle$ coincides with the diagonal element H_{kk}. The spread of the unperturbed energies due to the diagonal elements of the interaction can be characterized by the rms deviation Δ_E of the centroids $\bar{E}_k$ from the energy center; for the 2^+0 states $\Delta_E \approx 8$ MeV. The energy dispersion of basis states due to the off-diagonal interaction is

$$\sigma_k^2 \equiv \sum_\alpha (E_\alpha - \bar{E}_k)^2 W_k^\alpha = \sum_{l(\neq k)} (H_{lk}')^2. \qquad (3)$$

The energy dispersion (3) of individual 2^+0 basis states turns out (Fig. 1) to be uniform, $\sigma_k \approx \bar{\sigma} \approx 10$ MeV over the entire space. The remnants of the partition structure are visible at the low edges of partitions. This might be caused by a random choice of the initial simple states in the projection procedure. The dispersion σ_k is closely related to the spreading width (in the "strong coupling" case (Lauritzen et al. 1995, Frazier et al. 1996) $\Gamma \simeq 2\bar{\sigma}$). The uniformity of the dispersion supports the idea of saturation (Lauritzen et al. 1995) which has important consequences for damping of giant resonances.

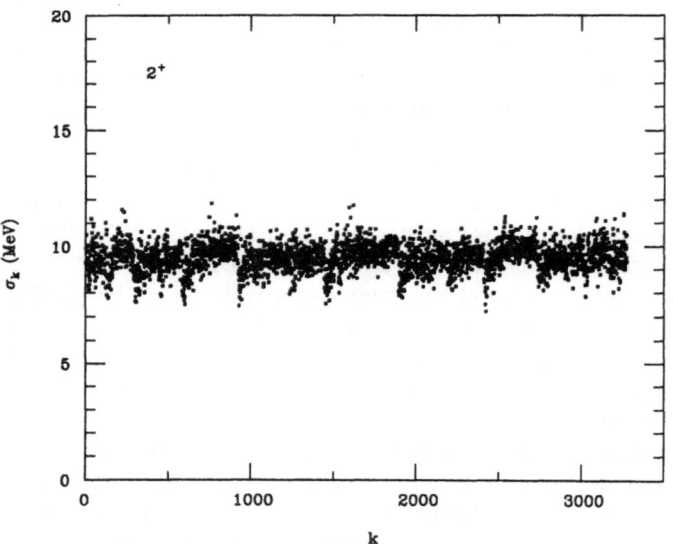

Fig. 1. Energy dispersions σ_k for the basis 2^+0 states.

Since variations of the effective spreading (3) of the basis states around the mean value $\bar{\sigma}$ are small, one can work out (Horoi et al. 1994) a simple truncation method to reduce a huge shell-model hamiltonian matrix to a manageable size. The method was tested for the sd and fp shells and was proven to be very efficient. In the middle of the fp shell (JT dimensions of the order of a few million) the size of the matrix is effectively reduced to a few thousand.

The primary characteristic of the spectrum is the level density

$$\rho(E) = \sum_\alpha \delta(E - E_\alpha). \tag{4}$$

It is normalized to the total number of states, $\int dE\rho(E) = N$. The total dispersion of energy

$$\sigma_E^2 \equiv \frac{1}{N} \int dE(E - \bar{E})^2 \rho(E) = \bar{\sigma}^2 + \Delta_E^2 \tag{5}$$

consists of (added in quadratures) the spread of centroids defined by the diagonal part of the hamiltonian and the fragmentation width (3) due to the off-diagonal part. With the above-mentioned values of $\bar{\sigma}$ and Δ_E for the 2^+0 states we get $\sigma_E = 13$ MeV.

We have investigated in some detail how the shell-model basis states become spread over the eigenstates (Frazier et al. 1996). In the "standard" model, the shape of this distribution has a Breit-Wigner shape with a width given by $\Gamma = 2\pi < V >^2 /D$. It is derived under the assumption that the mean energy spacing D is constant and the average coupling matrix elements $< V >^2$ are constant over the region of spreading. We show in Fig. 2 the actual results for the shape of the spreading as a function of the off-diagonal hamiltonian strength λ, $H = H_{\text{diag}} + \lambda H_{\text{off-diag}}$. The results for the $(0d1s)^8$, $J^\pi T = 0^+0$ states and 0^+0 classes are obtained by adding together the strength functions for 400 states in the middle spectrum. The distributions for each of the states are added after shifting the energy scale for each state so that the centroid is at zero. For small λ the shape is indeed Briet-Wigner. But by the time the interaction reaches its full strength ($\lambda = 1$) the shape deviates very much from Breit-Wigner and is actually closer to Gaussian. Although it is not clear from Fig. 2, the tails of the distribution are neither Breit-Wigner nor Gaussian but fall off exponentially. One reason for the deviation from the standard model at large λ is that the level density is not constant. In the Breit-Wigner region the FWHM increases quadratically in λ but in the Gaussian region it changes over to a linear dependence. These results are discussed in more detail in Frazier et al. 1996.

The random matrix theory usually considers matrix elements of the hamiltonian as random normally distributed variables. Canonical Gaussian ensembles with no regularly increasing diagonal elements have nothing to do with the evolution along the spectrum ("secular" behavior) and can properly account for the local correlations and fluctuations only. Moreover, the actual distribution of off-diagonal matrix elements in the "natural" shell-model basis is quite different from Gaussian. The analysis (Flambaum et al. 1994) of the similar problem for a heavy atom showed that the actual distribution can be written in the form analogous to the Porter-Thomas distribution,

$$P_\kappa(\tilde{H}_{kl}) = \frac{1}{2}[(2\tilde{H})^{\kappa+1}\Gamma(\kappa+1)]^{-1}|\tilde{H}_{kl}|^\kappa e^{-|\tilde{H}_{kl}|/2\tilde{H}} \tag{6}$$

where Γ stands for the Γ-function and κ is a numerical parameter. The distribution of matrix elements found in the shell-model calculations (Wambach

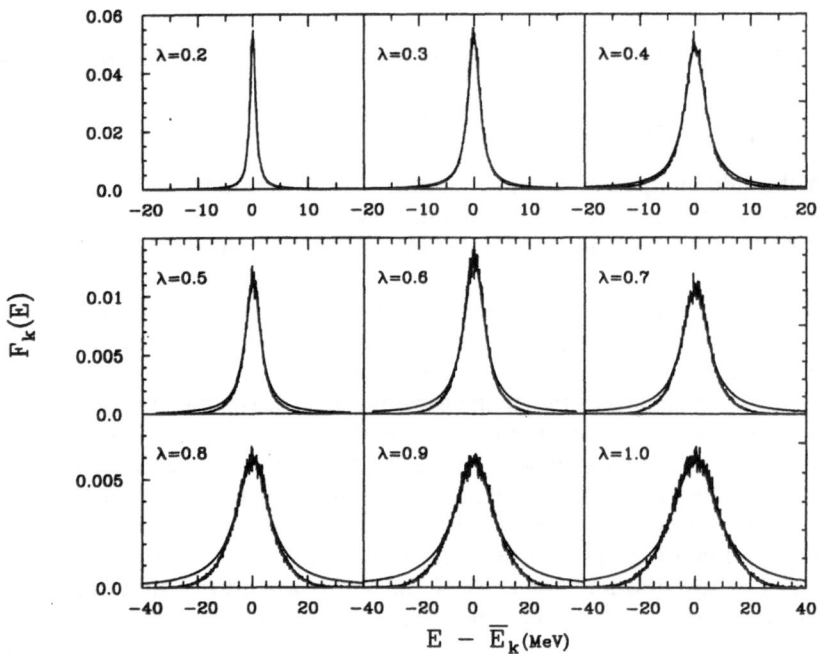

Fig. 2. Strength distribution for the shell-model basis states as a function of λ (histograms). They are compared with a Breit-Wigner shape (solid lines). The y-scale changes from panel to panel.

private communication) and in the interacting boson model (Kusnezov private communication) also agree with (6). Since our study reveals a very similar picture, the conclusion is plausible that this class of distributions is generic for the many-body interactions in heavy atoms or nuclei. For the distribution (6) taken literally for all values of $\tilde{H}_{kl}$, the mean absolute value of $\tilde{H}_{kl}$ is $2(\kappa + 1)\tilde{H}$. The power κ found in Flambaum et al. 1994 is close to the Porter-Thomas value -1/2.

The Porter-Thomas distribution for the reduced widths γ of the resonances follows from the Gaussian distribution for the decay amplitudes A if the proportionality $\gamma \propto |A|^2$ is assumed. Therefore eq.(6) implies that the normally distributed quantities in the realistic case are not the off-diagonal matrix elements themselves, as would be the case in Gaussian random matrix ensembles, but rather some quantities resembling square roots of them. The reason might be the domination of multipole-multipole forces. This is by construction the case in the interacting boson model. The Coulomb interaction in atoms is actually determined by a small number of low multipoles. The specific role of the pairing and quadrupole interactions in nuclei is also well known.

The distribution (6) diverges at small values of the matrix elements if $\kappa \leq -1$. In the actual analysis it is difficult to make a precise fit to this region. Fitting the rest of the histogram we allow all values of κ. The distribution function for 5.36×10^6 off-diagonal matrix elements between the 2^+0 states is shown in Fig. 3. Except for the region around zero and extreme wings corresponding to the exceptionally big elements, the distribution is in good agreement with expression (6) for $\kappa = -2$. The fit covers a change of matrix elements by four orders of magnitude. The off-diagonal matrix elements for the 0^+0 states agree with the distribution (6) at $\kappa = -1$. The origin of the apparent difference in the preexponent factor for different classes of states is not clear at this point.

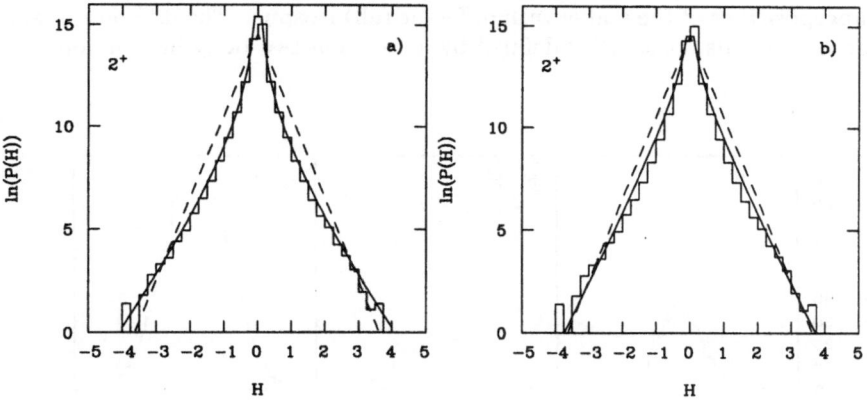

Fig. 3. (a) The distribution of off-diagonal matrix elements (in MeV) between the 2^+0 states (histogram), dashed line - pure exponential fit, eq.(6) with $\kappa = 0$, solid line - fit with $\kappa = -2$. (b) The same as (a) except the solid line is with $\kappa = -1$.

3 Level statistics

The total density $\rho(E)$ of states with given values of exact integrals of motion vanishes at boundaries of the finite spectrum, being maximum in the middle. The GOE predicts the Wigner semicircle rule for the level density. On the other hand, due to the two-body character of interaction there is a noticeable number of vanishing matrix elements in the truncated configuration space, and all many-body matrix elements are determined by a small number of the two-body matrix elements. However the realistic interaction is not strong enough to destroy completely the partition structure. In such cases, we should expect the level density $\rho(E)$ to be closer to the Gaussian shape (Brody et al. 1981). The transition from Gaussian to semicircle level density occurs (French and Wong 1971, Mon and French 1975, Brody et al. 1981) when many-body forces are introduced, lifting the selection rules for interactions between the partitions. The two-body matrix elements are the same for all classes of states with various J and T which can induce the correlations between the classes.

Depicting the level densities for 0^+0 and 2^+0 states as histograms we can fit both of them, Fig. 4, by the Gaussians with the same values of the centroid $E_0 = -90$ MeV and the dispersion $\sigma_E = 13$ MeV predicted above. The Gaussian shape of the level density, with the smaller dispersion $\Delta_E \approx 8$ MeV, is formed entirely due to the combinatorial nature of the fermionic excitation spectrum already for the unperturbed energies H_{kk} with no off-diagonal interaction. In the $p - n$ formalism, we get practically the same results for the level density of the superpositions of the states with different (all) isospins. The different isospin sectors have similar properties defined by a common two-body interaction.

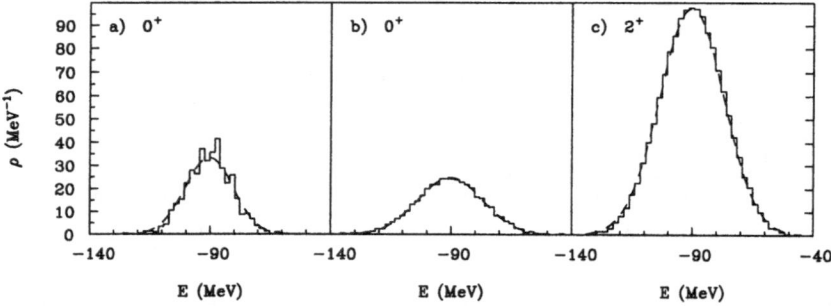

Fig. 4. Level densities for the 0^+0 with no off-diagonal interaction (panel a) and for the 0^+0 and 2^+0 states with the realistic hamiltonian, panels b and c, respectively. The results of calculations are shown by the histograms and Gaussian fits by the dashed lines.

We can also generate the random matrix ensemble defined by the actual exponential distribution of the matrix elements (6). Then the many-body matrix elements are uncorrelated, and the level density agrees with the semicircle law. The difference between the empirical Gaussian level density and the semicircle should be ascribed to the correlations within the many-body hamiltonian determined by a small number of two-body matrix elements regardless of their regularity or randomness. The initial part of the spectrum agrees also with the Fermi-gas level density (Bohr and Mottelson 1969) but at high excitation energy the Fermi-gas approach breaks down due to limitations of the finite Hilbert space.

The degeneracies caused by the shell structure in the spherical mean field are lifted by the residual interaction. In the stochastic limit, the mixing by the off-diagonal hamiltonian leads to the level repulsion and to a more uniform level spacing distribution. It results in the nearest level spacing distribution close to the Wigner surmise

$$P_W(s) = \frac{\pi}{2} s e^{-(\pi/4)s^2}. \tag{7}$$

The linear repulsion and Gaussian tail are the distinctive features of chaotic level statistics in contrast to the Poisson distribution of random events $P_P(s) = e^{-s}$ characteristic for integrable systems. Here $s = (E_{\alpha+1} - E_\alpha)/\bar{D} \equiv \mathcal{E}_{\alpha+1} - \mathcal{E}_\alpha$ is the nearest neighbor spacing in units of the local average spacing $\bar{D}$. This rescaling, or unfolding (Brody et al. 1981), $E_\alpha \to \mathcal{E}_\alpha$, is important to separate local level correlations from the global secular behavior.

Fig. 5 shows the nearest level spacing distribution $P(s)$ for 0^+0 states and the variable residual interaction. The relative intensity of the off-diagonal matrix elements is equal to $\lambda = 0.0, 0.1$ and 0.2, respectively. The noninteracting case reveals the Poisson-like distribution. The transition to the Wigner distribution occurs at $\lambda \approx 0.2$ of the realistic value, when typical off-diagonal matrix elements are of the order of the mean level spacing. The level spacing distribution is universal as can be seen for the other $J^\pi T$ classes. An analysis shows that the interpartition interaction is responsible for establishing the Wigner level spacing distribution. Diagonalization within the single largest partition alone both for 0^+0 and for 2^+0 states, results in a distribution with an excess of small and large spacings. Qualitative features of the Wigner distribution appear after the diagonalization is performed in the model space of the three largest partitions. The $p-n$ formalism reveals the intermediate level spacing distribution due to the absence of mixing and repulsion between levels of different isospin. It is known that a superposition of many independent level sequences leads to the Poisson limit (Gurevich and Pevsner 1957).

An interesting theoretical problem is related to the precise form of the level repulsion at small distances s. The GOE and regular dynamics predict the behavior $P(s) \propto s^\beta$, $s \to 0$, with $\beta = 1$ and $\beta = 0$, respectively. There is no consistent theory explaining how the Poisson level spacing evolves into the Wigner distribution as the stochastization occurs and levels repel each other. Various scenarios include (i) a decreasing finite value of $P(s = 0)$ determined (Berry and Robnik 1984, Berry 1985) by the regular and chaotic volume fractions of the classical phase space (the similar change was found (Sokolov and Zelevinsky 1988, Sokolov and Zelevinsky 1989, Sokolov and Zelevinsky 1992, Mizutori and Zelevinsky 1993) if the decay channels are open and the levels acquire the finite lifetime); (ii) the fractional power law $P(s) \sim s^\beta$, $0 < \beta < 1$ (Prosen and Robnik 1993), used with variable success in interpolation formulae (Brody et al. 1981, Izrailev 1990) (the correlation of β with the localization of wave functions was pointed out in Izrailev 1990); (iii) the linear repulsion in a narrow region of

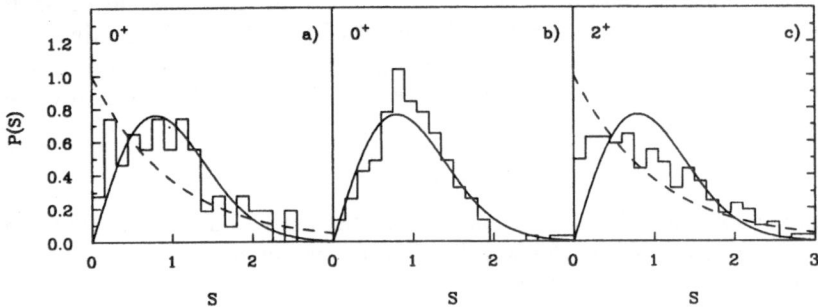

Fig. 5. Nearest level spacing distribution (histograms) compared to the Wigner surmise (solid lines) and the Poisson distribution for the 0^+0 states at different interaction strength, $\lambda = 0$, 0.1 and 0.2 (panels a,b and c, respectively).

spacings comparable to the magnitude of perturbation as implied by perturbation theory; (iv) the logarithmic singularities $\sim \ln(1/s)$ due to the presence of strictly forbidden (Molinari and Sokolov 1989) or exponentially small (Altland and Fuchs 1995) matrix elements. The perturbative arguments do not take into account the abundance of small off-diagonal matrix elements which seems to be a generic feature of realistic systems. In the semiclassical domain it can occur if the classical phase space consists of separated parts. The degeneracy of the states localized in different areas would lead to the Poisson distribution of level spacings. However the quantum tunneling restores the communication between those areas. To study the region of $s \ll 1$ with high precision, one needs much better statistics. It can be achieved combining properly the data for different classes of states and for the variable interaction strength.

It is known (Brody et al. 1981, Bohigas et al. 1984, Berry 1985) that chaotic dynamics lead to rather rigid spectra. The level repulsion creates a sequence of levels which "crystallize"; the fluctuations are suppressed in comparison with a pure random sequence. As an appropriate quantitative measure, the spectral rigidity $\Delta(L)$ is used,

$$\Delta(L) = \langle \min_{A,B} \int_x^{x+L} \frac{d\mathcal{E}}{L} [\mathcal{N}(\mathcal{E}) - A\mathcal{E} - B]^2 \rangle_x. \tag{8}$$

Here the average is taken of integral deviations of the cumulative unfolded level number $\mathcal{N}(\mathcal{E})$ from the best linear fit over various (overlapped) segments of length L. For a random level sequence with the Poisson nearest level spacing distribution, the deviation grows linearly, $\Delta(L) = L/15$. For the chaotic case and the Wigner distribution (7), the spectra are rigid. Starting at small L with the same linear behavior, the deviation grows logarithmically at $L \gg 1$,

$$\Delta(L) \approx \frac{1}{\pi^2} \ln L - 0.007. \tag{9}$$

In the semiclassical limit (Berry 1985), $\Delta(L)$ is expected to saturate at a nonuniversal value of $L \simeq L_{max} \simeq 2\pi\hbar/t_{min}D$ determined by the shortest periodic orbits with a period t_{min}. The number L_{max} measures the Weisskopf recurrence time of a wave packet, $2\pi\hbar/D$, in natural units of the shortest period. At $L > L_{max}$ one expects pseudooscillatory behavior with constant $\Delta(L)$.

Using our 2^+0 states we trace the behavior of $\Delta(L)$ up to very large L. The results, Fig. 6, display an agreement with the GOE prediction with no evidence of saturation up to $L \simeq 150$. At higher L, the spectral rigidity decreases revealing the upbend from the GOE curve. Such behavior is known in "one-body chaos" (anisotropic Kepler problem (Wintgen and Marxer 1988), Sinai billiard (Arve 1991) or the experiment with a superconducting stadium billiard (Gräf et al. 1992)) where the deviations start at much smaller L. According to Mottelson's conjecture (Mottelson unpublished), the remnants of regular behavior determine dynamics at times too short to resolve the signatures of chaos. For the stadium billiard (Gräf et al. 1992) the effect is due to the marginally stable "bouncing ball" orbits. In Arve 1991 the upbend point L_d was associated with the inverse Lyapunov exponents which give the time scale for the development of classical chaos. The upbend of the curve $\Delta(L)$ starting at $L \approx 200$ corresponds to the energy interval $\delta\epsilon \simeq 7$ MeV. This behavior lasts up to the highest values of $L \approx 2000$ attainable for our computations.

An additional analysis has to be performed to pin down the factors responsible for the upbend. In many-body dynamics within one major shell, the available

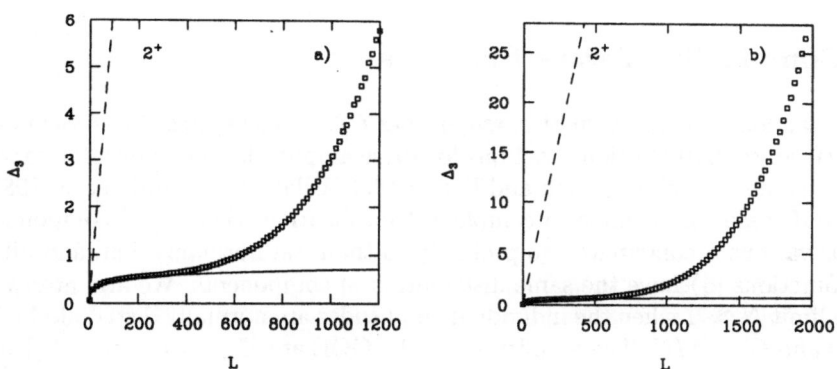

Fig. 6. The spectral rigidity $\Delta(L)$ for the 2^+0 states in the limit of large L, $L \leq 1200$ (panel a) and $L \leq 2000$ (panel b); squares (calculations), dashed line (Poisson statistics) and solid line (the GOE prediction).

regular energy parameters are the single-particle level spacings $\delta e \simeq 3$ to 7 MeV which would give $L_d \simeq 100$ to 200. One could also think of the "scars" related to quasiperiodic motion induced by the coherent two-body matrix elements as pairing. In this case one would expect the saturation of $\Delta(L)$ at $L_d \simeq 100$ to 150. The inverse lifetime of the simple configurations can be estimated by the fragmentation width (3), $\bar{\sigma} \simeq 10$ MeV. One can argue that at times shorter than $t_d \simeq 1/\bar{\sigma}$ the chaotic component of the evolution is still of minor importance. It would lead to the upbend of the spectral rigidity at $L_d \simeq \bar{\sigma}/\sqrt{2}D \simeq 200$ which approximately agrees with the observation. However, the time interval t_d cannot in general be identified with the inverse Lyapunov exponent. In contrast to the exponential decay, the survival amplitude decreases as a Gaussian function. The decrease takes place even in the case of regular dynamics if the initial state is not an eigenstate of the hamiltonian.

Important information on the stochastization process can be obtained from the level and wave function dynamics as a function of external parameters (Haake 1991, Szafer and Altshuler 1993). An example of such behaviour is shown in Fig. 7 where the levels in the middle part of the spectrum for the $(0d1s)^8$, $J^\pi T = 0^+0$ class of states (dimension 325) is shown as a function of the off-diagonal hamiltonian strength λ, $H = H_{\text{diag}} + \lambda H_{\text{off}-\text{diag}}$. $\lambda = 1$ (100 percent in Fig. 7) corresponds to the usual Wildenthal hamiltonian. Apparently, decorrelation of observables under the change of parameters obeys, in the stochastic regime, universal laws where only the scale factors are specific for a given system (Kusnezov and Lewenkopf 1996, Kusnezov et al. 1996). The strength λ of the residual interaction can be taken as a natural control parameter in many-body dynamics. The curvature distribution of unfolded energy levels $\mathcal{E}_\alpha(\lambda)$ in the sd-shell model takes the GOE form (Zakrzewski and Delande 1993, von Oppen 1994) at $\lambda \approx 0.3$, in parallel to the level spacing distribution (Zelevinsky et al. 1996).

4   Complexity of wave functions

In the random matrix ensembles, amplitudes C_k^α of eigenstates (5) become random variables. Distribution functions for these amplitudes are known for canonical Gaussian ensembles (Ullah and Porter 1963, Ullah 1967, Brody et al. 1981). The GOE case corresponds to complete delocalization when all N components C_k^α for various k contribute equiprobably to the total normalization, and all N eigenfunctions $|\alpha\rangle$ have the same distribution of components. We are interested in the limit $N \gg 1$ when the individual amplitudes are normally distributed with $\bar{C} = 0$ and $\overline{C^2} = 1/N$. The amplitudes in the GOE are slightly correlated (Ullah and Porter 1963, Ullah 1967, Brody et al. 1981) due to the orthonormalization. Thus, in the limit $N \gg 1$

$$\overline{W_k^\alpha W_l^\beta} = \frac{1}{N^2}(1 + 2\delta^{\alpha\beta}\delta_{kl}), \tag{10}$$

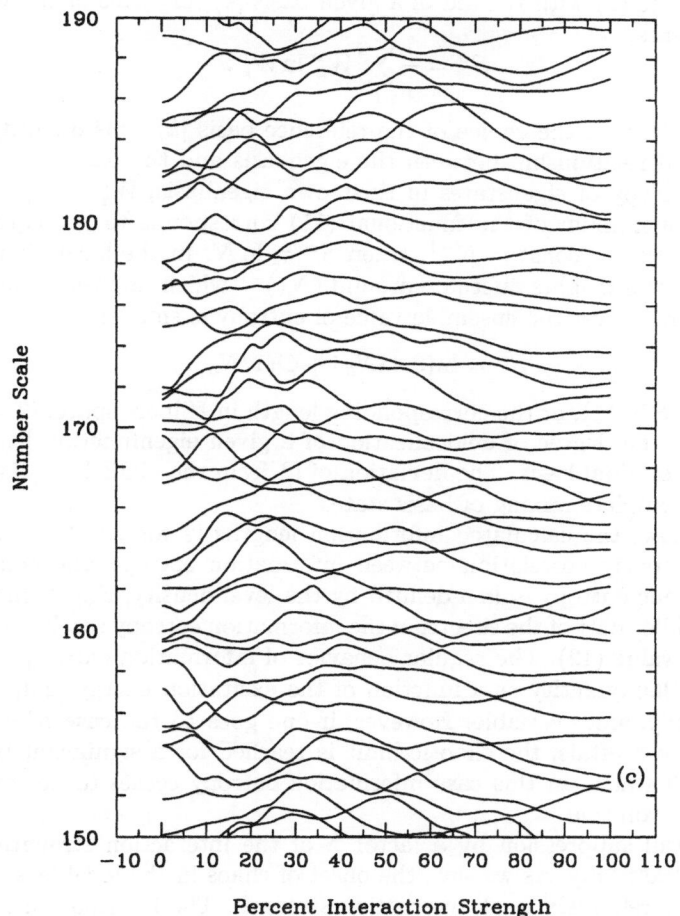

Fig. 7. Information length of the 2^+0 states as a function of energy, part a, and in the α-scale, part b. The horizontal solid line corresponds to the GOE value 1578.

In a gradual transition from regular to stochastic dynamics, the number N_α of principal components of the stationary state $|\alpha\rangle$ increases along with excitation energy towards the limit of complete delocalization. In a given energy range, $E \approx E_\alpha$, the distribution of components C_k^α presumably is similar to the Gaussian but with the local width $\overline{(C_k^\alpha)^2} = 1/N_\alpha$.

   Information entropy (Yonezawa 1980, Reichl et al. 1988, Izrailev 1990, Zelevinsky 1993) is a suitable candidate for measuring the degree of complexity of individual wave functions. It is defined for a given normalized function $|\alpha\rangle$, expanded as in eq.(2) with the aid of a given basis $|k\rangle$, in terms of the weights of the components,

$$S^\alpha = -\sum_k W_k^\alpha \ln W_k^\alpha. \tag{11}$$

Being dependent on the choice of the reference basis $|k\rangle$, this quantity reflects a complicated relationship between the eigenbasis and the basis of representation. The entropy of eigenstates in their own eigenbasis, $W_k^\alpha \to \delta_k^\alpha$, vanishes. The formal maximum of the functional (11) corresponds to the equiprobable distribution, $W_k^\alpha = \text{const} = N^{-1}$, when $S^\alpha = \ln N$. In the local Gaussian approximation, the weights fluctuate around $(N_\alpha)^{-1}$ which implies (Izrailev 1990) that the average over the ensemble value of entropy is smaller,

$$\overline{S^\alpha} = \ln(0.48 N_\alpha) + \mathcal{O}(1/N_\alpha). \tag{12}$$

The entropy S^α (11), or the corresponding length in Hilbert space, $l_S^\alpha = \exp S^\alpha$, characterizes the degree of delocalization of a given eigenfunction $|\alpha\rangle$ with respect to the original basis. The deviation of l_S^α from the GOE limit $0.48N$ indicates the incomplete mixing of basis states.

Fig. 8 shows the calculated information lengths l_S^α for all 3276 states 2^+0. We see the strong correlation between information entropy and conventional thermodynamic entropy $\sim \ln \rho$ defined by the level density, Fig. 4. In the most chaotic (middle) part of the spectrum the information entropy reaches about 90% of the GOE value (12). The regular behavior of information entropy allows one to consider this quantity as a function of the excitation energy and, therefore, as a thermodynamic variable. However, if one goes to the case of degenerate single-particle orbitals, the chaotic limit is reached for a significant portion of 0^+0 and 2^+0 states. In this case information entropy ceases to be sensitive to the spectral evolution.

The overall suppression by a factor λ of the interaction strength changes the results drastically. As we saw, the onset of chaos in the local level statistics occurred at a relatively weak interaction strength. The information entropy of the 0^+0 states at $\lambda = 0.4$ evolves regularly as a function of energy but the localization length is strongly diminished roughly in proportionality to λ.

The behavior of information entropy found in the above example is generic. The calculations were also performed for $N = 1183$ 0^+0 states in ^{12}C in a model space of the first four oscillator shells, taking into account $(0+2)\hbar\omega$ excitations, with the interaction (Warburton and Brown 1992) which contains a cross-shell part. The results (shown in Zelevinsky et al. 1996) clearly differentiate the states with the lowest center-of-mass energy from those with the excited center-of-mass motion. They have the same degree of internal complexity.

Two important features of our results are to be stressed. (i) The information entropy was calculated in the original basis for all individual eigenfunctions with no averaging elements. The eigenfunctions, adjacent in energy, could have different structure and localization properties which would lead to strong fluctuations. Instead, the degree of complexity regularly evolves along the spectrum

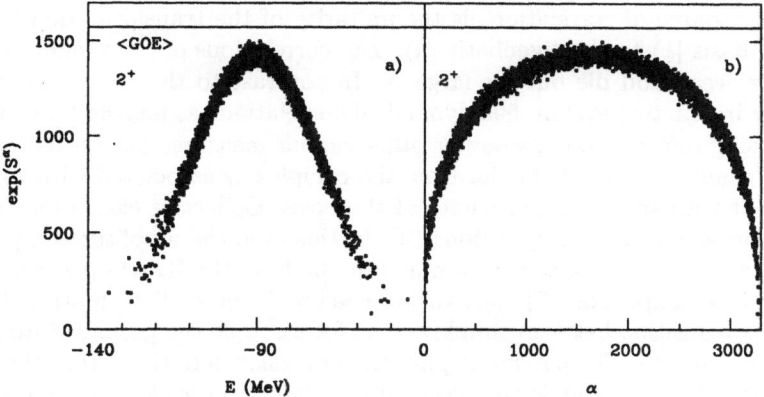

Fig. 8. Information length of the 2^+0 states as a function of energy, part a, and in the α-scale, part b. The horizontal solid line corresponds to the GOE value 1578.

reflecting common features of neighboring eigenstates. Being a function of excitation energy only, the degree of complexity can therefore be considered as a thermodynamic variable. (ii) The localization of the eigenfunctions in the shell model basis depends strongly on the strength of the interaction relative to the stabilizing influence of the mean field. This confirms the general trend of the mean field to quench the chaotic signatures of many-body dynamics (Zelevinsky 1993, Bauer et al. 1994, Bauer et al. 1995).

The results for the "natural" mean field basis can be compared with calculations using different representations. The $\mathcal{SU}(3)$ model (Elliott 1958) explains from the group-theoretical viewpoint the appearance of quadrupole deformation and rotational bands. In the basis of the $\mathcal{SU}(3)$ eigenfunctions, almost all eigenvectors are completely delocalized. This basis with the degenerate single-particle levels turns out to be almost random with respect to stochastization, analogously to a pure random basis which can also be used as a representation basis for comparison. A certain self-consistency between the representation basis and the residual interaction is necessary to achieve a meaningful description of stochastization. The mean field basis which separates in an optimal way local fluctuations from the global evolution is the most appropriate for this purpose (Zelevinsky 1993).

In the Gaussian BRM ensemble (Fyodorov and Mirlin 1991) the localization length of eigenstates is proportional to the square of the local level density $\rho(E)$. We found that, except for the edges of the spectrum, the localization length l_S is approximately proportional to $\rho(E)$ rather than to its square. This means the strong correlation between thermodynamic entropy $\sim \ln \rho$ and information entropy.

In the GOE the components of the eigenvectors are dynamically independent. The only source of correlations is the unitarity of the transformation from the original basis $|k\rangle$ to the eigenbasis $|\alpha\rangle$. The correlations of the weights W_k^α, eq. (10), are weak and die out for large N. In contrast to the GOE, the realistic strongly interacting system has dynamical correlations as pairing built in.

In using information entropy or other similar measures one needs to distinguish a genuine chaotic behavior from the complexity associated with collective motion or with the improper choice of the basis. Collective excitations $|c\rangle$ also can be presented by superpositions (2). In this case the amplitudes C_k^c are coherent with respect to a certain simple (one-body in the RPA) operator Q. The phases of the amplitudes C_k^c are synchronized with those of the matrix elements Q_{k0} for a transition between simple states, for example the ground state $|0\rangle$ and a $1p-1h$ state $|k\rangle$. The partial amplitudes add constructively so that the transition probability $|0\rangle \to |c\rangle$ is enhanced, as compared to the elementary transition $|0\rangle \to |k\rangle$, by a factor N^c which is a number of coherent components contributing to the wave function $|c\rangle$. If N^c is large, our measures of complexity will signal an appearance of a complicated state which has nothing to do with chaos.

However (i) the fraction of collective states is small ($\sim 1/N^c$), (ii) the degree of collectivization in nuclei N^c is small compared to the degree of complexity of typical complicated states (in heavy nuclei $N^c \simeq 10^2$ for low lying excitations of vibrational or rotational type but $N \simeq 10^6$ for neutron resonances), and (iii) in a realistic many-body system only the lowest collective states are approximately stationary, but in that energy domain there is no chaos anyway. Collective states at higher excitation energy are strongly mixed (damped), so the only surviving signature of collectivity ("scar") could be a nonstatistical excess of specific $1p-1h$ basis components concentrated at a certain energy in the interval of the spreading width and manifested by a peak of the strength function of the operator Q.

The problems associated with an inappropriate choice of the basis can be more dangerous. In many cases there exists a smooth evolution or phase transition of the mean field along with increasing energy. Even if the new shape supports regular single-particle motion, this can be misinterpreted as onset of chaos due to the complexity of new eigenstates expressed in the old representation. But in such cases the invariant measures of chaos connected to the level statistics unequivocally indicate absence of chaos. As an example, we studied (Zelevinsky et al. 1996) the pairing correlations as a function of excitation energy. This specific probe allows one to trace the behavior similar to the second order phase transition through the properties of individual eigenfunctions.

5   Chaoticity *vs* thermalization

We saw that the stationary states display the conventional signatures of quantum chaoticity. Local level correlations reveal Wigner repulsion. The spectra are rigid with no pronounced contribution from periodic orbits. The partition structure is smeared, the eigenstates are delocalized and their information entropy (11) in the shell-model basis is close to the GOE limit. These signatures give clear evidence that our system, at least in the middle of the energy range, is near the stochastic limit. Along with this, there exists a noticeable change of complexity as a function of excitation energy. The thermodynamic picture is frequently used in the description of excited states at high level density. What is the relation between the complicated structure of eigenstates and statistical mechanics? To address this question we first discuss the notion of statistical equilibrium as applied to an isolated mesoscopic system like a nucleus.

Statistical properties of a closed equilibrated system with a sufficiently high number of degrees of freedom are determined by the statistical weight $\Omega(E) = \rho(E)\delta E$ of states with given values of exact integrals of motion. Since the density of states $\rho(E)$ grows fast with energy E, the interval δE can still contain many levels which makes the concept of the smooth level density meaningful. The exact value of δE is not important as long as it is small compared to the energy interval where the macroscopic properties of the system change considerably. Now one can define thermodynamic entropy $S^{th}(E) = \ln \Omega(E)$ and temperature T according to

$$\frac{\partial S^{th}}{\partial E} = \frac{1}{T}. \tag{13}$$

Such a description corresponds to the minimum information available. Our knowledge of the microscopic state of the system at equilibrium is limited to what is given by exact integrals of motion.

The accuracy of the statistical approach implies that the results are insensitive to the actual microscopic state of the system. Average over the equilibrium statistical ensemble should give the same outcome as an expectation value for a typical stationary wave function at the same energy (Landau and Lifshitz 1958). The main underlying assumption is that of similarity of generic wave functions in a given energy region. At the same time, equilibrium statistical averaging discards possible phase relationships between the components of wave functions. This is justified if the phase coherence can appear with a very low statistical weight only. The similarity of close eigenstates is ensured by the mixing resulting from the chaotic many-body dynamics. The pioneering paper on the compound nucleus by Niels Bohr (Bohr 1936) already gives an equal footing to elements of both patterns, chaos and thermalization. The definition by Percival (Percival 1973) of chaotic wave functions goes along the same line. It was shown by van Hove (van Hove 1955-59) that a broad class of systems displays quantum ergodicity: a random initial wave function evolves with time into a state which gives the same values of observables as the microcanonical thermodynamic ensemble, see also Srednicki 1994.

We can compare statistical properties of eigenfunctions of the Fermi system with strong interaction (although in a truncated space) with those of the equilibrium statistical ensemble. The degree of complexity measured by information entropy of individual functions in the shell-model basis is the same for many states close in energy. Within small fluctuations, it changes smoothly with excitation energy and can be treated as a thermodynamic variable. To compare the global thermodynamic behavior with the features of the individual eigenfunctions, we calculate the evolution of single-particle occupation numbers n_{lj}^α along the spectrum of many-body states $|\alpha\rangle$,

$$n_{lj}^\alpha = \frac{1}{2} \sum_{m\tau} \langle \alpha | a_{ljm\tau}^\dagger a_{ljm\tau} | \alpha \rangle. \tag{14}$$

The results are shown in Fig. 9 where the panels a, b and c correspond to 0^+0, 2^+0 and 9^+0 ($N = 657$) states, respectively. All three classes of states exhibit an identical smooth behavior. It suggests that one can associate with each eigenstate $|\alpha\rangle$ a single-particle "temperature" T_{s-p}^α defined by the Fermi distribution

$$f_{lj}^\alpha = \{\exp[(e_{lj}' - \mu)/T_{s-p}^\alpha] + 1\}^{-1}. \tag{15}$$

In the center (infinite temperature), all occupancies $f_{lj}^\alpha = n_{lj}^\alpha/(2j + 1)$ indeed become equal to each other the common value being $1/2$ for our case of 12 particles in the sd-shell of the total capacity 24.

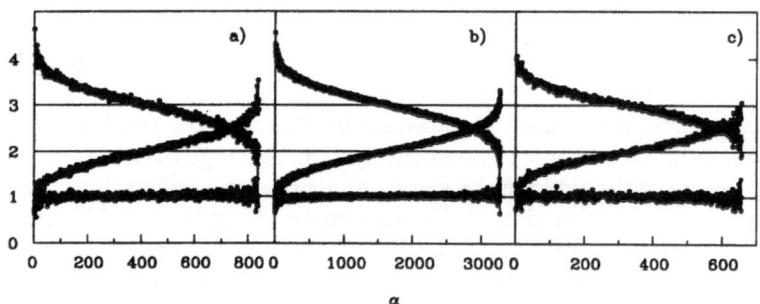

Fig. 9. Single-particle occupation numbers for states 0^+0 (panel a), 2^+0 (panel b), and 9^+0 (panel c). For all panels the three sets of points refer to $s_{1/2}, d_{3/2}$ and $d_{5/2}$ orbitals, from bottom to top on the left-hand side.

T_{s-p}^α changes smoothly with energy, being almost the same for all states within the narrow energy interval as it should be for an intensive thermodynamic quantity. It becomes infinite simultaneously with the thermodynamic temperature when the memory of the initial single-particle energies is lost. The effective energies $e_{lj}' - \mu$ can be found (Horoi et al. 1995, Zelevinsky et al. 1996) from the fit or directly from the slopes of the lines in Fig. 9; they are close to the bare values. Using these energies, one can extract the effective temperature T_{s-p}^α and check that, despite the strong interaction, the "single-particle thermometer" on average measures the same temperature as T obtained from the level density.

For the Gaussian level density $\rho(E)$ with the centroid at E_0 and variance σ_E^2, the temperature (13), see Fig. 10 (solid lines), is

$$T = \sigma_E^2/(E_0 - E). \tag{16}$$

The right half of the spectrum, $E > E_0$, is associated with decreasing entropy and negative temperature.

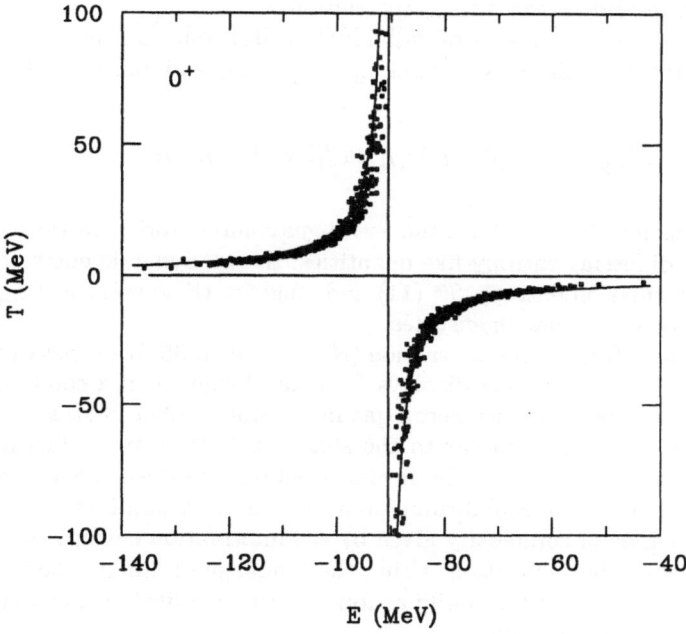

Fig. 10. Temperature calculated from the global fit to the level density of the 0^+0 states (solid line) and found from the occupation numbers of Fig. 9 (dots).

These results imply that the system can be considered as an equilibrated Fermi-liquid, and its properties can be expressed in terms of occupation numbers for a gas of interacting quasiparticles. We do not perform any ensemble averaging. The eigenfunctions individually show the distribution of occupancies expected from statistical mechanics of the equilibrium thermal ensemble. Thermodynamics of the system are determined mainly by the stabilizing action of the mean field. Using the mean field basis we segregate the incoherent processes leading to stochastization and chaos from the regular evolution along the spectrum. The stochastic part of the dynamics is responsible for the complexity of the eigenfunctions and their similarity, which can be interpreted in terms of thermal equilibrium. The regular (mean field) features allow us to use a simple language of average occupation numbers for quasiparticles in a heated Fermi liquid.

We mentioned that a certain level of self-consistency between the mean field and the residual interaction is necessary for the optimal separation of local and global features. In the shell-model calculations it is ensured by the semiempirical hamiltonian. In this case we can expect a strong correlation between the thermodynamical entropy (lack of knowledge about the precise microscopic state of the system) and information entropy (disorder of a given microscopic state computed in the mean field basis of simple quasiparticle configurations). A direct comparison indeed reveals such a correlation.

Using the occupancies f_{lj}^α of individual orbitals one can calculate the single-particle entropy of the quasiparticle gas (Landau and Lifshitz 1958) for each state $|\alpha\rangle$,

$$S_{s-p}^\alpha = - \sum_{lj\tau}(2j + 1)[f_{lj}^\alpha \ln f_{lj}^\alpha + (1 - f_{lj}^\alpha)\ln(1 - f_{lj}^\alpha)]. \tag{17}$$

The expression (17) comes from the Fermi-gas combinatorics. Now we have three, apparently different, entropy-like quantities: thermodynamic entropy $S^{th}(E) \sim \ln\rho(E)$, information entropy S^α (11) and single-particle entropy S_{s-p}^α (17), the latter two for individual eigenstates.

For a weak off-diagonal interaction (Horoi et al. 1995, Zelevinsky et al. 1996), the thermodynamic entropy displays Gaussian behavior of a combinatorial nature typical for an imperfect Fermi-gas in a finite number of states. Within the fluctuations, it is quite similar to the single-particle entropy. The information entropy in this case is low; only at high level density does one see some effects of mixing. This is an equilibrium picture of almost non-interacting particles where the degree of complexity given by the information entropy is only weakly correlated with thermalization. Using the language of kinetic theory, collisions (mixing) are necessary for equilibration but the equilibrium properties do not depend on the collision rate.

In the opposite case of too strong off-diagonal interaction, all states are strongly mixed and the information entropy is near the GOE maximum. The memory of the mean field is lost and S_{s-p}^α is also at the maximum corresponding to the equiprobable population of orbitals so that the response to thermal excitation cannot be expressed in terms of quasiparticles. Within the fluctuations, S^α and S_{s-p}^α coincide. The interaction is too strong, almost all wave functions "look the same" regardless of level density, and the quasiparticle "thermometer" cannot resolve the spectral regions with different temperatures. However, the system still has normal thermodynamic properties governed by the level density. In this case only the microcanonical description is possible.

For the realistic mean field and self-consistent residual interaction, all three entropies (correspondingly normalized) turn out (Horoi et al. 1995) to be identical within fluctuations except for the edges of the spectrum. Near the ground state the Fermi surface is already smeared due to two-body correlations so that the single-particle occupation numbers and information entropy show deviations from the frozen Fermi-gas. The difference between low thermodynamic temperature and single-particle temperature, as measured for instance in particle knockout experiments near the ground state, was discussed in Zelevinsky 1993. For the majority of states and for the mean field consistent with residual interactions, the thermodynamic entropy (defined either via the global level density or in terms of occupation numbers) behaves similarly to information entropy.

6   Conclusions

The nuclear shell model provides a realistic, exactly solvable example of a many-fermion system with strong interaction. The available dimensions are sufficiently high to allow for statistically reliable studies. The construction of the complete set of states with the given values of exact integrals of motion is an important prelude to the analysis. The role of this premixing and "geometrical chaoticity" is to be further studied. The exponential distribution of the off-diagonal matrix elements of the residual interaction appears a generic feature of the realistic many-body systems.

As excitation energy and level density increase, the local level statistics quickly reveal signatures of chaotic dynamics predicted by the GOE. This occurs for an interaction strength much lower than its actual value. The problem of the transitional nearest-level spacing distribution in the region of onset of chaos and the fractional power law for level repulsion remains to be solved.

The degree of complexity of stationary wave functions can be measured by the information entropy and the moments of the distribution function of the components. These measures depend on the representation which can be used to gain additional knowledge on the structure of the eigenvectors. The mean field (shell-model) basis appears to be the preferred representation which allows for the optimal separation of local spectral properties from global secular dynamics. The structure of the wave functions presented in the mean field basis evolves in a regular way along the spectrum. The measures of complexity can be considered as functions of the excitation energy. The distribution of the components of the eigenvectors in the shell model basis is close to Gaussian although the correlational analysis reveals deviations. The GOE limit of the complete delocalization can be reached by the majority of the eigenvectors only with an artificially suppressed stabilizing action of the mean field.

The single-particle occupation numbers of the shell-model orbitals regularly evolve along the spectrum, being nearly the same for different classes of states. They can be described by the Fermi distribution with effective energies close to the bare ones. The similarity of the wave functions and occupation numbers of the states close in energy can be interpreted in terms of statistical equilibrium. In spite of the presence of strong interactions, the system behaves at high excitation energy as a heated Fermi gas of fermionic quasiparticles. This indicates the possibility of using the thermal ensemble for calculating matrix elements between the compound states (Flambaum and Vorov 1993). The apparent decoherence emerges here as a property of individual complicated wave functions in a closed mesoscopic system, with no heat bath involved. Different definitions of temperature, related to the thermal microcanonical ensemble, single-particle occupancies, and information entropy, practically coincide for the mean field representation used in the last two cases, where the temperature scale is extracted for each individual eigenstate. It gives new arguments for understanding the foundations of quantum statistical mechanics and its relationship to quantum chaos.

Note to be added in proof: We have recently found that the results for the spectral rigidity $\Delta(L)$ at the large L are very sensitive to the details of the unfolding procedure. In particular, we have found a new unfolding procedure for which the results for $L < 150$ are not changed, but the results for high L become in better agreement with the GOE prediction. We plan to present details of this new analysis elsewhere.

Acknowledgments

The authors acknowledge support from the NSF grant 94-03666.

References

Alhassid Y., Novoselsky A. (1992): Phys. Rev. **C45**, 1677

Altland A., Fuchs D. (1995): Phys. Rev. Lett. **74**, 4269

Arve P. (1991): Phys. Rev. **A44**, 6920

Auerbach N., Brown B.A. (1994): Phys. Lett. **B340**, 6

Bauer W., McGrew D., Zelevinsky V., Schuck P. (1994): Phys. Rev. Lett. **72**, 3771

Bauer W., McGrew D., Zelevinsky V., Schuck P. (1995): Nucl. Phys. **A583**, 93

Berry M.V. (1985): Proc. Roy. Soc. London, Ser. A **400**, 229

Berry M.V., Robnik M. (1984): J. Phys. **A17**, 2413

Berry M.V., Robnik M. (1986): J. Phys. **A19**, 649

Bertulani C.A., Zelevinsky V. (1994): Nucl. Phys. **A568**, 931

Bohr N. (1936): Nature **137**, 344

Bohr A., Mottelson B. (1969): *Nuclear Structure*, vol. 1 (Benjamin, New York)

Bohigas O., Giannoni M.J., Schmit C. (1984):, Phys. Rev. Lett. **52**, 1

Bohigas O., Giannoni M. (1984): *Mathematical and Computational Methods in Nuclear Physics*, ed. J.S. Dehesa, J.M.G. Gomez and A.Polls *Lecture Notes in Physics* **209** (New York: Springer) p. 1

Bohigas O., Weidenmüller H.A. (1988): Ann. Rev. Nucl. Part. Sci. **38**, 421

Brenneisen J., et al. (1995): Z. Phys. **A352**, 149, 279, 403

Brody T.A., Flores J., French J.B., Mello P.A., Pandey A, Wong S.S.M (1981): Rev. Mod. Phys. **53**, 385

Brown B.A., Bertsch G. (1984): Phys. Lett. **148B**, 5

Brown B.A., Etchegoyen A., Rae W.D.M., Ormand W.E., Winfield J.S., Zhao L.(1988): *OXBASH code*, MSUNSCL Report **524**

Brown B.A., Wildenthal B.H. (1988): Ann. Rev. Nucl. Part. Sci. **38**, 29

Casati J., *et al.* (1990): Phys. Rev. Lett.**64**, 1

Døssing T., *et al.* (1996): Phys. Rep., *in press*

Elliott J.P. (1958): Proc. Roy. Soc. **A245**, 128, 562

Elyutin P.V. (1988): Usp. Fiz. Nauk **155**, 397 [Sov. Phys. Usp. **31**, 597 (1988)]

Feingold M., Leitner D.M., Wilkinson M. (1991): Phys. Rev. Lett. **66**, 986

Flambaum V.V., Gribakin G.F. (1995): Prog. Part. Nucl. Phys., **35**, 423

Flambaum V.V., Vorov O.K. (1993): Phys. Rev. Lett. **70**, 4051

Flambaum V.V., Gribakina A.A., Gribakin G.F., Kozlov M.G. (1994): Phys. Rev. **A50**, 267

Frazier N., Brown B.A., Zelevinsky V. (1996): *to be published*

French J.B., Wong S.S.M. (1971): Phys. Lett. **B35**, 5

French J.B., Ratcliff K.F. (1971): Phys. Rev. **C3**, 94

Fyodorov Y.V., Mirlin A.D. (1991): Phys. Rev. Lett. **67**, 2405

Gaardhøje J.J. (1992): Annu. Rev. Nucl. Part. Sci. **42**, 483

Garrett J.D., et al. (1991): *Future Directions with 4π Gamma Detection Systems of the New Generation*, ed. J.Dudek and B.Haas (AIP) p.345

Gräf H.-D., Harney H.L., Lengeler H., Lewenkopf C.H., Rangacharyulu C., Richter A., Schardt P., Weidenmüller H.A. (1992): Phys. Rev. Lett. **69**, 1296

Gurevich I.I., Pevsner M.I. (1957): Nucl. Phys. **2**, 575

Haake F. (1991): *Quantum Signatures of Chaos* (Springer, New York)

Harney H.L., Richter A., H.A.Weidenmüller H.A. (1986): Rev. Mod. Phys. **58**, 607

Horoi M., Brown B.A., Zelevinsky V. (1994): Phys. Rev. **C50**, R2274

Horoi M., Zelevinsky V., Brown B.A. (1995): Phys. Rev. Lett. **74**, 5194

Izrailev F.M. (1990): Phys. Rep. **196**, 299

Kuś M., Lewenstein M., Haake F. (1991): Phys. Rev. **A44**, 2800

Kusnezov D., Lewenkopf C.H. (1996): *in press*

Kusnezov D., Brown B.A., Zelevinsky V. (1996): *to be published*

Kusnezov D., private communication

Landau L.D., Lifshitz E.M. (1958): *Statistical Physics* (Pergamon Press)

Lauritzen B., Døssing T., Broglia R.A. (1986): Nucl. Phys. **A457**, 61

Lauritzen B., Bortignon P.F., Broglia R.A., Zelevinsky V. (1995): Phys. Rev. Lett. **74**, 5190

Lewenkopf C.H., V.Zelevinsky V. (1994): Nucl. Phys. **A569**, 183

Lopac V., Brant S., Paar V. (1990): Z. Phys. **A337**, 131

Matsuo M., et al. (1993): Nucl. Phys. **A557**, 211c

Mitchell G.E., et al. (1988): Phys. Rev. Lett. **61**, 1473

Mizutori S., Zelevinsky V. (1993): Z. Phys. **A346**, 1

Molinari L., Sokolov V.V. (1989): J. Phys. **A22**, L999

Mon K.K., J.B.French J.B. (1975): Ann. Phys. **95**, 90

Mottelson B., *unpublished*.

Ormand W.E., Brown B.A. (1989): Nucl. Phys. **A491**, 1

Ormand W.E., Broglia R.A. (1992): Phys. Rev. **C46**, 1710

Percival I.C. (1973): J. Phys. **B6**, L229

Porter C.E. (editor) (1965): *Statistical Theories of Spectra: Fluctuations*, (Academic Press, New York)

Prosen T., Robnik M. (1993): J. Phys. **A26**, 2371

Raman S., et al. (1991): Phys. Rev. **C43**, 521

Ratcliff K.F. (1971): Phys. Rev. **C3**, 117

Reichl J. (1988): Europhys. Lett. **6**, 669

Schuster H.G. (1989): *Deterministic Chaos* (VCH Verlagsgesellschaft, Weinheim)

Sokolov V.V., Zelevinsky V. (1988): Phys. Lett. **B202**, 10

Sokolov V.V., Zelevinsky V. (1989): Nucl. Phys. **A504**, 562
Sokolov V.V., Zelevinsky V. (1992): Ann. Phys. **216**, 323
Sushkov O.P., Flambaum V.V. (1980): Pis'ma Zh. Exptl. Teor. Fiz. **32**, 377 [JETP Lett. **32**, 353 (1980)]
Sushkov O.P., Flambaum V.V. (1982): Usp. Fiz. Nauk **136**, 3 [Sov. Phys. Usp. **25**, 1 (1982)]
Srednicki M. (1994): Phys. Rev. **E50**, 888
Shriner J.F., *et al.* (1990): Z. Phys. **A335**, 393
Szafer A., Altshuler B.L. (1993): Phys. Rev. Lett. **70**, 587
Ullah N., Porter C.E. (1963): Phys. Lett. **6**, 301
Ullah N. (1967): *Int. Nucl. Phys. Conf.*, ed. R.L.Becker (Academic, New York) p. 812
van Hove L.: Physica **21** (1955), 517; **23**, 441 (1957); **25**, 268 (1959)
von Oppen F. (1994): Phys. Rev. Lett. **73**, 798
Wambach J., private communication
Warburton E.K., Brown B.A. (1992): Phys. Rev. **C46**, 923
Wigner E.P. (1955): Ann. Math. **62**, 548
Wintgen D., Marxer H. (1988): Phys. Rev. Lett. **60**, 971
Yonezawa F.J. (1980): Non-Cryst. Solids **35 & 36**, 29
Zakrzewski J., Delande D. (1993): Phys. Rev. **E47**, 1650
Zelevinsky V., von Brentano P. (1991): Nucl. Phys. **A529**, 141
Zelevinsky V. (1993): Nucl. Phys. **A555**, 109
   Zelevinsky V.: Nucl. Phys. **A553**, 125c (1993) **A570**, 411c (1994)
Zelevinsky V., Horoi M., B.A.Brown B.A. (1995): Phys. Lett. **B350**, 141
Zelevinsky V., Brown B.A., Frazier N., M.Horoi M. (1996): Phys. Reports, *in press*

Large scale shell model calculations: The physics in and the physics out

Andrés P. Zuker

[1] Physique Théorique. Bât 40/1 Centre de Recherches Nucléaires, CNRS, IN2P3, Université Louis Pasteur. BP28 F-67037 Strasbourg Cedex-2 , France
[2] IBERDROLA visiting professor during march-september 1996 at Departamento de Física Teórica C-XI, Universidad Autónoma de Madrid, E–28049 Madrid, Spain

Abstract. After giving a few examples of recent results of the $(SM)^2$ collaboration *, the monopole modified realistic interactions to be used in shell model calculations are described and analyzed. Rotational motion is discussed in some detail, and some introductory remarks on level densities are made.

1 Introduction and some results

To do a shell model (SM) calculation one needs an interaction and a code.

The choice of a realistic interaction is conceptually satisfactory, but there are some problems.

- The realistic interactions saturate poorly. This problem can be bypassed efficiently in a SM context by modifying H_m, the monopole part of the Hamiltonian (i.e., the one that contains number and isospin operators). The modifications amount to force the correct unperturbed energies on closed shells and the particle and hole states built on them. (We call this set $cs\pm1$.). Section 2.1 is devoted to this point.
- There are many realistic interactions and they are complicated. True mostly for H_m. The residual, multipole Hamiltonian H_M is very much the same for all the forces that fit well the NN phase shifts at moderate energies because it is central. Furthermore, it is very simple: A *normalized* version of the standard pairing plus multipole forces. Section 2.2 describes H_M.
- For use in a finite space the interaction has to be renormalized: A capital subject that has kept many people busy for years. In Section 2.3 we shall see that a Gamow Teller strength function provides direct clues about the effective interactions by pointing clearly to what is achieved and what is missed in a SM calculation.

About the code(s) I will say little. ANTOINE [1] is currently capable of dealing with m-scheme dimensionalities of 10^6 (comfortably), 10^7 (not so comfortably), and 7×10^7 (very hard, the largest calculation so far [2]) on an HP715 (192 Mbytes, some 30 Mflops). There is also a coupled version of the code [3]

---

* $(SM)^2=$ Strasbourg Madrid Shell Model.

that has diagonalized matrices of dimensionality 10^6 at fixed J. Memories of 2 Gbytes and speeds of 100 Mflops would lead to gains of one order of magnitude. Moreover, the codes lend themselves naturally to parallel processing, and hence to further potential gains of one, or perhaps two orders of magnitude. A very schematic explanation of the algorithmic idea behind ANTOINE can be found in Ref. [4].

The prospects of diagonalizing exactly larger and larger matrices may be quite encouraging, but we know that sooner or later the fun will stop. And it will stop at dimensionalities that will be far short of 10^{40}, say, that would perhaps guarantee the solution of most nuclear structure problems without further thinking. It is to be hoped that one can learn about these truly large matrices by studying smaller ones. How big should these be? In other words: When does the fun start? In Sections 3 and 4, I will try to convince you that it has started already. Let me give here the general idea.

In many branches of physics one tries to learn about big systems by studying smaller ones. For instance by calculating in spin lattices of 32×32. The dimensionality (10^{100}) is substantial, but the system does not represent any existing object. It is a model to be used, through scaling arguments, to extrapolate to larger lattices. In nuclear physics much smaller matrices give realistic descriptions of existing systems. But then: Why not treat them also as models, so as to extrapolate, through scaling arguments, to larger systems?

There are two subjects, so far, where scaling arguments seem promising: rotational motion and level densities and their study will provide helpful hints about new computational strategies. The one that seems most promising at the moment is the Shell Model Monte Carlo (SMMC) method. The exact ANTOINE results have already been useful benchmarks for SMMC, as mentioned by Steve Koonin in his talk and it is my pupose to show that they could also be useful in other respects.

As Michel Vallières' review covers some of the $(SM)^2$ results, I will give only some examples taken from very recent work. They are meant to illustrate that the pf shell is a frontier region, where the SM can demonstrate both its unique capacity to cope with spectroscopic detail and its possibilities to provide fun in the sense mentioed above.

In Fig. 1 we have the spectrum of ^{49}Ti [5]. The agreement with experiment is good, and representative of the results in $A = 47\text{-}49$, although some are better. (Unless noted the experimental information is from Burrow's compilations [6].) One of the reasons to pick this nucleus, is that it is among the ones in which the pure $f_{7/2}^n$ model is less successful [7], [8], in that it misses several of the low lying states. Since this model space is definitely a good one, we may have here a realistic testing ground for the effect of intruders in the renormalization process. Apart from that, ^{49}Ti does not seem to be of particular use in raising broader issues.

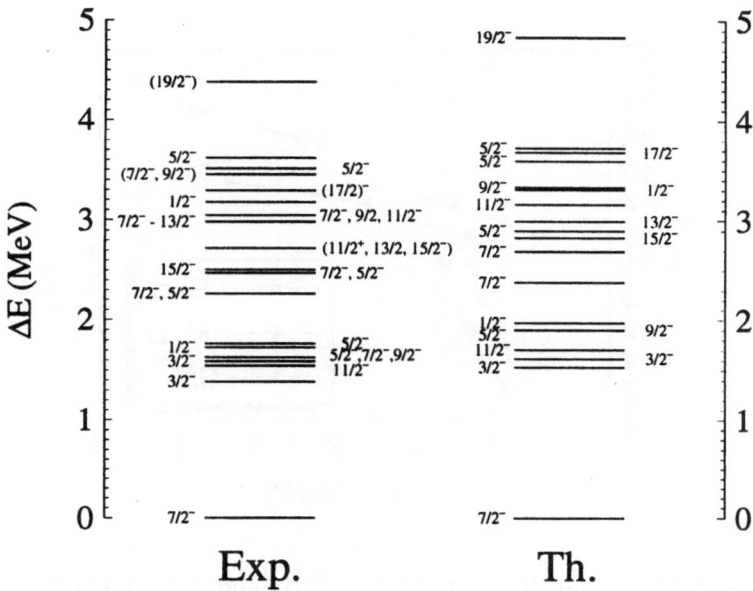

Fig. 1. Theoretical and experimental energy levels of ^{49}Ti.

In Table 1, we have the static moments of the $A = 47$ isobars. They are shown mainly to give an idea of typical agreements with measures for quantities other than the energies. For our present purpose we shall catalogue the discrepancy in the quadrupole moment of ^{47}Ca as a detail, but it may also be a remainder that there are renormalization effects we do not control well.

The first two examples correspond to what is expected of a SM calculation: good spectroscopy, but limited insight into general problems. We move now to Fig. 2, which shows the yrast γ energies of ^{50}Cr: A backbending rotor, with a predicted second backbend to boot [12]. The known points agree nicely with the calculations, and the unknown ones are been investigated [13]. Backbenbing bands were thought to occur only in much heavier nuclei, but there are several of them in the neighborhood of ^{48}Cr. Now we are faced with a problem whose interest goes well beyond the pf shell. Section 3 will be devoted to collect and organize what we have learned so far about rotors in a SM description.

Fig. 3 shows the calculated quenching factor for Gamow Teller (GT) transitions [9] in the pf shell. The extracted value is practically the same as the one in the sd shell [10] (0.744 vs 0.77). The interesting point is that according to Ref. [11], this number is a direct measure of the occupancy of the active particles (pf or sd) in the exact wavefunction. If ^{16}O or ^{40}Ca were perfect closed shells the value would be one. Therefore, deep correlations are rather strong, and very

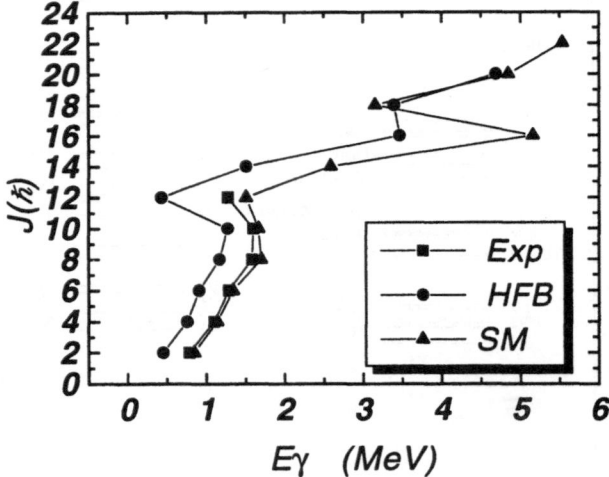

Fig. 2. Theoretical (triangles, SM; circles HFB) and experimental (squares) gamma ray energies versus the angular momentum J for ^{50}Cr.

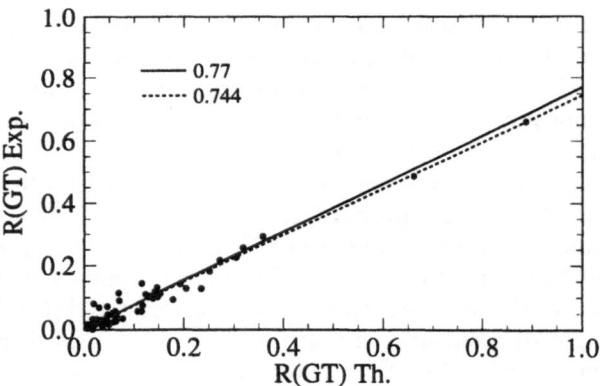

Fig. 3. Comparison of the experimental matrix elements $R(GT)$ with the theoretical calculations based on the "free-nucleon" Gamow-Teller operator. Each transition is indicated by a point in the x-y plane, with the theoretical value given by the x coordinate of the point and the experimental value by the y coordinate.

Table 1. Dipole magnetic moments and quadrupole electric moments of the
$A = 47$ isobars.

Nucleus	State	μ (μ_N)		Q (e fm^2)	
		Expt.	Theor.	Expt.	Theor.
^{47}Ca	$7/2^-$ (g.s.)	$-1.380(24)$	-1.41	$2.1(4)$	6.7
^{47}Sc	$7/2^-$ (g.s.)	$5.34(2)$	5.12	$-22(3)$	-21
^{47}Ti	$5/2^-$ (g.s.)	$-0.78848(1)$	-0.97	$30.3(24)$	22.7
	$5/2^-$	$-1.9(6)$	-1.16		8.16
^{47}V	$3/2^-$ (g.s.)		2.14		19.9
^{47}Cr	$3/2^-$ (g.s.)		-0.47		20.6

constant. They also tend to be difficult to detect, by hiding beyond an overall
normalization, as in the case of the $\sigma\tau$ operator. Again, we are faced with a
truly general problem. The results in Ref. [11] were obtained by analyzing the
GT strength function that will be revisited in Section 2.3 (Fig. 6). It will suggest ways of attacking the problem of level densities, which will be the subject
of Section 4.

2 The interaction

When used directly in a SM calculation, realistic matrix elements ([14](KB),
[15](KLS), [16]) give results that deteriorate rapidly as the number of particles
increases. One way of dealing with the problem is to use the matrix elements
as input for a fit to the data. The most ambitious project of this kind lead to
Hobson Wildenthal's W interaction for the sd shell [10]. An alternative consists
in identifying which matrix elements are wrong and change only these ([17], [18],
[19]). This approach lead to a fairly succesful perturbative calculation in the pf
shell [8], but it took some time to demonstrate conclusively that the trouble
with the realistic interactions is due to their bad saturation properties [20]. The
quality of the $(SM)^2$ calculations bears witness of this fact, but it is not clear
that the very simple mechanisms involved are clearly understood by the experts,
so I will say a few words about them in Section 2.1. Then I'll move to the part
of the interaction that works well, and to some aspects of the renormalization
process.

   NOTATIONS Some equations will be necessary and I will use Bruce French's
product notation (and many of his ideas [21]).

   Γ stands for JT. Then $(-)^\Gamma = (-)^{J+T}$, $[\Gamma] = (2J+1)(2T+1)$, and in general
$F(\Gamma) = F(J)F(T)$. Orbits are called r, s, etc., and we use $(-)^r = (-)^{j_r+1/2}$,
$[r] = 2(2j_r + 1)$. The beauty of the notation is that all the expressions are valid
in neutron-proton formalism (i.e., when isospin is not introduced explicitly) by

simply dropping all reference to the isospin part. Then $(-)^r = (-)^{j_r}$, $[r] = (2j_r + 1)$, and so on.

n_r is the number of particles in orbit r, and $\bar{n}_r = [r] - n_r$ is the number of holes. T_r is used for both the isospin and the isospin operator.

$Z^\dagger_{rs\Gamma}$ is an operator of type $a^\dagger_r a^\dagger_s$ coupled to good spin and isospin JT. S^γ_{rs} is an operator of type $a^\dagger_r a_s$ coupled to good spin and isospin $\lambda\tau$.

V^Γ_{rstu} is a two-body matrix element. W^Γ_{rstu} is used after the monopole part has been subtracted.

p is the principal oscillator quantum number.

2.1   The monopole Hamiltonian

The separation of the Hamiltonian into an "unperturbed" and a "residual" part , $H = H_0 + H'$, is at the heart of many-body physics, and the idea that H_0, could be represented by some central - i.e. single particle - field is of heuristic and qualitative value. However, it is formally unsatisfactory because H is a two body operator (the kinetic energy can be written as two body by subtracting the center of mass part), and it would be more natural for H_0 to be a two body operator.

The proper procedure is to extract from H everything that is diagonal in the representation defined by the central field. It amounts to calculate the traces of H over configurations i;e., states of fixed number of particles in each orbit. When working in an isospin representation, the traces are better calculated at both fixed n_r and T_r. The thing to do then is to calculate the centroids

$$V_{rs} = \frac{\sum_\Gamma V^\Gamma_{rsrs}[\Gamma]}{\sum_\Gamma [\Gamma]} = \frac{\sum_\Gamma V^\Gamma_{rsrs}[\Gamma](1 - (-)^\Gamma \delta_{rs})}{[r]([s] - \delta_{rs})}, \tag{1}$$

and

$$V^T_{rs} = \frac{\sum_J V^{JT}_{rsrs}[J]}{\sum_J [J]} = \frac{4\sum_J V^{JT}_{rsrs}[J](1 - (-)^{J+T}\delta_{rs})}{[r]([s] + 2\delta_{rs}(-)^T)}. \tag{2}$$

The sums run over Pauli allowed values, and the second equality in each equation gives the form in which the Pauli restrictions are explicitly incorporated. Introducing now $b_{rs} = V^1_{rs} - V^0_{rs}$, we can write the French trace equivalent Hamiltonian as

$$H_m = H_{sp} + \sum \frac{1}{(1 + \delta_{rs})}\left[V_{rs}\, n_r(n_s - \delta_{rs}) + b_{rs}\left(T_r \cdot T_s - \frac{3n_r \bar{n}_r}{4([r] - 1)}\delta_{rs}\right)\right]. \tag{3}$$

H_{sp} is the single particle term generated by the core orbits. The result is adapted from [22], and it has has been given a form that makes the terms in b_{rs} traceless at fixed n_r. The advantage of this choice is that these terms can be simply dropped if one is interested only in isoscalar averages.

The expectation value of H for any basic state is the average energy (i.e. the trace) of the configuration to which it belongs. For the $cs \pm 1$ states there is only one state per configuration, and the trace reduces to the expectation value. It means that $H_M = H - H_m$ does not contribute to the unperturbed energies of the $cs \pm 1$ set. Hence the statement that H_m is fully in charge of saturation properties. The set of orbits should result in principle from a selfconsistent calculation. Under Hartree Fock variation, the n_r operators will go into non diagonal ones of the form S_{st}^{00}. Therefore H_m should be generalized to the full monopole Hamiltonian, $\mathcal{H}_m$, containing such objects (or better, all two body quadratic forms in the $S_{st}^{0\tau}$ operators, see [23], [24]). In practice the orbits are chosen from the prescription [25]

$$\hbar\omega = 40A^{-1/3}, \tag{4}$$

which forces the correct value for the radius, but if we are not careful to have also the correct energies, trouble will ensue, as I'll show next.

Let me write Eq. 3 for two shells, keeping only the isoscalar part for simplicity:

$$H_m = \varepsilon_p n_p + \varepsilon_h n_h + \frac{1}{2}V_{pp}n_p(n_p - 1) + \frac{1}{2}V_{hh}n_h(n_h - 1) + V_{ph}n_p n_h, \tag{5}$$

and calculate the excitation energy of kp-kh states with respect to the cs in which orbit h is full

$$\epsilon(k) = \left[(\varepsilon_p - \varepsilon_h) + \frac{1}{2}(|V_{pp}| - |V_{hh}|)([h] - 1)\right] k$$
$$+ (\frac{1}{2}|V_{pp} + V_{hh}| - |V_{ph}|)k([h] - k). \tag{6}$$

(I have used the fact that all centroids are attractive, i.e., negative.)

First let me illustrate the phenomenal importance of the two body terms in H_m. In ^{16}O one can infer from binding energies that the lowest ph states should come at around the value of the "gap" i.e., BE(^{17}O) +BE(^{15}O)−2BE(^{16}O)=11.5 MeV. Experimentally, there is a doublet near 6 MeV, consisting of a 3^- ph state and a 0^+ state of well established 4p-4h nature. To bring down the 3^- from 11.5 to 6 Mev is fairly easy (mainly an monopole isovector effect), but to bring down the 0^+ from $4 \times 11.5 = 46$ to 6 Mev seems impossible. To understand what happens, I shall borrow numbers from the calculation that first gave a consistent picture of the low lying states in the region [26]. Four particles were allowed to move on top of a ^{12}C core in a space of $p_{1/2}$, $s_{1/2}$ and $d_{5/2}$ orbits. For simplicity I treat the last two as as single one (p), and call the first h.

We have then $[h] = 4$, $[p] = 16$, and the single particle splitting and the centroids are (in MeV)

$$\varepsilon_p - \varepsilon_h \approx 3, \qquad \frac{1}{2}|V_{pp} + V_{hh}| - |V_{ph}| \approx 2.5, \qquad |V_{pp}| - |V_{hh}| \approx 0.$$

For $k = 1$, the 3 MeV coming from the single particle term in Eq. (6) is boosted to some 10.5 MeV by the contributions in the second line. For $k = 4$, these contributions vanish and we are left with an excitation energy of 12 MeV. Now it becomes quite plausible that correlation effects could detach a coherent 4p-4h state down to 6 MeV.

These estimates are very rough, but still sufficiently precise to indicate that quadratic effects may be spectacular, and that H_m provides energies that are close enough to the observed ones to be an excellent unperturbed Hamitonian.

Equation (6) can be of help in many other situations. If we consider the pf shell, the single particle spectrum can be obtained from ^{41}Ca. Using p and h for the $p_{3/2}$ and $f_{7/2}$ orbits respectively we have $\varepsilon_p - \varepsilon_h \approx 2$ MeV, a splitting that in ^{56}Ni, becomes BE(^{55}Ni)+BE(^{57}Ni)-2BE(^{56}Ni)=6.3 Mev, which demands $([h] - 1)(|V_{hh}| - |V_{ph}|)| \approx 4.3$ MeV, again a huge non linear effect.

The monopole trouble of the realistic interactions is basically that $|V_{ph}|$ is too strong. In the case of the pf shell, instead of the well established closure in ^{56}Ni they lead to a ground state dominated by 4p-4h excitations. The only corrective action that seems indispensable, amounts to forcing good single particle gaps and spectra around the closures. The point is analyzed in detail in Ref. [5] for the neighborhood of ^{48}Ca, where correlation effects can be treated exactly. A fairly good H_m that gives all the observed $cs \pm 1$ energies from ^{5}He to ^{209}Pb with a precision of some 300 keV, using a dozen parameters, has been constructed and will be soon available [27]. It should provide a reliable guide for calculations in any region.

Let me enter now a word of caution and a request.

Word of caution. Quite a while ago ^{28}Si produced a thrilling experience [28]. Using the KB interaction in a truncated space (up to 4p-4h on top of the $d_{5/2}$ cs) we obtained a very decent spectrum (including transitions). However, a deformed HF calculation produced a ground state well below ours. Then we redid the SM work—in as big spaces as were possible at the time—the results were equally satisfactory, but the wavefunctions had next to zero overlap with the original ones. (I think in Hsua Feng's talk there is a remark about something similar happening for competing symmetries.) Had we calculated occupancies, they would have shown that the truncated ones were closer to the truth. Hence, a truncated calculation with an uncorrected realistic interaction *can* be quite efficient, but it is quite dangerous. A good recent example comes from ^{56}Fe, where KB produces again a magnificent spectrum and transitons [29] in a truncated space. In enlarging it, trouble will come sooner or later (we checked). In Aldo Covello's talk we have another example of beautiful results in the Sn isotopes that will suffer if the whole space is included (we also checked). The irony is that bad monopole behaviour in truncated calculations *helps*, by inducing extra configuration mixing that mocks well the effect of the neglected configurations. If would be certainly interesting to bring this behaviour under control, but in the meantime, caution is in order since the monopole problem is quite real.

Request. The monopole behaviour is related to all that is complicated in the forces (tensor, spin orbit and other relativistic effects). However, it is unlikely that the modern potentials—that fit the phase shifts as well as Bob Wiringa showed—produce the *same* trouble as the older forces. In the light nuclei in particular, they may be doing fairly well. Since the radial (i.e., monopole!) behaviour in the p shell is quite strange, it would be quite interesting to see what a HF variation says. In the context of the no core SM Bruce Barrett told us about, the task is quite feasible. A very encouraging hint comes from Tom Kuo's talk in which he showed that at the point in the two well description that corresponds to the correct radius of ^{6}Li, core polarization vanishes. Since this is precisely the indication of HF stability, it would mean that ^{6}Li is *predicted* to have the right radius. Perhaps I am being over enthusiastic by missing something about this result, but still, I think it would be nice to have HF calculations in the region, with the best potentials.

The bad saturation properties of the realistic forces is in my mind the most fundamental problem in nuclear physics. The introduction of density dependent potentials for HF use was the first major step in bypassing it, and Skyrme forces are still going strong: we heard from Karlheinz Langanke that one of the recent variants (presented in Jacek Dobaczewski's talk) seems capable of solving a long standing riddle in r-processes. However, one thing is to bypass the problem, as we all try, and another, to solve it. I have always had the feeling that a bit of relativistic mustard in the menu of our Schroedinger staple would help. Peter Ring's spectacular results may be taken to indicate that more than mustard is needed.

2.2   The multipole Hamiltonian

The multipole Hamiltonian H_M is what is left of H when it has been made monopole free. In all probability it must contain "something" very close to the pairing plus quadrupole Hamiltonian, that was postulated on physical grounds by Bohr, Mottelson and Elliott and has been of enormous service [30], [31]. We also know that "something" must differ from the standard definitions of pairing plus quadrupole, which produce collapse if used in the full Hilbert space. A massive forthcoming paper [32] deals in detail with the problem of extracting the collective part of H_M, and here I shall extract from the paper what is strictly necessary to show that "something" is a *normalized* form of pairing plus quadrupole.

There are two standard ways of writing H_M:

$$H_M = \sum_{r \le s, t \le u, \Gamma} W_{rstu}^{\Gamma} Z_{rs\Gamma}^{+} \cdot Z_{tu\Gamma}, \quad \text{or} \tag{7}$$

$$H_M = \sum_{rstu\Gamma} [\gamma]^{1/2} \frac{(1+\delta_{rs})^{1/2}(1+\delta_{tu})^{1/2}}{4} \omega_{rtsu}^{\gamma} (S_{rt}^{\gamma} S_{su}^{\gamma})^0, \tag{8}$$

with the matrix elements related through

$$\omega_{rtsu}^{\gamma} = \sum_{\Gamma} (-)^{s+t-\gamma-\Gamma} \left\{ \begin{matrix} r & s & \Gamma \\ u & t & \gamma \end{matrix} \right\} W_{rstu}^{\Gamma}[\Gamma], \tag{9}$$

$$W_{rstu}^{\Gamma} = \sum_{\gamma} (-)^{s+t-\gamma-\Gamma} \left\{ \begin{matrix} r & s & \Gamma \\ u & t & \gamma \end{matrix} \right\} \omega_{rtsu}^{\gamma}[\gamma]. \tag{10}$$

Replacing pairs by single indeces $rs \equiv x$, $tu = y$ in eq. (7) and $rt \equiv a$, $su = b$ in eq. (8), we bring the matrices W_{xy}^{Γ} and $f_{ab}^{\gamma} = \omega_{ab}^{\gamma} \sqrt{(1 + \delta_{rs})(1 + \delta_{tu})}/4$, to diagonal form through unitary transformations $U_{xk}^{\Gamma}, u_{ak}^{\gamma}$:

$$U^{-1}WU = E \implies W_{xy}^{\Gamma} = \sum_{k} U_{xk}^{\Gamma} U_{yk}^{\Gamma} E_{k}^{\Gamma} \tag{11}$$

$$u^{-1}fu = e \implies f_{ab}^{\gamma} = \sum_{k} u_{ak}^{\gamma} u_{bk}^{\gamma} e_{k}^{\gamma}, \tag{12}$$

and then,

$$H_{M} = \sum_{k,\Gamma} E_{k}^{\Gamma} \sum_{x} U_{xk}^{\Gamma} Z_{x\Gamma}^{+} \cdot \sum_{y} U_{yk}^{\Gamma} Z_{y\Gamma}, \tag{13}$$

$$H_{M} = \sum_{k,\gamma} e_{k}^{\gamma} \left(\sum_{a} u_{ak}^{\gamma} S_{a}^{\gamma} \sum_{b} u_{bk}^{\gamma} S_{b}^{\gamma} \right)^{0} [\gamma]^{1/2}, \tag{14}$$

which we call the E and e representations.

For the diagonalizations we used the KLS force in two contiguous major shells (The KLS force is used because Paul Lee gave me the code 28 years ago. As explained in Ref. [32] the realistic H_M has changed little in the meantime). For the E representation the density of eigenvalues is asymmetric with a long tail to the left, which is what to expect of an attractive force. For the e representation the result is shown in Fig. 4. When the contributions to H_M of the five largest peaks (in absolute value) are eliminated, the E histogram in Fig. 5 becames symmetric, the residual skewness being associated with the three isolated peaks. French and Mon have shown that a symmetric E distribution is all that is needed to characterize a random Hamiltonian [33]. Therefore, we conclude that the five peaks extracted from the e distribution, plus the three big ones left in the E distribution design the candidates to a realistic collective Hamiltonian. Using the fact that a generic matrix element goes with the oscillator parameter as [20]

$$W_{xy}^{\Gamma}(\omega) \cong \frac{\omega}{\omega_0} W_{xy}^{\Gamma}(\omega_0), \tag{15}$$

we find that the contributions to H_M of the $\Gamma = 01$, and $\gamma = 20$ terms can be writtten as (E^{01} and e^{20} are the eigenvalues corresponding to a single shell, which are half of the two shell ones in the figures)

$$H_{\bar{P}} = -\frac{\hbar\omega}{\hbar\omega_0}|E^{01}|(\overline{P}_p^+ + \overline{P}_{p+1}^+) \cdot (\overline{P}_p + \overline{P}_{p+1})$$

$$H_{\bar{q}} = -\frac{\hbar\omega}{\hbar\omega_0}|e^{20}|(\bar{q}_p + \bar{q}_{p+1}) \cdot (\bar{q}_p + \bar{q}_{p+1}), \qquad (16)$$

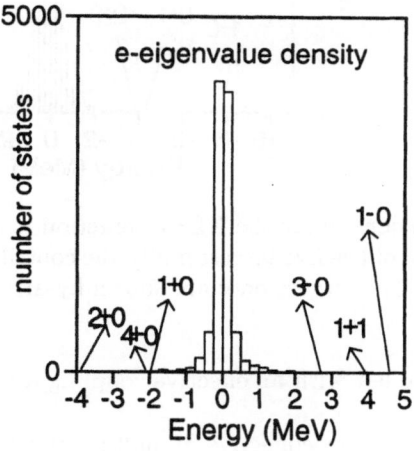

Fig. 4. e-eigenvalue density for the KLS interaction in the pf+sdg major shells. Each eigenvalue has multiplicity $[\gamma]$. The largest ones are shown by arrows.

which are the usual pairing plus quadrupole Hamiltonians, except that the operators for each major shell of principal quantum number p are affected by a normalization, which is (the square root of)

$$\Omega_p = \sum_r \Omega_r = \frac{1}{2}(p+2)^{(2)} \qquad \mathcal{N}_p^2 = \Sigma q_{rs}^2 \cong \frac{1}{8}(p+3/2)^4, \qquad (17)$$

for the pairing and quadrupole operators respectively. ($q = (4\pi/5)^{1/2}r^2Y^2$.) The effect of the normalization is to transform $|E^{01}|/\hbar\omega_0 = g' = 0.32$ and $|e^{20}|/\hbar\omega_0 = \kappa' = 0.216$ in Eqs. 16 into universal coupling constants, and in their new form the forces do not lead to collapse. We know that the bare values of g' and κ' are boosted through renormalizations. For the pairing case the process is delicate, and here we only estimate the effective quadrupole constant κ which we can do with some precision [32]. Core polarization contributes 30%, and if only the quadrupole part of the Hamiltonian is kept in a schematic calculation a further boost of 15% is necessary to account for the effect of the neglected

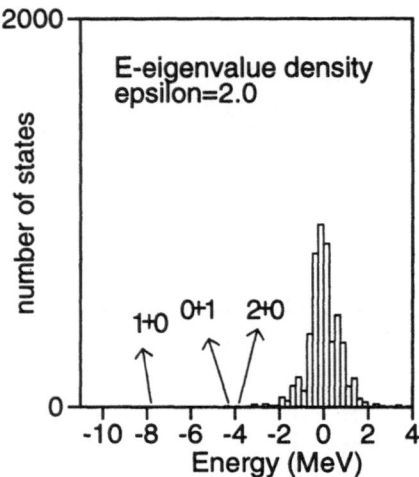

Fig. 5. *E*-eigenvalue density for the KLS interaction in the pf+sdg major shells $\hbar\omega = 9$, after removal of the five largest multipole contributions. Each eigenvalue has multiplicity $[\Gamma]$. The largest ones are shown by arrows.

contributions. We are left with an effective coupling $\kappa = 0.33$. This number will be used in Section 3.

The other terms in the collective Hamiltonian are very much what they should: The very large $\Gamma = 10$ one is the $ST = 10$ part of a pairing force in LS scheme, and then we have octupole, hexadecapole and $\sigma\tau$ forces. The simplest we could hope for, and realistic at that. The preeminence of pairing and quadrupole is due to their association with exact symmetries. It is important to note that the quadrupole terms involving $2\hbar\omega$ jumps are strongly suppressed in the the realistic force: They happen to be exactly what is needed to produce the correct effective charges in perturbation theory. Therefore we are left with a $q \cdot q$ force of Elliott's type, associated with the SU(3) symmetry [34]. All details in [32].

2.3   Remark on renormalization

To do SM calculations, the interaction has to be adapted to the model space we are prepared to tackle. The theory of effective interactions may be a rather complicated affair, but its practice *must* be simple: if second order perturbative estimates are not basically sound, we are in trouble, because we soon end up with strong state dependence and non Hermitian effective many body forces. To understand how something potentially complicated may turn out to be simple it must be known that all theories start with the ambitious project of decoupling exactly the model states from the rest of the Hilbert space. Now: We may well start with a set of integral equations that guarantee the decoupling (Eq. (37)

in Ref. [36]), but we never solve it, since it would be as difficult as an exact diagonalization. What is done is explained in Ref. [37], and amounts to all intents and purposes to enforce the decoupling for a *single* state. The resulting effective interaction may turn out to decouple, at least approximately, some other states and that's it.

To see how this works in practice, let us examine Fig. 6 adapted from [11].

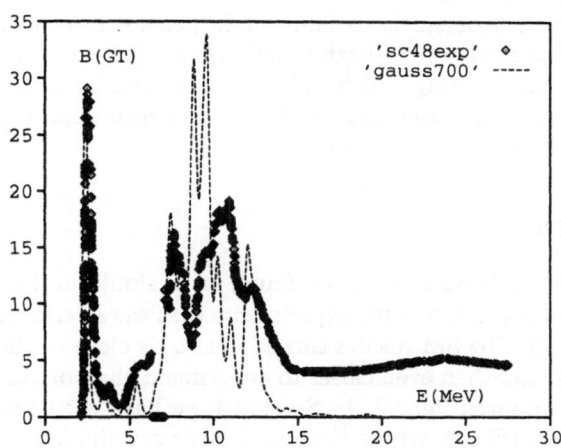

Fig. 6. Experimental GT strength in ^{48}Ca$(p, n)^{48}$Sc from [35], after elimination of the Fermi peak at around 6 MeV, compared with the calculated peaks after 700 Lanczos iterations. The peaks have been smoothed by gaussians having the instrumental width of the first measured level.

The peaks have all $J = 1$, $T = 3$. In the pf shell there are 8590 of them and we have gone to 700 iterations to ensure fully converged eigenstates below 11 MeV. Of these eigenstates 30 are below 8 MeV: They are at the right energy and have the right GT strength. At higher energies our peaks are much too narrow compared with experiment. *It means that they may well be eigenstates of our effective Hamiltonian in the pf shell, but not eigenstates of the full system.* Therefore, they should be viewed as doorways, subject to further mixing with the background of intruders which dominates the level density after 8 MeV, as corroborated by the experimental tail that contains only intruders and can be made to start naturally at that energy.

The message is that the effective interaction is doing a very good job, but it is certainly not decoupling 8590 pf states from the rest of the space. If the fact is not explicitly recognized we end up with the often raised [16] "intruder problem": Decoupling cannot be enforced perurbatively when intruders are energetically close to model states. Figure 6 points to the way out: Few model eigenstates are well decoupled, but it is possible to make sense of many others if one interprets them as doorways.

A recent review [16] of effective interactions confirms the validity of the original KB renormalizations, and in [32] further arguments are given in favour of the simplest second order estimates, but Fig. 6 also suggests that the physics does not stop at 8 MeV, and there are two questions to examine if we want to move higher up. One is related to level densities and thermodynamic properties. Alex Brown dealt with some facets of it in his talk, and Section 4 will deal with some others. The second problem is to find out in general how a doorway becomes an eigenstate. The study of strength functions (see [38] for an overview) can tell us much about the "mixing width" ($\Gamma_\downarrow$), but we still have to learn about the "escape width" ($\Gamma_\uparrow$), i.e., the shell model in the continuum, not a easy task, as we learned from Steve Cotanch's talk.

3 Rotations

In Ref. [39] the yrast band of ^{48}Cr was found to be similar to that of a backbending rotor, in good agreement with experiments that were performed very much at the same time [40]. The SM results turned out to be close to those of mean field calculations [41], and then even closer to experiment than originally thought [42] (see [38] for an updated figure). In Section 1 we have seen that ^{50}Cr is also a backbending rotor (Fig. 2, where the trends are very similar to those in ^{48}Cr),

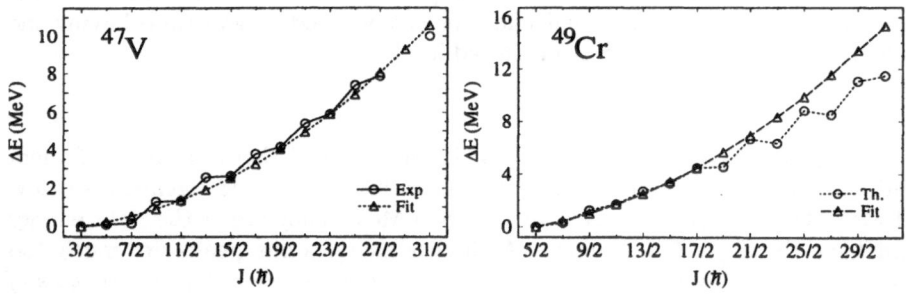

Fig. 7. Comparison between the spectrum of ^{47}V and ^{49}Cr and the predictions of the rotational model.

and in Fig. 7 (from [5]), we find again ^{47}V and ^{49}Cr behaving as good rotors at low spin, and even high spin for the former. In fact—to within some intersting staggering effects—both follow the predictions of the rotational model [43] up to some critical J, and then there is a change in regime, as made evident in

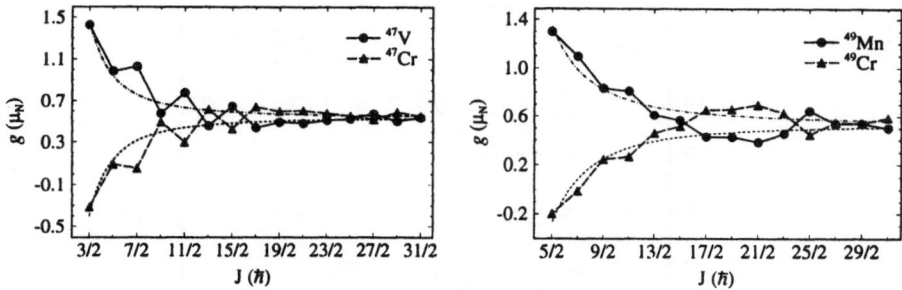

Fig. 8. Gyromagnetics factors of the nuclei ^{47}V, ^{47}Cr, ^{49}Cr and ^{49}Mn as a function of the angular momentum. The curves represent the predictions of the particle plus rotor model.

Fig.8 comparing the gyromagnetic factors predicted by the particle plus rotor model to the SM ones [5]. Therefore, the region $A = 47$-50 can be described as "rotational", and naturally we would like to find out what a SM description of rotational nuclei (i.e., a spherical description of deformation) might mean in general.

The aim of this section is to give a view of what we have learned so far: the main ideas and most of the figures come from Ref. [44], but there are some significant additions and clarifications from Refs. [5], [41], that call for a reorganized presentation. The idea is to show that the exact calculations are telling us an extremely simple story: In a rotational regime, the wavefunctions are determined by the interplay of the quadrupole force with the monopole field. Other parts of the interaction enter as perturbations.

Let us start by considering the quadrupole force alone, taken to act in the p-th oscillator shell. It will tend to maximize the quadrupole moment, which means filling the lowest orbits obtained by diagonalizing the operator $Q_0 = 2q_{20} = 2z^2 - x^2 - y^2$. Using the cartesian representation, $2q_{20} = 2n_z - n_x - n_y$, we find eigenvalues $2p$, $2p-3,\ldots$, etc., as shown in the left panel of Fig. 9, where spin has been included. By filling the orbits orderly we obtain the intrinsic states for the lowest SU(3) [34] representations: $(\lambda, 0)$ if all states are occupied up to a given level and (λ, μ) otherwise. For instance: putting two neutrons and two protons in the $K = 1/2$ level leads to the $(4p,0)$ representation. For four neutrons and four protons, the filling is not complete and we have the (triaxial) $(8(p-1),4)$ representation for which we expect a low lying γ band.

Next consider the influence of the monopole field, which I take to be well represented by a spin-orbit splittting that will separate the subshells into two $\Delta j = 2$ groups. For even p we have the sequences $j = p + 1/2 \ldots 1/2$ below and

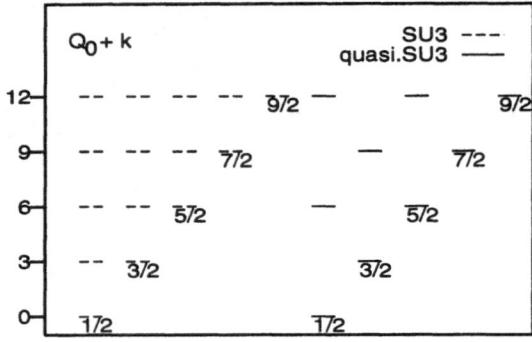

Fig. 9. Nilsson orbits for SU(3) ($k = 2p$) and quasi-SU(3) ($k = 2p - 1/2$).

$j = 3/2 \ldots p - 1/2$ above, while for odd p we have the sequences $j = p + 1/2 \ldots 3/2$ below and $j = 1/2 \ldots p - 1/2$ above. Let us keep only the lowest sequence and diagonalize again the quadrupole operator $2q_{20}$. The task is simplified if we refer to Table 2. Since we are keeping only $\Delta j = 2$ sequences, the $\Delta j = 1$ matrix elements do not contribute. Note that they are strongly suppressed both for large and small m, i.e., which lead to the orbits with largest oblate and prolate deformations respectively. The $\Delta j = 2$ matrix elements are practically identical to those in LS scheme if we make the identifications

$$l \longrightarrow j = l + 1/2 \quad m \longrightarrow m + 1/2 \times \text{sign}(m).$$

The correspondence is one-to-one and the resulting "quasi SU(3)" quadrupole operator respects SU(3) relationships, *except* for $m = 0$, where the correspondence breaks down. Still, it suggests that sequences $j = 1/2, 5/2, 9/2 \ldots$ or $3/2$, $7/2, 11/2 \ldots$, must have a behaviour close to that of the sequences $l = 0, 2, 4 \ldots$ or $1, 3, 5 \ldots$ that span the one particle representations of SU(3). The resulting spectrum for quasi-$2q_{20}$ is shown in the right panel of Fig 9. The result is not exact for the $K = 1/2$ orbits but a very good approximation.

The way to use the approximate quasi-SU(3) symmetry is simply to reason with the right panel of fig. 9 as we would do with the left one. Then, both the four and eight particle "representations" for $T = 0$ will be axial, while the ten particle $T = 1$ ones would be triaxial. At this point we note a positive indication in the absence of a γ band in ^{48}Cr (its counterpart in the sd shell, ^{24}Mg, is triaxial). For ^{50}Cr we could expect a γ band, but we shall see that it is not very robust, as it depends quite critically on the single particle splittings.(Experiment and calculations give no clear answer in this case [12].)

So far we have introduced quasi-SU(3) following the arguments of [44], which rest on the fact the two lowest $\Delta j = 2$ sequence of orbits, separated from the

Table 2. The matrix elements of r^2 and C_{20} in jj and LS coupling

$$< pl|r^2|pl >= p + 3/2$$
$$< pl|r^2|pl + 2 >= -[(p - l)(p + l + 3)]^{1/2}$$

$$< jm|C_2|jm >= \frac{j(j + 1) - 3m^2}{2j(2j + 2)}$$

$$< lm|C_2|lm >= \frac{l(l + 1) - 3m^2}{(2l + 3)(2l - 1)}$$

$$< jm|C_2|j + 1m >= -\frac{3m[(j + 1)^2 - m^2]^{1/2}}{(2j + 4)(2j + 2)(2j)}$$

$$< jm|C_2|j + 2m >= \frac{3}{2}\left\{ \frac{[(j + 2)^2 - m^2][(j + 1)^2 - m^2]}{(2j + 2)^2(2j + 4)^2} \right\}^{1/2}$$

$$< lm|C_2|l + 2m >= \frac{3}{2}\left\{ \frac{[(l + 2)^2 - m^2][(l + 1)^2 - m^2]}{(2l + 5)(2l + 3)^2(2l + 1)} \right\}^{1/2}$$

rest by the spin-orbit splitting, are sufficient to ensure quadrupole coherence. It does not mean that the higher orbits can simply be neglected. To study their influence, instead of reasoning simply in terms of maximizing $2q_{20}$, we want to diagonalize the schematic Hamiltonian [5]

$$H_{mq} = \hbar\omega \left(\sum \epsilon_i n_i - \frac{8\kappa}{(p + 3/2)^4} q \cdot q \right), \tag{18}$$

where we we have borrowed from Eqs. (16) and (17), and the discussion that follows them, the normalized form of the quadrupole force that emerges naturally when it is extracted from a realistic interaction. (Remember we use $q = (4\pi/5)^{1/2}r^2 Y^2$, r is the dimensionless coordinate, p the principal quantum number.) Since we are interested in situations of permanent deformation, q_{20} is expected to be a good approximate quantum number. Therefore, we could obtain the intrinsic state by linearizing H, which amounts to a mean field calculation. (Note here that we want a Hartree, *not* a Hartree-Fock variation, so as to guarantee the exact SU(3) solution for vanishing single particle splittings.) The operation amounts to replacing $q \cdot q$ by $q_{20}q_{20}$, and demands some care since q_{20} is a sum of neutron and proton contributions $q_{20} = q_{20}^\nu + q_{20}^\pi$. The correct linearization for the neutron operators, say, is then

$$q_{20}\, q_{20} \longrightarrow q_{20}^\nu \langle q_{20}^\nu + 2q_{20}^\pi \rangle \approx \frac{3}{2}q_{20}^\nu \langle q_{20} \rangle,$$

where we have assumed $\langle q_{20}^{\nu} \rangle \approx \langle q_{20}^{\pi} \rangle$. Therefore we are left with (H_{sp} is the single particle contribution, and ε a parameter)

$$H_{mq0} = \hbar\omega \left(\varepsilon H_{sp} - \frac{3\kappa}{(p+3/2)^4} \langle 2q_{20} \rangle 2q_{20} \right), \tag{19}$$

which is a Nilsson problem [45] with the coefficient of $2q_{20}$ under a new guise. In the usual formulation it is taken to be one third of the deformation parameter δ:

$$\frac{\delta}{3} = \frac{1}{4} \frac{\langle 2q_{20} \rangle}{\langle r^2 \rangle} = \frac{\langle 2q_{20} \rangle}{(p+3/2)^4}. \tag{20}$$

By equating with the coefficient of $2q_{20}$ in eq. (19) we find

$$\kappa = \frac{4}{12} = 0.33.$$

which can be interpreted as a "derivation" of the value of the quadrupole coupling constant. The agreement with the value of κ estimated after Eq. (17) seems too good to be true, but the reader is referred to [32] to check that there is no cheating.

Nilsson diagrams for the pf shell ($p = 3$) are shown in Fig. 10. The right panel corresponds to the usual representation with levels given as a function of the deformation parameter δ. In eq. (19) we have set $\varepsilon = 1$ and $\hbar\omega H_{sp} =$ the single particle spectrum as given in ^{41}Ca (basically equidistant single particle orbits $f_{7/2}, p_{3/2}, p_{1/2}, f_{5/2}$ with a splitting of 2 MeV). In the figure the centroid of the spectrum is made to vanish.

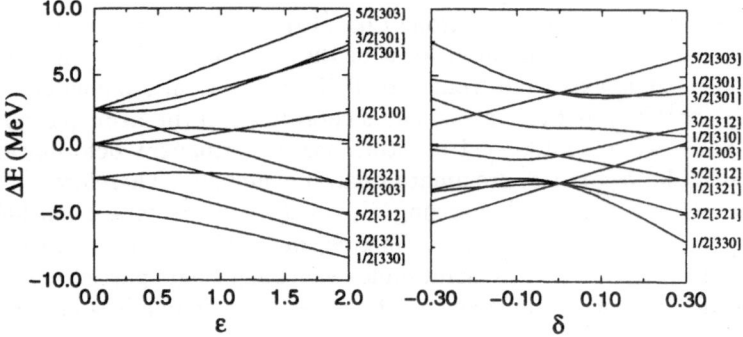

Fig. 10. Nilsson diagrams in the pf shell. Energy vs. single particle splitting ε (left panel), energy vs. deformation δ (right panel).

In the left panel we have turned the representation around: since we are interested in rotors, we start from perfect ones (SU(3)) and study what happens under the influence of an increasing single particle splitting. Here we have used $\langle 2q_{20} \rangle \approx 32-36$, obtained for 4 neutrons and 4 protons filling the lowest orbits in either side of fig. 9. In principle this number should be obtained self consistently, but we shall see that it varies little as a function of ε. For $p = 3$, and $\hbar\omega \approx$ 10 MeV, this means in round numbers $\delta = 0.25$.

The figures suggest that quasi-SU(3) operates in full at $\varepsilon \approx 0.8$ where the four lowest orbits are in the same sequence as the right side of fig. 9 (Remember here that the real situation corresponds to $\varepsilon \approx 1.0$). The agreement even extends to the next group, although now there is an intruder (1/2[310] orbit). The suggestion is confirmed by an analysis of the wavefunctions: For the lowest two orbits, the overlaps between the pure quasi-SU(3) wavefunctions calculated in the restricted $\Delta j = 2$ space (*fp from now on*) and the ones in the full *pf* shell exceeds 0.95 *throughout the interval* $0.5 < \varepsilon < 1$. More interesting still: the contributions to the quadrupole moments from these two orbits vary very little, and remain close to the values obtained at $\varepsilon = 0$ (i.e., from fig. 9).

These results are quite useful in giving qualitative interpretations of the exact calculations, in reducing them to simpler ones, and in suggesting reasonable estimates of the deformations. Let me give some examples.

For ^{48}Cr we would have $Q_0 \approx 2\langle q_{20}\rangle A^{1/3} \equiv 116e$ fm^2, not far from the exact $Q_0 \approx 100e$ fm^2 [39]

Then we can understand from the diagrams why ^{48}Cr is *not* triaxial, and why the expected γ band in ^{50}Cr fails to materialize: calculations in the $(fp)^{10}$ space indicate that its presence depends on the near degeneracy of the 5/2[312] and 1/2[321] orbits, which is broken because the effective δ is likely to be closer to 0.2 than 0.3. As a consolation we are left with a weaker prediction: in $A = 49$, there should be a low lying $K = 1/2$ band, and calculations in the $(fp)^9$ space predict unambiguously a fairly good excited rotational sequence, that persists well after the degeneracy between the $K = 1/2$ and 5/2 levels is broken. Details are given in Ref. [5]

Another useful thing we can do with the diagrams is is to explore what happens at the backbend. In Ref.[12], exact results for the yrast band in ^{50}Cr were compared with those obtained by projecting Nilsson states with $\delta = 0.22$ and -0.1, and then calculating their energies using the full Hamiltonian. *The two bands cross at $J = 10$, precisely were the backbend occurs.* The spectroscopic quadrupole (Q_s) moments for the $J = 2\hbar$ and $4\hbar$ states *reproduce to the decimal place the exact results.* Then the agreement deteriorates as ^{50}Cr does not rotate as well as the projected band. The oblate band hardly describes a rotor, but for $J = 10\hbar$, $12\hbar$ and $14\hbar$ yields (in e fm^2) $Q_s = 19.2, 11.7, 8.3$ respectively, not too far from the exact numbers $(23.7, 12.3, 7.3)$. (This example suggests that looking for SM basis constructed with few determinants is a good idea: see Takaharu Otsuka's talk for a possible strategy.)

We have learned that calculations in the restricted $(fp)^n$ spaces account remarkably well for the results in the full major shell $((pf)^n)$. Let us move now to larger spaces. In Fig. 11 we have yrast transition energies for different configurations of 8 particles in $\Delta j = 2$ spaces. The force is KLS, $\hbar\omega = 9$ MeV, the single particle splittings uniform at $\varepsilon = 1$ MeV, and gds, say, is the lower sequence in the sdg, $p = 4$, shell. Rotational behaviour is fair to excellent at low J. As expected from the normalization property of the realistic quadrupole force [Eq. (17)] the moments of inertia in the rotational region go as $(p + 3/2)^2$ $(p' + 3/2)^2$, i.e. if we multiply all the E_γ values by this factor the lines become parallel. The intrinsic quadrupole moment obtained from

$$Q_0(J) = \frac{(J+1)(2J+3)}{3K^2 - J(J+1)} < JJ|3z^2 - r^2|JJ >,$$

remains constant to within 5% up to a critical J value at which the bands backbend.

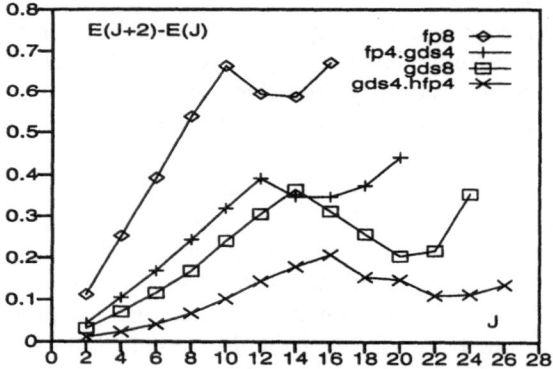

Fig. 11. Yrast transition energies $E_\gamma = E(J+2) - E(J)$ for different configurations, KLS interaction.

Two points are worth making:

- For $(fp)^8$ the Q_0 values are very much those of the full $(pf)^8$ calculation with a renormalized interaction and $\varepsilon = 2$ (Fig. 10 of ref.[39]) and the backbend occurs at the same J. The moments of inertia differ enormously, as they are very sensitive to ε and to the pairing force. However, the effect is perturbative as we shall see soon.
- All the spaces behave in the same way: a very good example of scaling.

Because of last item we specialize to $(gds)^8$ in what follows, and since we expect the quadrupole force to be responsible for most of what we see, we turn

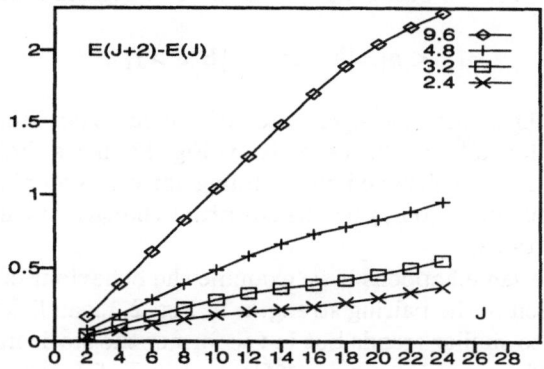

Fig. 12. Yrast transition energies $E_\gamma = E(J+2) - E(J)$ for the $(gds)^8$ configuration with an $-|e^{20}|\bar{q}\cdot\bar{q}$ force.

to Fig. 12 showing the results of diagonalizing $-|e^{20}|\,\bar{q}_p\cdot\bar{q}_p$ $(p=4)$. At $|e^{20}|=9.6$ the single particle splittings are overwhelmed and we have a nearly perfect rotor. Q_0 stays practically constant up to $J=16-18$ and then decreases slowly. At $|e^{20}|=4.8, 3.2$ and 2.4 the rotational behaviour remains very good below $J=14$. Then there is a break and the upper values are again aligned. Superficially, Fig. 11 and Fig. 12 are telling different stories, but it is not so.

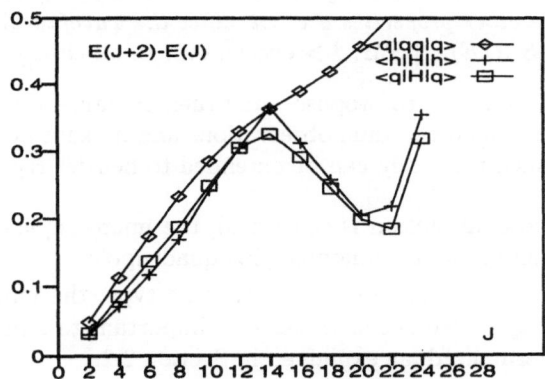

Fig. 13. Yrast transition energies for $< h|H|h > \equiv (gds)^8$ in Fig. 11 and $< q|qq|q > \equiv 3.2$ in Fig. 12 compared with $< q|H|q >$. See text.

At $|e^{20}|=3.2$ the overlap of each state with the one obtained with the full

KLS interaction is always better than 0.95, which suggests that

$$< h|H|h >_J \approx < q|H|q >_J,$$

where $|h >$ and $|q >$ are the eigenstates of the full Hamiltonian H and the quadrupole force for $|e^{20}| = 3.2$, respectively. Fig. 13 shows that this is the case indeed. It means that the observed backbending pattern is obtained by doing first order perturbation theory on $|q >$: the spectrum changes but not the *structure* (i.e. the wavefunctions).

Very much the same happens if we examine the behaviour of the moments of inertia as a function of the pairing strength. In Fig. 2, mean field (HFB) and SM calculations lead to similar trends but in the former the static moment of inertia is much larger. The same is true for ^{48}Cr, and in Ref. [41] it was shown that the origin of the discrepancy can be understood by redoing the SM calculations reducing the $JT = 01$ two-body matrix elements

$$W_{rrtt}^{01} \longrightarrow W_{rrtt}^{01} + \alpha\sqrt{(j_r + 1/2)(j_t + 1/2)}, \qquad (21)$$

which amounts to subtracting a standard pairing term. The resulting E_γ pattern for an exact calculation with $\alpha = 0.165$ is called SM(E) in Fig. 14. It amounts to a parallel displacement of the original SM points in the direction of the mean field ones (CHFB). Now note that SM(P)—obtained by taking expectation values of the modified interaction (21) using the SM wavefunctions—is very close to SM(E) except at the first point. The procedure is as in Fig. 13 and the conclusion is the same:

Although the energetics of the yrast band are strongly affected by the pairing modifications, the other properties are not, since the wavefunctions change little. (The overlaps $< SM(E), J| SM, J >$ exceed 0.97 in all cases).

At this point, allow me to propose—and then explain—two hypotheses that are consistent with the preceeding observations and make it possible to simplify calculations so much that they can be extended to heavier regions.

1. As far as rotational motion is concerned, the microscopic equivalent to the collective Hamiltonian is monopole plus quadrupole.
   - Since the quadrupole force is of Elliott's type, the number of particles in each major shell is conserved. It is important to rememeber that this force is normalized, as explained in Section 2.2.
   - It is also important to be careful about the monopole Hamiltonian. At fixed configuration (i.e., number of particles in each major shell) it could be approximated by a central field *which may be a sensitive function of the configuration* (refer to discussion in Section 2.1).
2. Once we are in a rotational regime the intrinsic quadrupole moment in each major shell remains close to its theoretical maximum as extracted directly from the diagrams in Fig. 9.

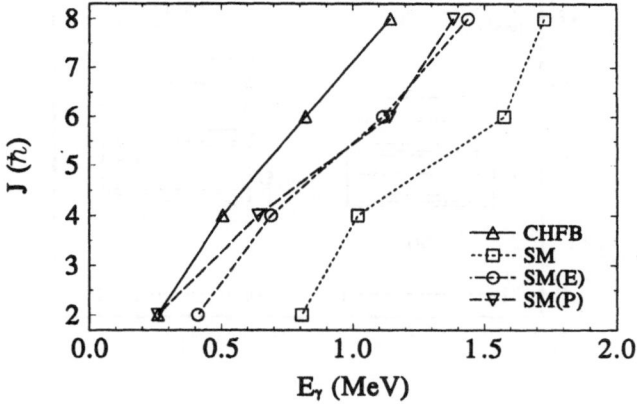

Fig. 14. Influence of the pairing strength on the moment of inertia. See text.

The main difference with conventional views of deformation is that Elliott forces build quadrupole coherence by aligning spherical oscillator quanta, which leads to clustering, rather than uniform distributions in elongated ellipsoids. Because of the second assumption we are always close to the maximum deformation attainable in each block (i.e.,major shell). It should be noted here that the more blocks we have, the largest will be the effective quadrupole strength within each block. (Refer to Eqs.(18,19), and generalize to several blocks.)

We have seen that quasi-SU(3) is a variant of SU(3) that obtains for moderate spin-orbit splittings. For other forms of single particle spacings, the pseudo-SU(3) scheme of Arima, Draayer, Harvey and Hecht [46] will be favoured (in which case we have to use the left panel of Fig. 9, with pseudo-$p = p - 1$). Other variants of SU(3) may be possible and are well worth exploring. In cases of truly large deformation SU(3) itself may be valid in some blocks.

To see how this works, consider Fig. 15 giving a schematic view of the single particle energies in the space of two contiguous major shells—in protons (π) and neutrons (ν)—adequate for a SM description of the rare earth region.

We want to estimate the quadrupole moments for nuclei at the onset of deformation. We shall assume quasi-SU(3) operates in the upper shells, and pseudo-SU(3) in the lower ones. The number of particles in each shell for which the energy will be lowest will depend on a balance of monopole and quadrupole effects, but Nilsson diagrams [45] suggest that when nuclei acquire stable deformation, two orbits $K=1/2$ and $3/2$—originating in the upper shells of Fig. 9—become occupied, i.e., the upper blocks are precisely the 8-particle configurations we have studied at length. Their contribution to the electric quadrupole moment is

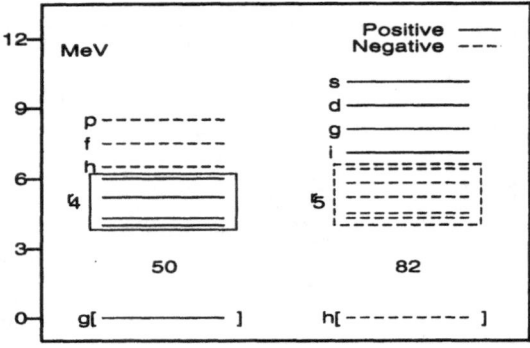

Fig. 15. Schematic single particle spectrum above ^{132}Sn. r_p is the set of orbits in shell p excluding the largest. For the upper shells the label l is used for $j = l+1/2$

then

$$Q_0 = 8[e_\pi(p_\pi - 1) + e_\nu(p_\nu - 1)], \tag{22}$$

with $p_\pi = 5$, $p_\nu = 6$; e_π and e_ν are the effective charges.

Consider even-even nuclei with Z=60-66 and N=92-98, corresponding to 6 to 10 protons with pseudo-$p = 3$, and 6 to 10 neutrons with pseudo-$p = 4$ in the lower shells. From the left part of Fig. 9 we obtain easily their contribution to Q_0, which added to that of Eq. (22) yields a total

$$Q_0 = 56e_\pi + (76 + 4n)e_\nu, \tag{23}$$

for $^{152+2n}$Nd, $^{154+2n}$Sm, $^{156+2n}$Gd and $^{158+2n}$Dy respectively. At fixed n, the value is constant in the four cases because the orbits of the triplet $K = 1/2, 3/2, 5/2$ in Fig. 9 have zero contribution for $p = 3$. Q_0 (given in dimensionless oscillator coordinates, i.e., $r \to r/b$ with $b^2 \approx 1.01A^{-1/3} fm^2$), is related to the $E2$ transition probability from the ground state by $B(E2) \uparrow = 10^{-5}A^{2/3}Q_0^2$. The results, using effective charges of $e_\pi = 1.4$, $e_\nu = 0.6$ calculated in [32] are compared in table 3 with the available experimental values . The agreement is quite remarkable and no free parameters are involved. Note in particular the quality of the prediction of constancy (or rather $A^{2/3}$ dependence) at fixed n, which does not depend on the choice of effective charges. The discrepancy in ^{152}Nd is likely to be of experimental origin, since systematics indicate, with no exception, much larger rates for a 2^+ state at such low energy (72.6 keV).

This section indicates that by careful analysis of exact results one may come to very simple computational strategies. In the last example on $BE2$ rates, the simplicity is such that the computation reduces to a couple of sums. In general, work may be a bit harder because we have to respect the fundamental specificity of nuclear states: They have good angular momentum. But then, dealing with

Table 3. $B(E2) \uparrow$ in e^2b^2 compared with experiment [47]

N	Nd	Sm	Gd	Dy
92	4.47	4.51	4.55	4.58
	2.6(7)	4.36(5)	4.64(5)	4.66(5)
94	4.68	4.72	4.76	4.80
			5.02(5)	5.06(4)
96	4.90	4.95	4.99	5.03
			5.25(6)	5.28(15)
98	5.13	5.18	5.22	5.26
				5.60(5)

a Hamiltonian as simple as monopole plus quadrupole should make it possible to push exact SM, SMMC, or other calculations very far, and then make them realistic by dealing with the rest of the Hamiltonian in perturbation theory.

4   Level densities[48]

Looking at Fig. 6 one may ask: Is it really necessary to go to 700 Lanczos iterations to obtain the exact strength function? At n iterations we have full information about the first $2n-1$ moments of the distribution. Therefore, something like 20 iterations should be quite sufficient. However, each iteration produces a new peak that acts as doorway, and to have the full information we must learn how this doorway mixes with the backgroung peaks. The idea is that one does not need to calculate exactly the background since the only thing we need to know about it is its level density. To understand what can be done we constructed the full tri-diagonal matrices for a few cases. In Fig. 16 we have plotted the diagonal $d \equiv H_{ii}$ and off diagonal $nd \equiv H_{i\,i+1}$ elements as a function of $x = 1/d$, (d =dimension of the matrix). Only one case is shown in full ($J = 2$ in ^{48}Ca) because it is representative of the universal curves on which all others fall, once nd is normalized so that its maximum value is unity. I have also included the 700 matrix elements for $J = 1$ in ^{48}Sc—used in Fig. 6—to give an idea of this universal behaviour.

By now we have a good answer to the question raised above: to obtain a reasonable strength function, calculate exactly a few terms of the exact matrix, and then model the other nd elements using the smooth function $g(x)$ shown in the figure, and set the diagonals to the adequate constant value. It works quite well, and I will say one word on the influence of fluctuations at the end.

What about level densities? In Fig. 17 the total area of the density curves and the spectrum spans are normalized to unity. The exact density and the one calculated with the model $g(x)$ off-diagonals in Fig. 16 are strikingly similar, and strikingly close to gaussians (actually binomials!) This is an old story [49],

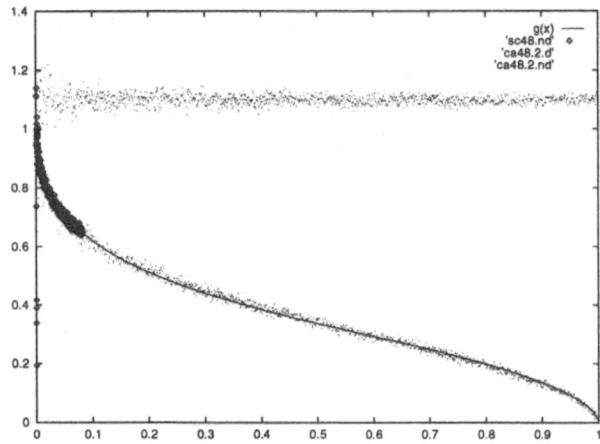

Fig. 16. Typical diagonal and off diagonal matrix elements

and one knows from Mon and French [33] that for random two body forces the densities should distribute normally. But we also know our forces are not random. So: Where is the difference? Simply in the lowest states, which are shifted to the left, but so little that in the figure the effect is hardly perceptible. To make it visible we calculate the specific heats—now using the density for all J states—and in Fig. 18 we see that the difference is now appreciable: The Schottky maximum, characteristic of finite spaces is now much higher. In Ref. [50] it was speculated that for an infinite system, this maximum would become a logarithmic singularity signalling a second order phase transition.

These very simple examples show that there is much of interest to be learned from SM matrices. Once their remarkable scaling properties are fully understood, they should make it possible to extend the studies to higher energies and larger systems.

One last word about fluctuations: On seeing the figures at the workshop, Steve Koonin asked about Anderson localization. Offhand my answer was that it depends on fluctuations, and that in our case fluctuations where too small to make a difference. Actually this is not quite true. The exact results tend to concentrate the strength around the big doorways while the model matrix leads to much more spreading. Adding fluctuations does not change the picture much. What seems to be happening is that *random* fluctuations can model the exact matrix elements only after some 70 iterations. Therefore, there seems to be some important information hidden in relatively high moments of the distribution that are responsible for the concentration of strength. I do not know whether this has anything to do with *Anderson* localization, but some localization there is.

I will stop here.

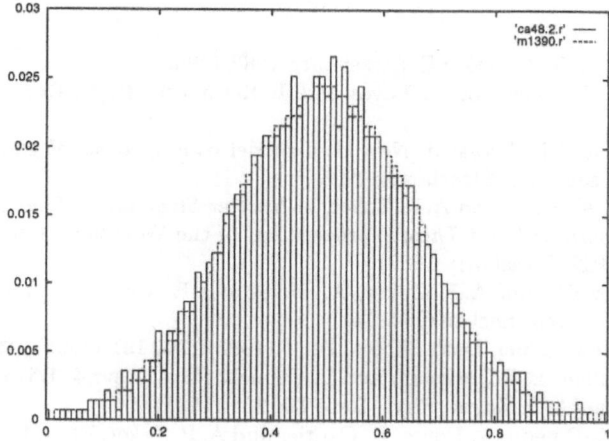

Fig. 17. Exact and model level densities for $J = 2$ in ^{48}Ca

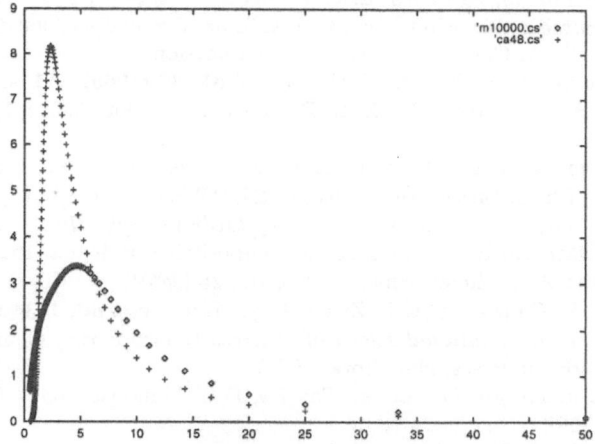

Fig. 18. Model and exact specific heat in ^{48}Ca (all states)

5   Acknowledgements

It goes without saying, but it is just as well to say it, that the enormous amount of work I have tried to summarize can only be teamwork. The $(SM)^2$ team is made of Etienne Caurier, Gabriel Martínez, Frèdèric Nowacki, Alfredo Poves, Joaquin Retamosa and myself. Marianne Dufour and Jean Duflo are my two partners in the monopole-multipole games.

References

[1] E. Caurier, code ANTOINE, Strasbourg 1989-1996.

[2] E. Caurier, F. Nowacki, A. Poves and J. Retamosa, Phys. Rev. Lett. **77**, 1954 (1996).

[3] E. Caurier and F. Nowacki, New shell model code in basis of good angular momentum and seniority. Strasbourg 1995.

[4] E. Caurier, A. Poves and A. P. Zuker, in *Nuclear Structure of Light Nuclei far from Stability. Experiment and Theory*, proceedings of the Workshop, Obernai, edited by G. Klotz, (CRN, Strasburg, 1989).

[5] G. Martínez-Pinedo, A. P. Zuker, A. Poves and E. Caurier, submitted to Phys. Rev. C. LANL arch. nucl-th/9608044.

[6] T. W. Burrows, Nucl. Data Sheets, **74**, 1 (1995); **76**, 191 (1995); **48**, 569 (1986).

[7] J. D. McCullen, B. F. Bayman, and L. Zamick, Phys. Rev. **4**, B515 (1964).

[8] A. Poves and A. P. Zuker, Phys. Rep. **70**, 235 (1981).

[9] G. Martínez-Pinedo, A. Poves, E. Caurier and A. P. Zuker, Phys. Rev. C **53**, R2602 (1996)

[10] B.A. Brown and B.H. Wildenthal, *Annu. Rev. Nucl. Part. Sci.* **38**, 29 (1988).

[11] E. Caurier, A. Poves and A. P. Zuker, Phys. Rev. Lett. **74**, 1517 (1995).

[12] G. Martínez-Pinedo, A. Poves, L. M. Robledo, E. Caurier, F. Nowacki, and A. P. Zuker, to de published in Phys. Rev. C, LANL archive nucl-th/9604039.

[13] S. Lenzi and J. A. Cameron, private communication.

[14] T. T. S. Kuo and G. E. Brown, Nucl. Phys. **A85**, 40 (1966); **A114**, 235 (1968).

[15] S. Kahana, H.C. Lee and C.K. Scott, Phys. Rev. **180**, 956 (1969); Phys. Rev. **185**, 1378 (1969).

[16] M. Hjorth-Jensen, T.T.S. Kuo and E. Osnes, Phys. Reps. **261**, 125 (1995).

[17] E. Pasquini, Ph. D. thesis, Report No. CRN/PT 76-14, Strasburg, 1976.

[18] E. Pasquini and A. P. Zuker in *Physics of Medium Light Nuclei*, Florence, 1977, edited by P. Blasi and R. Ricci (Editrice Compositrice, Bologna, 1978).

[19] A. Cortés and A. P. Zuker, Phys. Lett. **84B**, 25 (1979)

[20] A. Abzouzi, E. Caurier et A. P. Zuker, Phys. Rev. Lett. **66**, 1134 (1991).

[21] J.B. French, in *International school of physiccs Enrico Fermi*, Course XXXVI, C. Bloch ed. (Academic Press, New York, 1966).

[22] J.B. French in *Isospin in Nuclear Physics*, D.H. Wilkinson ed. (North-Holland, Amsterdam 1969)

[23] A.P. Zuker, *Nucl. Phys.* **A576**, 65 (1994).

[24] A. P. Zuker and M. Dufour, LANL archive nucl-th/9505012. Unpublished.

[25] A. Bohr and B. Mottelson, *Nuclear Structure* vol I (Benjamin, Reading, 1964).

[26] A. P. Zuker, B. Buck and J. B. Mc Grory Phys. Rev. Lett. **21**, 39 (1968).

[27] J. Duflo and A. P. Zuker, in preparation.

[28] M. Soyeur and A. P. Zuker, Phys. Lett. **41B**, 135 (1972).

[29] H. Nakada, T. Otsuka, T. Sebe, Phys. Rev. Lett. **67**, 1086 (1991).

[30] D.R. Bes and R.A. Sorensen, Adv. Nuc. Phys. **2**, 129 (1969).

[31] M. Baranger and K. Kumar, Nucl. Phys. **A110**, 490 (1968); Nucl. Phys. **A110**, 529 (1968).

[32] M. Dufour et A.P. Zuker, to appear in Phys. Rev. C **54**, xxxx (1996)

[33] K.K. Mon and J.B. French, Ann. Phys. (NY) **95**, 90 (1975).

[34] J. P. Elliott, Proc. R. Soc. London **245**, 128,562 (1956)

[35] B.D. Anderson *et al.*, Phys. Rev. C **31**, 1161 (1985).

[36] A. Poves, and A. P. Zuker, Phys. Rep. **71**, 141 (1981).

[37] A. P. Zuker in *Mathematical and Computational Methods in Nuclear Physics*, J.S. Dehesa et al eds. (Springer 1984).

[38] A. P. Zuker, in Nuclear dynamics at short and long ranges, edited by A. Gattone *et al.* (World Scientific, Singapore) (To be published.)

[39] E. Caurier, A. P. Zuker, A. Poves, and G. Martínez-Pinedo, Phys. Rev. C **50**, 225 (1994).

[40] J. A. Cameron, M. A. Bentley, A. M. Bruce, R. A. Cunningham, H. G. Price, J. Simpson, D. D. Warner, A. N. James, W. Gelletly and P. Van Isacker, Phys. Lett. **319B**, 58 (1993).

[41] E. Caurier, J.L. Egido, G. Martínez-Pinedo, A. Poves, J. Retamosa, L.M. Robledo and A.P. Zuker, Phys. Rev. Lett. **75**,2466 (1995).

[42] S. Lenzi. et al., Z. Phys. **A354**, 117 (1996)

[43] A. Bohr and B. R. Mottelson, *Nuclear Structure*, Vol. II (Benjamin, New York, 1975).

[44] A. Zuker, J. Retamosa, A. Poves and E. Caurier, Phys. Rev. C **52**, R1742 (1995).

[45] S.G. Nilsson, *Mat. Fys. Medd. Dan. Vid. Selsk.* **29**, 1 (1955)

[46] J.P. Draayer in *Future Directions in Nuclear Physics with $4\pi\gamma$ Detectors*, edited by J.Dudek and B. Haas (A.I.P. 1992)

[47] S. Raman, C.W. Nestor and K.H. Bhatt, *Phys. Rev.* **C37**, 805 (1988).

[48] L. Waha, E. Caurier, F. Nowacki and A. P. Zuker, in preparation.

[49] M.Chang and A.P.Zuker. Nuc. Phys.**A198**, 417 (1972).

[50] O.Civitarese, G. Dussel and A.P.Zuker. Phys. Rev. C **40**, 2900 (1989).

Realistic Shell-Model Calculations for Sn Isotopes

A. Covello, F. Andreozzi, L. Coraggio, A. Gargano, and A. Porrino

Dipartimento di Scienze Fisiche, Università di Napoli Federico II,
and Istituto Nazionale di Fisica Nucleare,
Mostra d'Oltremare, Pad. 20, I-80125 Italy

Abstract. We report on a shell-model study of the Sn isotopes in which a realistic effective interaction derived from the Paris free nucleon-nucleon potential is employed. The calculations are performed within the framework of the seniority scheme by making use of the chain-calculation method. This provides practically exact solutions while cutting down the amount of computational work required by a standard seniority-truncated calculation. The behavior of the energy of several low-lying states in the isotopes with A ranging from 122 to 130 is presented and compared with the experimental one.

1 Introduction

Shell-model calculations with realistic effective interactions have been mainly performed for light nuclei, in particular for those of the s-d shell [1]. Recently, however, work on extending this kind of calculations to medium- and heavy-mass nuclei has started. In this context, attention has been focused so far on the Sn isotopes, especially on the light ones [2–6] for which experimental data have become available down to ^{103}Sn in the last few years. Our own study of the light Sn isotopes [6] was undertaken as a part of a comprehensive program of calculations aimed at describing in a consistent way the spectroscopic properties of the whole chain of Sn isotopes and assessing the role of realistic effective interactions in this mass region. The results obtained in [6] were quite satisfactory encouraging us to pursue this program. In this paper we present some results of our calculations concerning the isotopes with A ranging from 122 to 130. They have been obtained by making use of a realistic effective interaction derived from the Paris free nucleon-nucleon potential [7].

In all of our calculations we assume that ^{100}Sn is a closed core and let the valence neutrons occupy the five single-particle orbits $0g_{7/2}$, $1d_{5/2}$, $2s_{1/2}$, $1d_{3/2}$, and $0h_{11/2}$. In this model space, however, a complete-basis diagonalization becomes very impractical if one wants to study nuclei with more than six or eight valence neutrons. As is well known, in single-closed-shell nuclei, such as the Sn isotopes and the $N = 82$ isotones, the seniority scheme provides a very valuable tool to reduce the model-space dimensions. There is in fact clear evidence that only states with low seniority play a significant role in the description of the low-energy spectra of these nuclei. One can therefore truncate the configuration space by retaining only those basis vectors for which the seniority $v = \sum_j v_j$ (v_j is the seniority of level j) does not exceed a certain value v_{max}. In standard shell-model calculations, however, this truncation scheme implies that the configuration space can be truncated only in rather large steps. As a consequence, the size of the energy matrices is still quite large if one wants to go beyond $v_{max}=2$. Aside from computational problems, the usefulness of describing the low-lying states of single-closed-shell nuclei by wave functions containing thousands of components is certainly questionable. Hence, it is desirable to have less computer demanding methods so that shell-model calculations can be performed with relative ease, thereby facilitating the confrontation between theory and experiment. In our study of the Sn isotopes we make use of a new approach to shell-model problems within the seniority scheme, the chain-calculation method (CCM), developed by ourselves over the past several years [8, 9]. Our approach is based on a chain calculation across nuclei differing by two in nucleon number and has proven to be an effective way for further reducing the dimensions of seniority-truncated shell-model spaces without significant loss in the accuracy of the results (we shall come back to this point in the next section). We can now handle model spaces including seniority-three and seniority-four states for odd and even nuclei, respectively.

Our presentation is organized as follows. In Sect. 2 we give a brief outline of the CCM and evidence its practical value by a specific example. In Sect. 3 we first give some details of our calculations and then present our results comparing them with experimental data. Sect. 4 presents some concluding remarks.

2   Outline of the Chain-Calculation Method

Let us consider the general shell-model Hamiltonian

$$H = \sum_j \epsilon_j \hat{N}_j + \frac{1}{4} \sum_{j_1 j_2 j_3 j_4 JM} G_J(j_1 j_2 j_3 j_4) A^\dagger_{JM}(j_1 j_2) A_{JM}(j_3 j_4) \ , \qquad (1)$$

where

$$\hat{N}_j = \sum_m a^\dagger_{jm} a_{jm} \ , \qquad (2)$$

$$A^\dagger_{JM}(j_1 j_2) = \sum_{m_1 m_2} \langle j_1 m_1 j_2 m_2 \mid JM \rangle a^\dagger_{j_1 m_1} a^\dagger_{j_2 m_2} \ . \qquad (3)$$

As already mentioned in the Introduction, our approach consists in solving the N-particle problem through a chain calculation involving only nuclei differing by two in nucleon number. In other words, the solution for the N-particle problem is built by starting from an initial value of N, say N_0, and then progressively adding pairs of particles up to the desired value of N. This stems from the fact that we write the wave function for a system of N (identical) particles and angular momentum J in the form

$$|N, \beta, J, M\rangle = \sum_{j\gamma} c^J_{j\beta\gamma}(N) A^\dagger_0(j) |N - 2, \gamma, J, M\rangle \ , \qquad (4)$$

where $A_0^\dagger(j) \equiv A_{00}^\dagger(jj)$, and β and γ specify the states with N and $N-2$ particles, respectively.

Since only zero-coupled pair operators are included in the expansion (4), the maximum seniority v_{max} in the states with N particles is that of the core states $|N-2, \gamma, J, M\rangle$. It is therefore clear that the initial value N_0, at which one starts the chain calculation, determines the seniority truncation, namely $v_{max} = N_0$ for any value of $N > N_0$. The practical value of this approach lies in the fact that it allows a further reduction of the dimensions of seniority-truncated spaces by restricting the number of core states in (4).

It should be noted that inherent in our formalism is the use of an overcomplete set of basis vectors $A_0^\dagger(j)|N-2, \gamma, J, M\rangle$. To single out a linearly independent set of states we analyze the metric matrix, whose elements are defined as

$$d^J_{j\gamma j'\gamma'}(N-2) = \langle N-2, \gamma, J \mid A_0(j)A_0^\dagger(j') \mid N-2, \gamma', J\rangle , \tag{5}$$

through the Cholesky decomposition of symmetric positive definite matrices [10]. An account of our procedure for removing the redundant states may be found in [11]. It is therefore to be understood that the sum in (4) is restricted to indices $(j\gamma)$ corresponding to linearly independent basis vectors. Making use of the expansion (4) in the Schrödinger equation,

$$H|N, \beta, J, M\rangle = E_{\beta J}(N)|N, \beta, J, M\rangle , \tag{6}$$

leads straightforwardly to the formulation of the eigenvalue problem in the form

$$\sum_{j'\gamma'}\left\{\sum_{j''\gamma''} [\bar{d}^J(N-2)]^{-1}_{j\gamma j''\gamma''} H^J_{j''\gamma''j'\gamma'}\right\} c^J_{j'\beta\gamma'}(N) = E_{\beta J}(N)c^J_{j\beta\gamma}(N) , \tag{7}$$

where

$$H^J_{j''\gamma''j'\gamma'} = \langle N-2, \gamma'', J|A_0(j'')HA_0^\dagger(j')|N-2, \gamma', J\rangle , \tag{8}$$

and $\bar{d}^J$ denotes the metric matrix obtained from the original one by removing rows and columns corresponding to redundant states according to the procedure described in [11].

The elements of the metric matrix (5) as well as those of the Hamiltonian matrix (8) are calculated by making use of the explicit expression of the wave function $|N-2, \gamma, J, M\rangle$ in the occupation-number representation,

$$|N-2, \gamma, J, M\rangle = \sum_{n_1...n_l q} K^J_{n_1...n_l\gamma q}(N-2)[A_0^\dagger(j_1)]^{n_1} ... [A_0^\dagger(j_l)]^{n_l}|N_0, q, J, M\rangle , \tag{9}$$

where $0 \leq n_i \leq \Omega_{j_i}$ and $n_1 + n_2 + \cdots + n_l = (N-2-N_0)/2$, l being the number of single-particle levels and $\Omega_j = j + 1/2$. The label q stands for all the quantum numbers specifying the states of the N_0-particle system, at which one starts the chain calculation. The coefficients K^J are related to the c^J's through

$$K^J_{n_1...n_l\beta q}(N) = \sum_{j_i\gamma} c^J_{j_i\beta\gamma}(N)K^J_{n_1...n_i-1...n_l\gamma q}(N-2)(1 - \delta_{n_i 0}) . \tag{10}$$

To sum up, the chain calculation is carried out as follows (more details may be found in [9]). Once the elements of the Hamiltonian matrix and of the metric matrix, $\langle N_0, q, J | A_0(j) H A_0^\dagger(j') | N_0, q', J \rangle$ and $\langle N_0, q, J | A_0(j) A_0^\dagger(j') | N_0, q', J \rangle$, respectively, are calculated, equation (7) can be solved together with the normalization condition,

$$\sum_{j\gamma j'\gamma'} \bar{d}^J_{j\gamma j'\gamma'}(N-2)\, c^J_{j\beta\gamma}(N)\, c^J_{j'\beta\gamma'}(N) = 1 \; , \tag{11}$$

for N_0+2. In this way one obtains $E_{\beta J}(N_0+2)$ and $c^J_{j\beta q}(N_0+2)$. At this first step the latter quantities are just equal to $K^J_{0\ldots n_j=1\ldots 0\beta q}(N_0+2)$, as can be seen from (10). This allows one to construct both the Hamiltonian and metric matrices for N_0+4. The chain calculation can then proceed up to the desired value of N.

As already pointed out above, the main advantage of this approach is that, within a given seniority truncation, a variety of approximations can be produced by reducing the number of core states $|N-2, \gamma, J, M\rangle$ in (4). We call kth-order theory the approximation in which the core states are restricted to the lowest k states. The dimensions of the matrices to be diagonalized at each step of the chain calculation are at most of order $k \times l$. By several numerical applications we have verified that the use of rather low orders of approximation suffices to produce practically exact results for at least the three or four lowest states for each value of the angular momentum. In this context, it should be stressed that an essential feature of the CCM is that it allows a stringent control of the accuracy of the approximations. Clearly, this makes it possible to cut down the amount of numerical work necessary to carry out a standard seniority-truncated shell-model calculation.

By way of illustration we give here some results of an application of our method in a shell-model space truncated at $v_{max} = 2$. As a test case we consider the nucleus ^{116}Sn. In this case we have 16 valence neutrons distributed over the five single-particle levels of the 50-82 shell, whose energies are taken to be (in MeV) $\epsilon_{5/2} = 0.0$, $\epsilon_{7/2} = 0.10$, $\epsilon_{1/2} = 2.20$, $\epsilon_{3/2} = 2.30$, $\epsilon_{11/2} = 3.15$. The residual interaction is the realistic effective interaction derived from the Paris free nucleon-nucleon potential which we are currently using in our study of the Sn isotopes. In Fig. 1 we show the energy of the lowest $0^+, 2^+, 4^+$ and 6^+ states obtained at various orders of approximation. The exact values are also reported for comparison. We see that a fifteenth-order calculation suffices to produce practically exact results for the seniority-two states while for the seniority-zero ground state only a fifth-order calculation is required. We note too how the increase in the accuracy of the results is fast and steady as the order of approximation increases. Regarding the reduction in the size of the energy matrices, the fifteenth-order calculation involves the diagonalization of 75×75 matrices, whereas the total number of seniority-two basis states with $J^\pi = 2^+, 4^+$ and 6^+ states is 529, 456 and 232, respectively. For the $J^\pi = 0^+$ ground state the number of seniority-zero basis states is 110 while the energy matrix is of order 25. It should be stressed that this reduction in the size of the energy matrices is obtained at no cost, namely without any significant loss in the accuracy of the results.

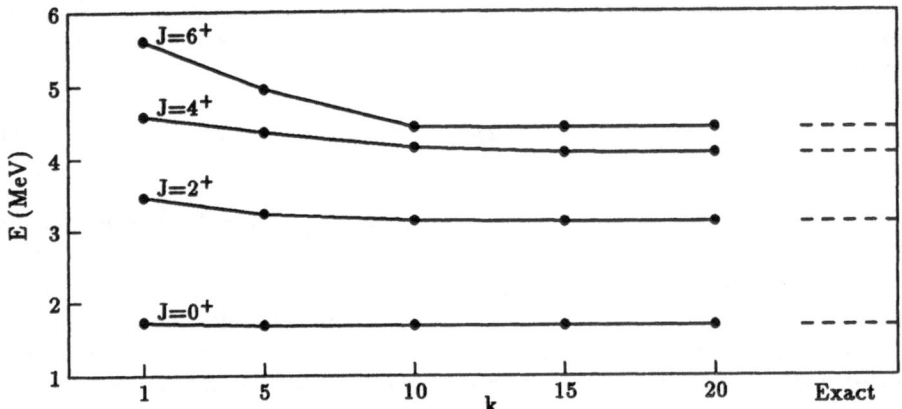

Fig. 1. Energy of the lowest $0^+, 2^+, 4^+$ and 6^+ states in ^{116}Sn. The results are plotted as a function of the order of approximation k. See text for details.

3   Calculations and Results

The results presented in this section concern the even and odd Sn isotopes with A ranging from 122 to 130. They have been obtained within the framework of the method described in the preceding section including states with seniority up to 4 and 3 for even and odd nuclei, respectively. As usual, the calculations have been performed considering holes rather than particles. As regards the order of approximation (see Sect. 2), we have carried out a 20th-order calculation for the odd nuclei while a 50th order has been used for the even ones. These levels of approximation are quite adequate to give extremely accurate results while providing a considerable reduction in the size of the energy matrices. We may mention, as an example, that for ^{123}Sn the total number of basis states with $J^\pi = \frac{5}{2}^+$ and $\frac{7}{2}^+$ is 739 and 763, respectively, while our energy matrices are of order 100.

A basic input to the calculation is clearly the model-space effective interaction V_{eff}. In the present work we have used a realistic V_{eff} derived from the Paris free nucleon-nucleon potential within the framework of a folded diagram method. A brief description of how this derivation is carried out is given in [6], where a list of relevant references can also be found.

As regards the single-hole (s.h.) energies, we take them from the experimental spectrum of ^{131}Sn [12, 13]. This means that our s.h. energies are (in MeV) $\bar{\epsilon}_{3/2} = 0.0$, $\bar{\epsilon}_{11/2} = 0.242$, $\bar{\epsilon}_{1/2} = 0.332$, $\bar{\epsilon}_{5/2} = 1.655$, $\bar{\epsilon}_{7/2} = 2.434$. The use of a unique set of s.h. energies for all of the considered isotopes is likely to be too restrictive, especially for the odd nuclei, in which the states of $v = 1$ nature are rather sensitive to changes in the mean field. We found it interesting, however, to see how far one can go in describing the low-energy properties of the considered nuclei without worrying about these changes. In Figs. 2–6 we compare, for A=122, 124, 126, 128 and 130, the calculated excitation energies of the first 2^+, 4^+, 6^+, 8^+ and 10^+ states with the experimental ones.

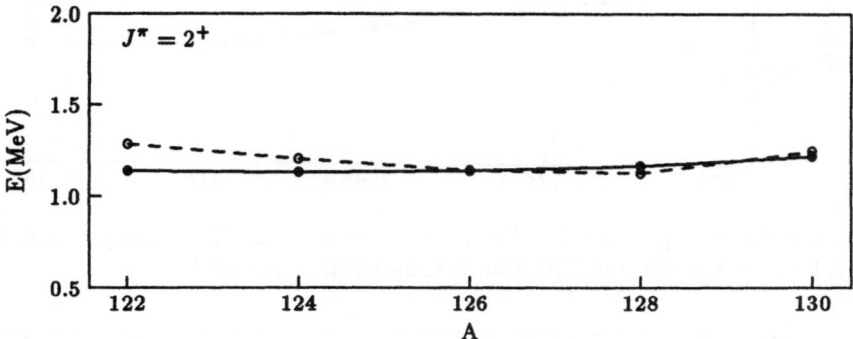

Fig. 2. Energy of the first 2^+ excited state in the even Sn isotopes with A ranging from 122 to 130. The theoretical results are represented by open circles, while the experimental data by solid circles. The latter are taken for A=122, 124, 126, 128 and 130 from [14, 15, 16, 17, and 18], respectively.

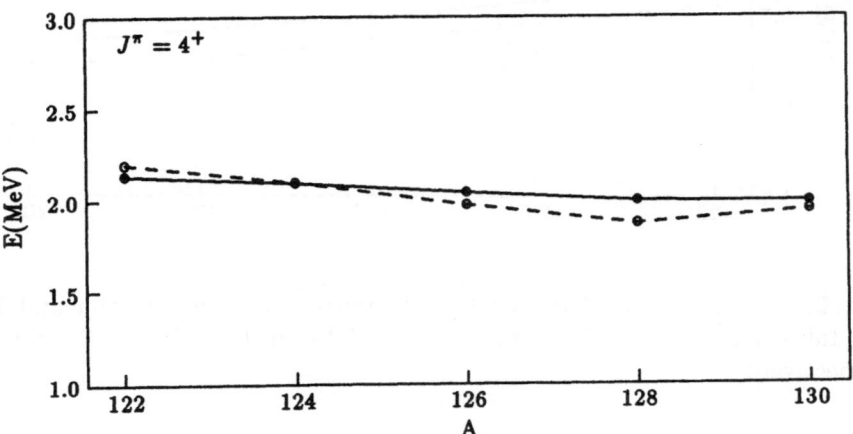

Fig. 3. Same as Fig. 2 but for the first 4^+ excited state. The experimental data are taken for A= 122, 124, 126, 128, and 130 from [14, 15, 16, 17, and 18], respectively.

We see that the behavior of the energy as a function of A is remarkably well reproduced in all cases. The quantitative agreement with experiment may also be considered more than satisfactory, the largest discrepancy being 186 keV for the 10^+ state in ^{128}Sn.

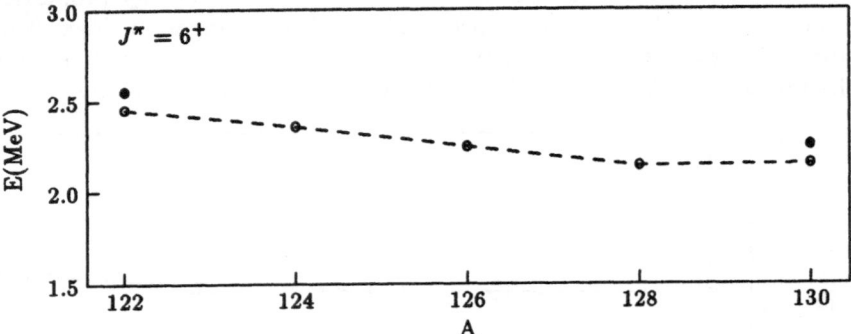

Fig. 4. Same as Fig. 2 but for the first 6^+ excited state. The experimental data are taken for $A=$ 122 and 130 from [14, and 18], respectively.

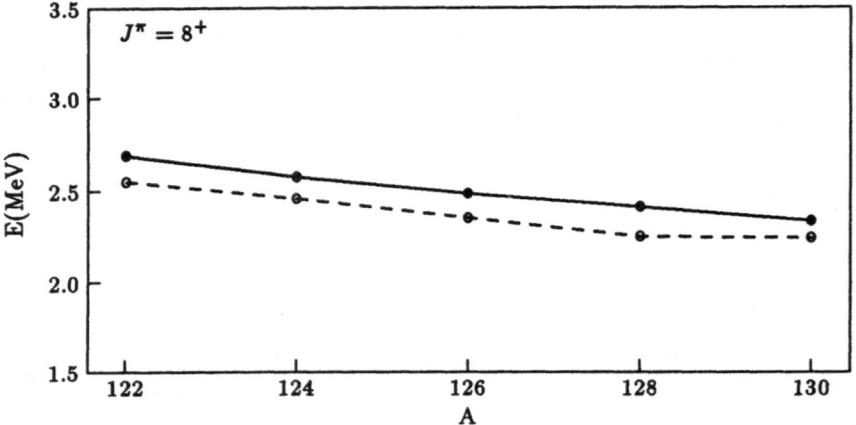

Fig. 5. Same as Fig. 2 but for the first 8^+ excited state. The experimental data are taken for $A=$ 122, 124, 126, 128, and 130 from [14, 19, 20, 21, and 18], respectively.

The results for the odd isotopes are presented in Figs. 7 and 8, where we report the energies of the lowest-lying states with angular momentum and parity corresponding to the valence s.h. orbits.

In Fig. 7 we compare, for $A=$123, 125, 127 and 129, the calculated excitation energies of the first $\frac{11}{2}^-$, $\frac{3}{2}^+$ and $\frac{1}{2}^+$ states with the experimental ones. These states, which are the three lowest ones in all of the four isotopes, are almost pure $v=1$ states (the minimum percentage of $v=1$ components is 95% for the $\frac{3}{2}^+$ state in ^{123}Sn). It appears that the behavior of these three states is very

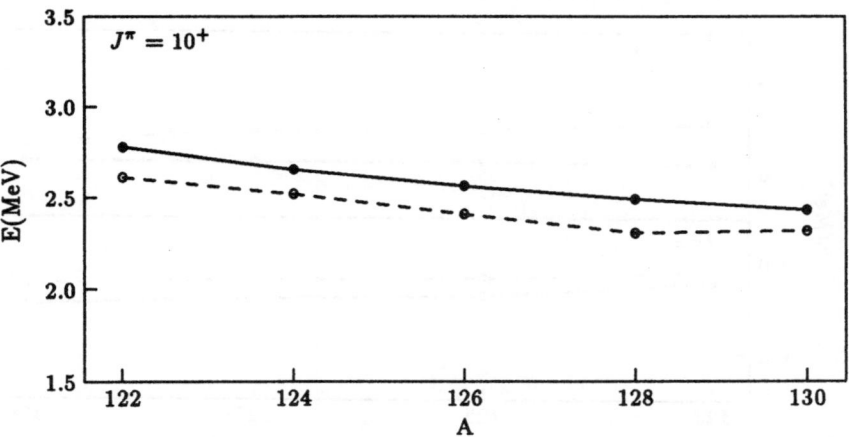

Fig. 6. Same as Fig. 2 but for the first 10^+ excited state. The experimental data are taken for $A=$ 122, 124, 126, 128, and 130 from [14, 19, 20, 21, and 18], respectively.

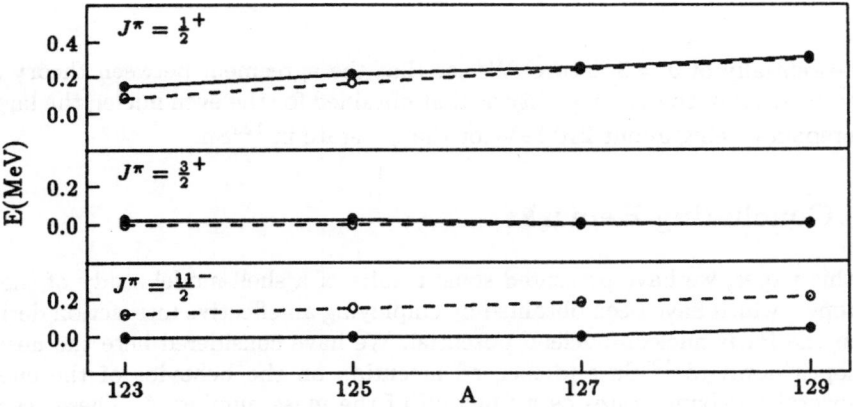

Fig. 7. Energies of the first $\frac{11}{2}^-$, $\frac{3}{2}^+$ and $\frac{1}{2}^+$ states in the odd Sn isotopes with A ranging from 123 to 129. The conventions of the presentation are the same as those used in Fig. 2. The experimental data are taken for $A=123$ and 125 from [22, and 23], respectively, and for $A = 127$ and 129 from [24].

well reproduced, the quantitative agreement for the $\frac{1}{2}^+$ and $\frac{3}{2}^+$ states being just excellent. For the $\frac{11}{2}^-$ states the discrepancy with experiment ranges from about 120 keV in ^{123}Sn to 180 keV in ^{129}Sn. As a consequence, the ordering of the close-lying levels $\frac{11}{2}^-$ and $\frac{3}{2}^+$ is inverted in ^{123,125,127}Sn. In this connection, we may mention that rather slight changes in the s.h. energies lead to right ordering [25].

In Fig. 8 the calculated excitation energies of the first $\frac{5}{2}^+$ and $\frac{7}{2}^+$ states are compared with experiment. We find that in all of the four isotopes both states

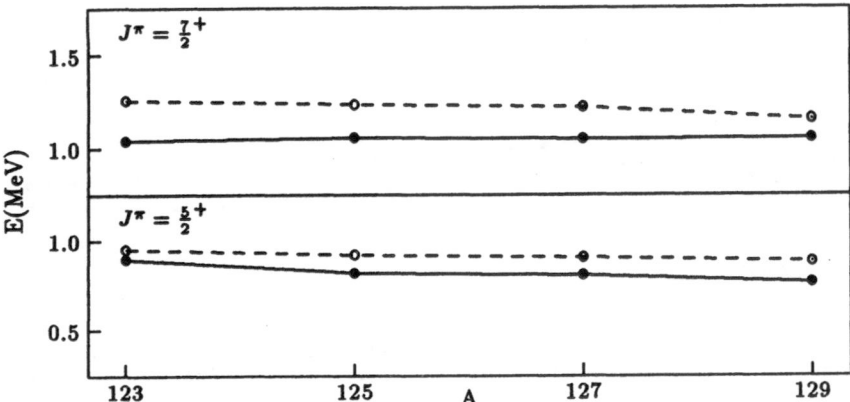

Fig. 8. Same as Fig. 7 but for the first $\frac{5}{2}^+$ and $\frac{7}{2}^+$ states. The experimental data are taken for $A=123$ and 125 from [22, and 26], respectively, and for $A=127$ and 129 from [24].

are essentially of $v = 3$ nature. We see that the agreement between theory and experiment is of the same quality as that obtained for the even nuclei, the largest discrepancy being about 210 keV for the $\frac{7}{2}^+$ state in ^{123}Sn.

4   Concluding Remarks

In this paper, we have presented some results of a shell-model study of the Sn isotopes, which have been obtained by employing an effective interaction derived from the Paris nucleon-nucleon potential. We have considered here the heavier isotopes down to ^{122}Sn and focused attention on the behavior of the energy of several low-lying states as a function of the mass number A. These results, as well as those reported in [6] for the light isotopes, are quite satisfactory. As mentioned earlier, we are currently carrying out a complete study of the whole isotopic Sn chain, where use is also made of the Bonn A and Argonne V_{18} nucleon-nucleon potentials. A companion study of the $N = 82$ isotones is also under way. This should permit to draw definite conclusions on the role of modern realistic interactions in nuclear structure calculations.

The results presented in this paper are part of a research project carried out in collaboration with T.T.S. Kuo. This work was supported in part by the Italian Ministero dell'Università e della Ricerca Scientifica e Tecnologica (MURST).

References

1. Jiang, M.F., Machleidt R., Stout D.B., Kuo, T.T.S.: Phys. Rev. C**46**, 910 (1992); earlier references are given there

2. Engeland T., Hjorth-Jensen M., Holt A., Osnes E.: Phys. Rev. **C48**, R535 (1993)
3. Holt A., Engeland T., Hjorth-Jensen M., Osnes, E.: Nucl. Phys. A **570**, 137c (1994)
4. Engeland T., Hjorth-Jensen M., Holt A., Osnes E.: Phys. Scr. **T56**, 58 (1995)
5. Hjorth-Jensen, M., Kuo T.T.S., Osnes, E.: Phys. Rep. **261**, 125 (1995)
6. Andreozzi, F., Coraggio, L., Covello, A., Gargano, A., Kuo, T.T.S., Li, Z.B., Porrino, A.:(submitted to Phys. Rev. C)
7. Lacombe, M., Loiseau, B., Richard, J.M., Vinh Mau R., Côté, J., Pires, P., de Tourreil, R.: Phys. Rev. **C21**, 861 (1980).
8. Covello, A., Andreozzi, F., Gargano, A., Porrino, A.: Proceedings of the Predeal International Summer School 1991. Raduta, A.A., Delion, D.S., Ursu, I.I.(eds.), p. 2. Singapore: World Scientific 1992
9. Covello, A., Andreozzi, F., Coraggio, L., Gargano, A., Porrino, A.: Proceedings of the Fifth International Spring Seminar on Nuclear Physics 1995. Covello, A.(ed.), p. 147. Singapore: World Scientific 1996
10. Wilkinson, J.H.: *The Algebraic Eigenvalue Problem* (Clarendon Press, Oxford, 1965), Chap. 4.
11. Andreozzi, F., Covello, A., Gargano, A., Porrino, A.: Proceedings of the International Symposium on Nuclear Shell Models, 1984. Vallieres, M., Wildenthal, B.H., p. 610. Singapore: World Scientific 1985
12. Fogelberg, B., Blomqvist, J.: Phys. Lett. **B137**, 20 (1984)
13. Fogelberg, B., Blomqvist, J.: Nucl. Phys. **A429**, 205 (1984)
14. Tamura, T.,: Nucl. Data Sheets **71**, 461 (1994)
15. Tamura, T., Miyano, K., Ohya, S.: Nucl. Data Sheets **41**, 413 (1984)
16. Miyano. K.: Nucl. Data Sheets **3**, 429 (1993)
17. Kitao, K., Kanbe, M., Matumoto, Z.: Nucl. Data Sheets, **38**, 191 (1983)
18. Sergenkov, Yu. V.,: Nucl. Data Sheets **58**, 765 (1989)
19. Broda, R., Mayer, R.H., Bearden, I.G., Benet, Ph., Daly, P.J., Grabowski, Z.W., Carpenter, M.P., Janssen, R.V.F., Khoo, T.L., Lauritsen, T., Moore, E.F., Lunardi, S., Blomqvist, J.: Phys. Rev. Lett. **68**, 1671 (1992).
20. Broda, R.: (private communication)
21. Fogelberg, B., Heyde, K., Sau, J.: Nucl. Phys. **A352**, 157 (1981)
22. Ohya, S., Tamura, T.: Nucl. Data Sheets **70**, 531 (1993)
23. Katakura, K., Oshima, M., Kitao, K., Iimura, H.: Nucl. Data Sheets **70**, 217 (1993)
24. De Geer, L.-E., Holm, G.B.: Phys. Rev. **C22**, 2163 (1980)
25. Covello, A., Andreozzi, F., Coraggio, L., Gargano, A., Porrino, A.: Proceedings of the XII International School on Nuclear Physics, Neutron Physics and Nuclear Energy 1995. Andrejtscheff, A., Elenkov, D.(eds.). Sofia: Institute for Nuclear Research and Nuclear Energy 1996
26. Fogelberg, B., De Geer, L.-E., Fransson, K., af Ugglas, M.: Z. Phys. **A276**, 381 (1976)

Shell Model Monte Carlo Methods

Steven E. Koonin

W.K. Kellogg Radiation Laboratory, California Institute of Technology, Pasadena, CA 91125 USA

Abstract. We review quantum Monte Carlo methods for dealing with large shell model problems. These methods reduce the imaginary-time many-body evolution operator to a coherent superposition of one-body evolutions in fluctuating one-body fields; the resultant path integral is evaluated stochastically. We first discuss the motivation, formalism, and implementation of such Shell Model Monte Carlo (SMMC) methods. There then follows a sampler of results and insights obtained from a number of applications. These include the ground state and thermal properties of pf-shell nuclei, the thermal behavior of γ-soft nuclei, and the calculation of double beta-decay matrix elements. Finally, prospects for further progress in such calculations are discussed.

This lecture covered material that can be found in Physics Reports (S.E. Koonin, D.J. Dean, and K. Langanke, to be published, 1996) or on the world wide web at the URL address http://xxx.lanl.gov/abs/nucl-th/9602006.

Shell Model for Large Systems and Quantum Monte Carlo Diagonalization Method

Takaharu Otsuka[1,2], Michio Honma[3] and Takahiro Mizusaki[1]

[1] Department of Physics, University of Tokyo, Hongo, Tokyo 113, Japan
[2] RIKEN, Hirosawa, Wako-shi, Saitama 351-01, Japan
[3] Center for Mathematical Sciences, University of Aizu
    Tsuruga, Ikki-machi Aizu-Wakamatsu, Fukushima 965, Japan

Abstract. We propose a Quantum Monte Carlo diagonalization method (QMCD) for solving the quantum many-body interacting systems. One-body fields dominating the structure of low-lying states are selected by an auxiliary field Monte Carlo method, which produce many-body basis states for diagonalizing the Hamiltonian consisting of one- and two-body terms. Not only the ground state but also low-lying excited states are obtained with their wave functions. The feasibility of shell-model calculations is radically extended by the QMCD method. It is shown that there are two crucial points for the realistic shell model calculations: one is to enhance the efficiency dynamically in the generation of shell-model basis vectors, and the other is to implement the restoration of symmetries such as angular momentum and isospin. Consequently the level structure of low-lying states can be studied with realistic interactions. After testing this method on ^{24}Mg, we present first results for energy levels and E2 properties of ^{64}Ge, indicating its large and γ-soft deformation.

1 Introduction

The nuclear shell model has been successful in the description of various aspects of nuclear structure, partly because it is based on a minimum number of natural assumptions. Although the direct diagonalization of the Hamiltonian matrix in the full valence-nucleon Hilbert space is desired, the dimension of such a space is too large in many cases, preventing us from performing the full calculations. Fig. 1 shows the m-scheme dimension of $N = Z$ even-even nuclei in the pf shell for both M=0 and full spaces. The direct diagonalization has been carried out up to ^{48}Cr (Caurier et al. 1994), for which the dimension of the M=0 space is 1,963,461. By adding only 2 protons and 2 neutrons, one obtains ^{52}Fe, for which the dimension increases from ^{48}Cr by a factor of about 50. Therefore conventional diagonalization of this nucleus without truncation of the Hilbert space is hopeless in the near future. Thus the applicability of the usual shell model is restricted due to the explosive increase of the dimension.

Recently, in order to relax this restriction drastically, active developments have been made in the stochastic approaches (Johnson et al. 1992, Varga and Liotta 1994, Horoi et al. 1994). For instance, the quantum Monte Carlo (QMC) method, which was first proposed for the condensed matter problem, has been

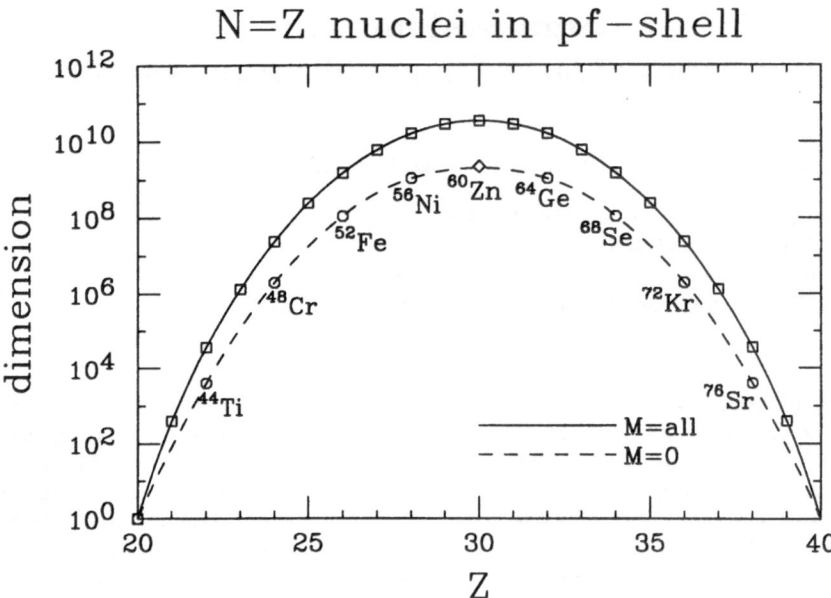

Fig. 1. The m-scheme dimension of $M=0$ (dashed line) and complete (solid line) spaces for $N = Z$ even-even nuclei in the pf shell.

applied to the nuclear shell model (Johnson et al. 1992, Ormand et al. 1994) by utilizing the auxiliary field Monte Carlo technique.

The auxiliary field Monte Carlo technique has been widely used in the solid state physics. This technique enables us to treat the many-body interacting system which spans so huge dimension of its Hilbert space that the direct diagonalization of the Hamiltonian matrix cannot be carried out. It has been successful in clarifying the zero temperature and thermal properties of the ground state for huge systems. In the nuclear structure physics, Johnson et al. has introduced this technique into the nuclear shell model, referring it to the Shell Model Monte Carlo (SMMC) (Koonin et al. 1996). In fact, ground-state (Langanke et al. 1995) and thermal properties (Dean et al. 1995) have been described well by the SMMC method (Koonin et al. 1996).

However, its application is rather restricted by the so-called *minus-sign* problem. In the SMMC method, the Hamiltonian must have specific properties with respect to the time-reversal transformation. Furthermore, the structure of excited states can be seen only through response functions. For spectroscopic investigations of the nuclear structure physics, it is a crucial step to remove or relax such a restriction.

For the diagonalization in shell model, we usually use the complete spherical bases of a given Hilbert space, although the amplitudes of wave functions distribute over almost all the bases, in general. Consequently, a huge model space is required. On the other hand, the mean field solutions (intrinsic states) have been known to give a good approximation, in particular, for low-lying states. Therefore, it is expected that, once appropriate intrinsic states are selected as many-body basis states, the Hamiltonian can be diagonalized in a reasonable approximation in a subspace spanned by such basis states. The number of basis states can then be made much smaller, so that the practical calculation becomes feasible. In the following we propose a method of choosing the many-body basis states in the form of intrinsic states by using the auxiliary field Monte Carlo technique, so as to diagonalize the Hamiltonian in a reasonable approximation. This method is referred to as the Quantum Monte Carlo Diagonalization (QMCD) method.

2   Formulation of QMCD

2.1   Auxiliary field Monte Carlo method

We consider the standard shell model Hamiltonian which consists of single particle energies and a two-body interaction:

$$H = \sum_{i,j=1}^{N_{sp}} \epsilon_{ij} c_i^\dagger c_j + \frac{1}{4} \sum_{i,j,k,l=1}^{N_{sp}} v_{ijkl} c_i^\dagger c_j^\dagger c_l c_k, \tag{1}$$

where $c_i^\dagger$ (c_i) denotes the creation (annihilation) operator of a nucleon in a single particle state i. The dimension of the single particle states is denoted as N_{sp}; $N_{sp}=24$ for the sd shell and 40 for the pf shell. This Hamiltonian can be rewritten in the quadratic form of one-body operators O_α:

$$H = \sum_{\alpha=1}^{N_f} (E_\alpha O_\alpha + \frac{1}{2} V_\alpha O_\alpha^2), \tag{2}$$

where the number of the O_α's, called N_f, can be at most N_{sp}^2 and usually appears to be much smaller. We consider the imaginary time evolution operator $e^{-\beta H}$, and divide the imaginary time β into N_t steps,

$$e^{-\beta H} = \prod_{n=1}^{N_t} e^{-\Delta\beta H}, \tag{3}$$

where $\Delta\beta = \beta/N_t$. According to Johnson et al. 1992, by applying the Hubbard-Stratonovich transformation (Hubbard 1959, Stratonovich 1957) at each time step, this operator can be expressed as an integral of one body evolution operators with respect to the auxiliary fields $\sigma_{\alpha n}$:

$$e^{-\beta H} \approx \int_{-\infty}^{\infty} \prod_{\alpha,n} d\sigma_{\alpha n} \left(\frac{\Delta\beta|V_\alpha|}{2\pi}\right)^{1/2} \cdot G(\sigma) \cdot \prod_n e^{-\Delta\beta h(\sigma_n)}, \tag{4}$$

where σ_n means a set of auxiliary fields of the n-th time step, $\sigma_n = (\sigma_{1n}, \sigma_{2n}, \cdots, \sigma_{N_f n})$, and σ denotes the assembly of the auxiliary fields over all the time steps, $\sigma = \{\sigma_1, \sigma_2, \cdots, \sigma_{N_t}\}$. The Gaussian weight factor $G(\sigma)$ is defined by

$$G(\sigma) = e^{-\sum_{\alpha,n} \frac{\Delta\beta}{2}|V_\alpha|\sigma_{\alpha n}^2}, \tag{5}$$

and the one-body Hamiltonian $h(\sigma_n)$ is defined by

$$h(\sigma_n) = \sum_\alpha (E_\alpha + s_\alpha V_\alpha \sigma_{\alpha n})O_\alpha, \tag{6}$$

where $s_\alpha = \pm 1 \ (= \pm i)$ if $V_\alpha < 0 \ (> 0)$.

If the $N_t \times N_f$-dimensional integral is treated with sufficient accuracy, we can obtain the ground state wave function by operating (4) with sufficiently large β on any initial state $|\Psi\rangle$ which is not orthogonal to the ground state. In numerical calculations the integral is evaluated by discretizing the $\sigma_{\alpha n}$ variables, and the integrand is computed for several specific sets σ. In the Monte Carlo integration, each set of σ is generated stochastically according to some weight functions. Then many sets of auxiliary fields are generated and the corresponding wave functions should be added with an equal weight. In many cases this integral does not converge with a tractable number of sets σ, since the variance of the integrand is in general too large. In the SMMC (Johnson et al. 1992) this difficulty can be avoided by considering only the expectation values. In this treatment, the so-called "minus sign" problem arises.

2.2 Basic Concepts of QMCD

In general, low-lying states of nuclei are described to a good extent in terms of static and/or dynamic mean fields and their fluctuations. The basic idea of the QMCD method is to diagonalize the shell model Hamiltonian, by using this property, in a subspace spanned by a small number of selected basis states obtained by stochastically generated one-body fields. Thus, the ground state and several excited states are expected to be obtained.

Based on this expectation, we propose a new method, consisting of the following steps:

1. We take an initial intrinsic state $|\Psi^{(0)}\rangle$ which can be determined, for example, by a variational method. Then the initial energy is calculated as $E^{(0)} = \langle\Psi^{(0)}|H|\Psi^{(0)}\rangle$.
2. A set of the auxiliary fields σ is given stochastically according to the Gaussian weight function (5). Note that it is practically easy and fast to generate random numbers obeying the Gaussian distribution.
3. We calculate a wave function $|\Phi(\sigma)\rangle$ for the present set σ:

$$|\Phi(\sigma)\rangle \propto \prod_{n=1}^{N_t} e^{-\Delta\beta h(\sigma_n)} |\Psi^{(0)}\rangle. \tag{7}$$

(We note that if an infinite number of auxiliary fields are generated and the corresponding wave functions of this type are all included for infinitesimal $\Delta\beta$, the exact time-evolved wave function is obtained.)

4. The state $|\Phi(\sigma)\rangle$ is ortho-normalized, by means of the Gram-Schmidt method, with respect to all other basis states obtained previously, and a new basis $|\Phi'\rangle$ is determined.

5. By including the new basis state obtained in step 4, we diagonalize the Hamiltonian H, and obtain an improved ground state energy E and its wave function $|\Psi\rangle$.

6. Steps from 2 to 5 are repeated until the the ground state energy E converges. We can also confirm the convergence by calculating the expectation value of angular momentum operator.

In order to accelerate the convergence, the following process can be added to step 4 according to De Readt and Frick 1993. The energy decrease ΔE which originates in the addition of the new basis $|\Phi'\rangle$ can be estimated by

$$\Delta E \sim \frac{1}{2}\left\{E - E_1 + \sqrt{(E - E_1)^2 + 4\,|E_2\,|^2}\right\}, \tag{8}$$

where E denotes the ground state energy obtained in the previous step, $E_1 = \langle\Phi'\,|\,H\,|\,\Phi'\rangle$, and $E_2 = \langle\Psi\,|\,H\,|\,\Phi'\rangle$. If ΔE is too small, for example, less than 10% in comparison to the energy decrease in the previous steps, the state $|\Phi'\rangle$ is discarded, and we return to step 2.

The number of adopted basis states is referred to as the QMCD basis dimension, which is increased as the above steps are repeated. We emphasize that energies and wave functions are determined by the diagonalization, and that we can obtain excited states as well as the ground state. Since certain basis states are commonly important for low-lying states, several lowest excited states are expected to be obtained with correlated accuracies using the same dimension of the basis states.

In practical shell model calculations, it is convenient to adopt basis states in the form of Slater determinants:

$$|\Phi\rangle = \prod_{\alpha=1}^{N} a_\alpha^\dagger\,|-\rangle, \tag{9}$$

where N denotes the number of valence nucleons, $|-\rangle$ is an inert spherical core, and $a_\alpha^\dagger$ represents the nucleon creation operator in a canonical single-particle state α, which is a linear combination of the spherical bases.

$$a_\alpha^\dagger = \sum_{i=1}^{N_{sp}} c_i^\dagger D_{i\alpha}. \tag{10}$$

We can specify the basis state $|\Phi\rangle$ in terms of an $N_{sp} \times N$ complex matrix D. Note that the operation of an exponential of any one-body operator $T = \sum_{ij} T_{ij} c_i^\dagger c_j$

on a state gives a new matrix $D'_{i\alpha} = \sum_j T_{ij} D_{j\alpha}$, while the form of a Slater determinant remains.

As another method, we can take condensed states of nucleon pairs in place of Slater determinants. This is a generalization of BCS/HFB approach.

3   Projection of Magnetic Quantum Number

Since a nucleus has rotational symmetry, the restoration of the total angular momentum, denoted as J, is quite crucial. The intrinsic state is not an eigenstate of the angular momentum or not even an eigenstate of the z-component of angular momentum. If we have no restriction on the angular momentum of the basis state, we should have degenerate eigenstates for the magnetic quantum number. In general, the full angular momentum projector can be expressed as the 3-dimensional integral involving the D-function. This projection is somewhat difficult for the numerical calculation unless the system has axial symmetry. However, the projection of z-component of the angular momentum is enough for selecting one component of degenerate eigenstates and is easily implemented in the QMCD method (Mizusaki et al. 1996) because it is carried out by inserting the one-dimensional numerical integration for evaluating overlaps and expectation values of the Hamiltonian.

In order to extract the component of a given magnetic quantum number, M, from the intrinsic state, we introduce the following state,

$$ | \Phi(\sigma, M)\rangle = P_M |\Phi(\sigma)\rangle = \frac{1}{2\pi} \int_0^{2\pi} d\phi\, e^{-i\phi(J_z - M)} |\Phi(\sigma)\rangle , \qquad (11) $$

where P_M is the projector onto the total magnetic quantum number M. Here J_z stands for the z-component of angular momentum operators. Note that J_z is an operator and M is its quantum number. The state $| \Phi(\sigma, M)\rangle$ will be referred to as the M-projected intrinsic state. The corresponding overlap and matrix element of the Hamiltonian are evaluated with one-dimensional integration over ϕ in eq.(11). The numerical calculation for the M-projection is easier in comparison to the full angular momentum projection.

If we consider the states with angular momentum J, we can use the M-projected intrinsic states with $M = J$. As this space certainly includes the eigenstates with angular momenta higher than J, we add the $J \cdot J$ term to the Hamiltonian, so as to push up the eigenstates with higher angular momenta. Hence, with this procedure, we can obtain low-lying states with the angular momentum J, separating them from other states with different angular momenta.

The present angular momentum projection is carried out only in the diagonalization process. We do not apply any modification to the Monte Carlo procedure. However, because the intrinsic state generated by the imaginary time evolution operator (7) contains various angular momenta, the M-projected basis can be considered to be good as basis for the subspace with a given magnetic quantum number. Moreover, the M-projection reduces the numerical burden of the diagonalization by decreasing the dimension.

4   Demonstration of QMCD for Simple Systems

4.1   Interacting Boson Model

As a demonstration of the QMCD method, we first consider the Interacting Boson Model (IBM) (Iachello and Arima 1987) as an easily accessible, yet still realistic, many-body system. The bosons are denoted as b_i $(i = 1, \cdots, N_{sp})$. In the IBM-1, $N_{sp}=6$ and $b_1 = s$, $b_2 = d_{-2}$, $\cdots$, $b_6 = d_2$. It is straightforward to apply the QMCD procedure mentioned above to the IBM. In this case we can take boson condensed states as basis states:

$$|\Phi\rangle = \frac{1}{\sqrt{N_B!}} \left(\sum_{i=1}^{N_{sp}} x_i b_i^\dagger \right)^{N_B} |0\rangle, \qquad (12)$$

where $|0\rangle$ is the boson vacuum and the (generally complex) amplitudes x_i specify the basis states.

The QMCD method is applied to the O(6) limit of sd-IBM and the SU(3) limit of the sdg-IBM in the following. The O(6) and SU(3) limits represent the γ-unstable and axially symmetric deformed nuclei, respectively. The sdg-IBM is an extension of the sd-IBM by including the hexadecapole degree of freedom. We start with the IBM Hamiltonian:

$$H = -\kappa Q \cdot Q + \kappa' J \cdot J, \qquad (13)$$

where the J represents the angular momentum operator and the quadrupole operator Q is defined by

$$Q = s^\dagger \tilde{d} + d^\dagger s + \chi [d^\dagger \tilde{d}]^{(2)}, \qquad (14)$$

for sd-IBM and in the case of the O(6) limit, χ is 0. For sdg-IBM,

$$Q = s^\dagger \tilde{d} + d^\dagger s + \chi [d^\dagger \tilde{d}]^{(2)} + \lambda [d^\dagger \tilde{g} + g^\dagger \tilde{d}]^{(2)} + \omega [g^\dagger \tilde{g}]^{(2)}. \qquad (15)$$

In the case of the SU(3) limit, those parameters are $\chi = -11\sqrt{10}/28$, $\lambda = 9/7$, $\omega = -3\sqrt{55}/14$.

First we consider the SU(3) limit of the sdg-IBM. Figure 2 shows the energies of the ground and the first excited states as a function of the QMCD basis dimension for two cases of $N_B=10$ and 20. The adopted values of other parameters are shown in the figure caption. The energies come down rapidly as more bases are included. Note that the energies of initial coherent states are -28.909 and -114.892 for these two cases. The dimension of the m-scheme basis in the exact diagonalization of the Hamiltonian matrix appears to be 92,123 and 39,180,981 for $N_B =10$ and 20, respectively, in the case of the total magnetic quantum number $M=0$. The exact numerical diagonalization is already very difficult in the former case, and is indeed practically impossible in the latter case. In Fig.2, the exact energies are calculated analytically. We can see that in the QMCD method the convergence is attained at about 400 and 900 basis states for these two cases. Although the number of QMCD basis states increases as N_B increases, this increase is much slower than that of the dimension of the entire Hilbert space.

The ground-state expectation values of the $J \cdot J$ operator are 0.003 and 0.037 for these two cases, respectively, while they are 6.002 and 6.017 for the first 2^+ state. These results confirm that the convergence of the angular momentum is fulfilled. Since these values are 74.474 and 154.641 for initial coherent states, we can see that the restoration of the angular momentum is significant. This feature is also found in the expectation value of the quadrupole operator $Q_{\mu=0}$. In one of the calculations with $N_B=10$, this value starts from -9.501 (intrinsic state value), and ends at 0.004 for the resultant ground state. We repeated the calculation using different sets of σ. The results are shown by the various lines in Fig.2. We find that all lines converge to the same value.

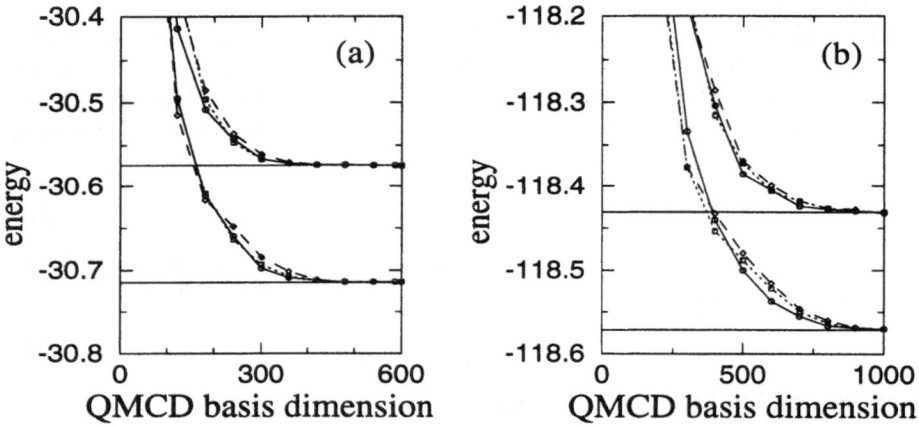

Fig. 2. Energies of the lowest two states for (a) $N_B=10$ and (b) $N_B=20$ as a function of QMCD basis dimension, in the SU(3) limit of the sdg-IBM. Used parameters are $\Delta\beta = 16$, $N_t = 20$, $\kappa = 0.1$ and $\kappa' = 0.01$. The solid, dashed and dotted lines stand for different sets of the σ's. The exact values are shown by horizontal lines.

Next we compare the efficiency in convergence between the results obtained with and without the M-projection. Figure 3 shows the convergence pattern of several low-lying levels as a function of the number of the QMCD basis dimension. We show the cases of (A) SU(3) and (B) O(6) limits. The boson number is taken to be 16. Other parameters are shown in the figure caption. The results of the QMCD method without the M-projection are shown by open diamonds for up to the 8th level. We observe the five fold degeneracy with regard to the magnetic quantum number of the 2_1^+ level.

The degeneracy with respect to magnetic quantum number means that we solve the Hamiltonian in a redundant way. As described above, the M-projection technique removes this redundancy by lifting the degeneracy. Closed symbols in Fig.3 indicate the convergence pattern of certain lowest levels obtained with the M-projection. In fact, we show the results of the QMCD calculation with the M-projection onto the $M = 0$ space. In order to compare the efficiency of convergence in the QMCD method with that in the M-projected QMCD method, we show the results using the same parameters of the Hamiltonian. One finds that the efficiency of the M-projected QMCD method for the O(6) and SU(3) limits is quite high compared to the QMCD method without the M-projection. In the QMCD method without the M-projection, since the eigenstates are degenerate with respect to magnetic quantum number, a larger QMCD basis dimension is needed to complete such multiplets of a given J. Moreover, in the QMCD method without the M-projection, it is difficult to obtain higher excited states, because basically all the eigenstates below the state of the interest have to be obtained. In turn, in the M-projected QMCD method, this difficulty is removed. As a consequence, we can easily handle many excited states with various angular momenta. From this comparison, one concludes that the M-projection plays a very important role in the practical calculations.

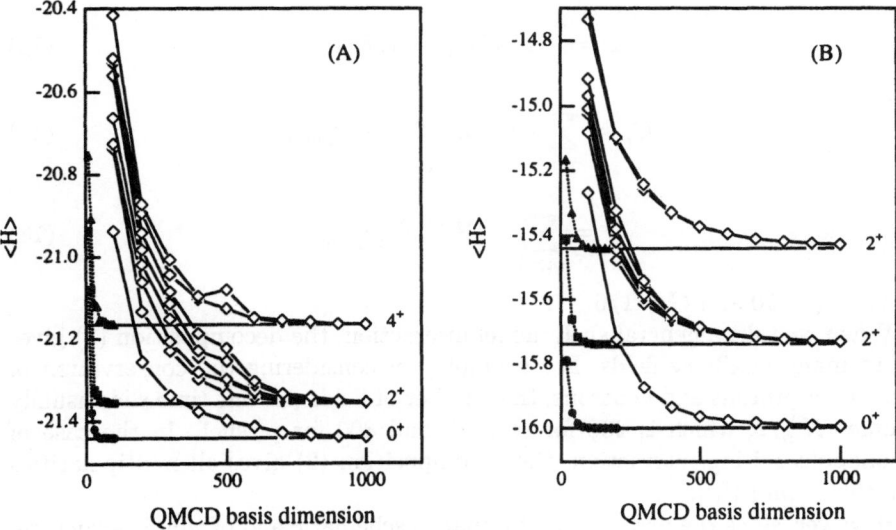

Fig. 3. Energies of the low-lying states for the results of QMCD method and spin-projected QMCD method as a function of the basis number, which correspond to the solid and dash lines, respectively. Used parameters are $\Delta\beta = 16$, $N_t = 20$, $\kappa = 0.1$ and $\kappa' = 0.01$. The exact values are also shown in horizontal lines.

The O(6) limit represents γ-unstable nuclei, and its eigenstates are formed by a superposition of an infinite number of intrinsic states (Ginocchio and Kirson 1980). Although this feature is remedied for small boson numbers due to quantum fluctuations, this feature becomes prominent for $N_B \gtrsim 10$ (Otsuka and Sugita 1987, Sugita et al. 1989). We can see a good agreement between the QMCD and the corresponding exact values. Note that the excitation energies of 2_2^+ states are still obtained correctly. It has been shown that the 2_2^+ excitation energy calculated from a single triaxial intrinsic state deviates much more than the present result. Thus, one finds the efficiency of the QMCD method also for the states without simple intrinsic structure.

We have compared the QMCD results with those obtained by the variation after the angular momentum projection (VAP) calculation where a single intrinsic coherent state is optimized with respect to the energy calculated by projecting this state onto a good angular momentum. The QMCD method always gives significantly better agreement with the exact results than the VAP.

4.2 Single-j Shell Model

As a simple example of the QMCD application to the shell model, we take a single j-orbit model with $j=23/2$ with one kind of nucleons. The Hamiltonian consisting of a normal-ordered quadrupole-quadrupole interaction and the monopole pairing interaction is assumed,

$$H = -\chi : Q \cdot Q : -GP^\dagger P, \tag{16}$$

where

$$Q_\mu = \sum_{mm'} (jmjm' \mid 2\mu) c_{jm}^\dagger \tilde{c}_{jm'}, \tag{17}$$

and

$$P^\dagger = \sum_{m>0} (-1)^{j+m} c_{jm}^\dagger c_{j-m}^\dagger. \tag{18}$$

We take $\chi = 10$ and $G = 1/6$.

When we take a general shell model interaction, the decomposition (2) gives rise to many auxiliary fields. For example, by considering the conservation of numbers of protons and neutrons, the number of fields per one time slice usually becomes $N_{sp}^2/2$, which is 288 for sd shell and 800 for pf shell. In the case of a monopole pairing interaction, the decomposition (2) gives all multipolarities with the equal $\mid V_\alpha \mid$.

However, since the present Hamiltonian is schematic and of simple structure, it is expected that we can generate favourable basis states for diagonalization by taking smaller number of one-body operators. In this case, the dominant term in the Hamiltonian is clearly the quadrupole-quadrupole interaction. Therefore, important basis states can be generated mainly by quadrupole fields. Note that the QMCD method is a method for generating efficient basis states for diagonalization, and there is no need to carry out the stochastic integration over all auxiliary fields. Thus the Hamiltonian which determines one-body fields for the basis generation can be different from that for calculation of matrix elements. By considering this point, the calculation can be much simpler.

Figure 4 shows (a) energies and (b) expectation values of $J \cdot J$ for several yrast states as a function of the QMCD basis dimension for the case of 8 nucleons. The exact values are shown by horizontal lines. In this case 5 quadrupole and 9 hexadecapole auxiliary fields are taken for generating the QMCD basis states. The strengths of these fields are determined from a Hamiltonian comprised of quadrupole-quadrupole and hexadecapole-hexadecapole interactions, with their strengths adjusted so as to reproduce approximately the matrix elements of the original Hamiltonian. It can be seen that good convergence is obtained at around 300 QMCD basis dimension by including only the above 14 fields out of 576 in total.

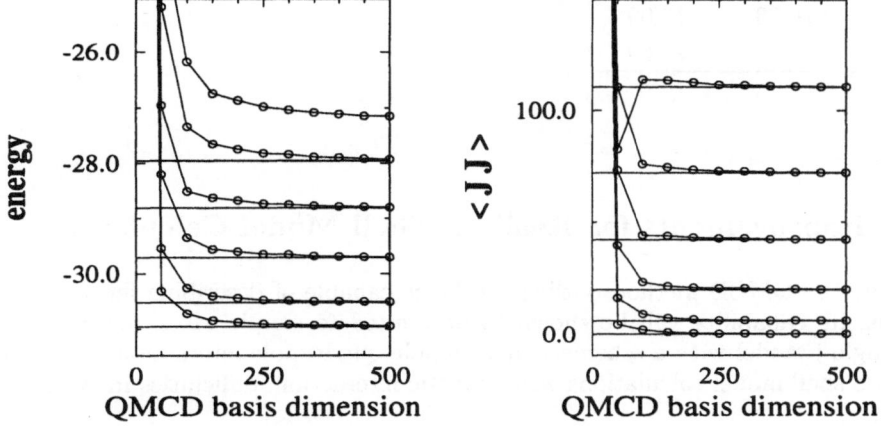

Fig. 4. (left) Energies and (right) expectation values of $J \cdot J$ of yrast states for a system of 8 nucleons as a function of QMCD basis dimension. The exact values are shown by horizontal lines.

4.3   Large-Scale Calculation

As an example of large-scale calculations, we present some results of full pf shell calculations. Table 1 includes the excitation energies of low-lying states of ^{52}Fe (6 protons and 6 neutrons) with the SU(3) interaction (Eliott 1958, Harvey 1968). In this case 17 fields out of 800 fields are selected which mainly correspond to quadrupole and hexadecapole one-body operators with relatively large strength $| V_\alpha |$. We mention that all possible fields are present with non-zero V_α in the original Hamiltonian. The maximum number of the QMCD basis is 320 in Table

1, while the m-scheme dimension of $M=0$ space is 109,954,620. The convergence is quite good. In fact, even in the total energy, the calculated value is -452.993 MeV, while the exact value is -453.000 MeV. We have also obtained a similarly good result for ^{56}Ni. These results suggest that the present QMCD method enables us to study the collective properties of heavier nuclei.

Table 1. Comparison between excitation energies (MeV) of the QMCD and the exact results for ^{52}Fe with SU3 interaction.

state	QMCD	exact
2^+_1	1.50	1.50
4^+_1	5.01	5.00
6^+_1	10.52	10.50
8^+_1	18.03	18.00
10^+_1	27.55	27.50

5   Improvements for Realistic Shell Model Calculations

While the QMCD method outlined so far is capable of describing fermion systems, its capability will be shown to be limited to simple cases, for instance, a single-j model with a schematic quadrupole-quadrupole interaction. In applying to shell model calculations with realistic interaction, difficulties in efficiency arise.

In Fig.5, calculated energies and expectation values of $J \cdot J$ of the lowest two states, 0^+_1 and 2^+_1, are shown as a function of the QMCD basis dimension for ^{24}Mg (4 protons and 4 neutrons in the sd shell). The USD interaction (Brown and Wildenthal 1988) is used. Solid lines show the results obtained by considering 50 fields with relatively large strength, $| V_\alpha |$, out of 288. We can see that the convergence is insufficient.

Thus, a further substantial improvement of the method is required for realistic shell model calculations. Such improvements are: (i) the sampling scheme is modified, and (ii) additional processes are introduced to restore symmetries.

5.1   Improvement in Sampling Processes

We start with the sampling. The original version of the QMCD method directly follows from the expression (4), where a rather naive sampling is performed. This sampling creates many unnecessary basis vectors in general. Therefore the actual sampling has to be modified for large-scale realistic shell model calculations so

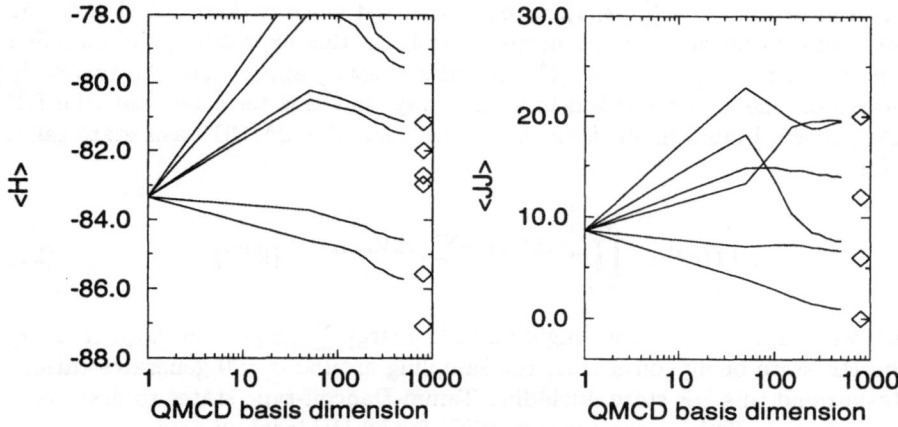

Fig. 5. (left) Energies and (right) expectation values of $J \cdot J$ of the lowest 6 states of ^{24}Mg plotted as a function of the QMCD basis dimension. The exact values are shown by diamonds.

that important basis vectors are generated still stochastically but more efficiently by considering the many-body dynamics.

The modification regarding the sampling consists of two parts. In the first part, the basis state generation is refined so as to make use of the local Hartree-Fock (HF) energy minima. In the QMCD calculation, one has to generate good basis states (*i.e.*, Slater determinants in deformed bases) which have (i) low values of diagonal matrix elements and/or (ii) large off-diagonal matrix elements of the Hamiltonian. The point (i) can be fulfilled by using, as $|\Psi^{(0)}\rangle$ in eq.(7), a deformed HF solution within the present shell model space. The QMCD process is comprised practically of several segments starting with different initial states, which are HF states at different local minima.

States around a minimum satisfy point (ii) in most cases. We then rearrange the one-body evolution process so that the basis states are sampled most frequently near the HF local minima, accelerating the generation of state vectors having larger overlap with eigenstates of interest.

For this purpose the Hamiltonian is rewritten, by introducing the constants c_α, as

$$H = \sum_\alpha (E_\alpha O_\alpha + \frac{1}{2} V_\alpha (O_\alpha - c_\alpha)^2 + V_\alpha c_\alpha O_\alpha), \qquad (19)$$

where a constant term is omitted. After the HS transformation, the one-body Hamiltonian becomes

$$h(\sigma_n) = \sum_\alpha ((E_\alpha + V_\alpha c_\alpha) O_\alpha + s_\alpha V_\alpha \sigma_{\alpha n} O_\alpha), \qquad (20)$$

where the c-number $-\sum_\alpha s_\alpha V_\alpha \sigma_{\alpha n} c_\alpha$ is omitted since it does not change the wave functions apart from the normalization. In this expression, the modified one-body term $\sum_\alpha (E_\alpha + V_\alpha c_\alpha) O_\alpha$ includes effects coming from the two-body interaction. The c_α's are taken in such a way that this term becomes the HF single-particle Hamiltonian, h_{HF}. With this h_{HF}, the QMCD basis state takes the form,

$$|\Phi(\sigma)\rangle \propto \prod_{n=1}^{N_t} e^{-\Delta\beta(h_{HF}+\sum_\alpha s_\alpha V_\alpha \sigma_{\alpha n} O_\alpha)} |\Psi^{(0)}\rangle. \qquad (21)$$

Thus, we simply replace the single particle energy $\sum_\alpha E_\alpha O_\alpha$ by h_{HF}. If $|\Psi^{(0)}\rangle$ is the HF state being considered, the sampling around $\sigma = 0$ generates various states around this HF state, including Tamm-Dancoff-type states to first order in σ, and so on. This treatment is possible for all HF local minima.

In cases of non-spherical nuclei, many Hartree-Fock local minima appear in the search for the initial state. By stochastically taking those local minima as the initial states, it is possible to take into account a wider variety of configurations.

The second major improvement on the sampling is the ordering of one-body fields according to their importance. As mentioned before, the QMCD method is a method for generating favorable basis states for diagonalization, and there is no need to carry out the stochastic integration over all auxiliary fields. In constructing the basis states, we start with the most relevant part of the Hamiltonian, which yields fewer auxiliary fields than the whole Hamiltonian. The calculation can then be performed more efficiently. After certain basis states are obtained, we take an enlarged portion of the Hamiltonian, so that other terms of the Hamiltonian can be properly included in constructing the basis states. Eventually, as in eq.(4), the completeness of the QMCD basis is guaranteed for the ground state by taking all fields.

In most cases, the auxiliary fields with large values of $|V_\alpha|$ in eq.(2) turn out to have quadrupole, hexadecapole or monopole nature. Therefore it is reasonable to arrange all fields in descending order of $|V_\alpha|$, and take them starting from the largest one. In addition, for a fixed initial state $|\Psi^{(0)}\rangle$, the total strength of each O_α changes due to the Pauli principle and to collective effects. Therefore we consider an excitation sum-rule;

$$S_\alpha = \langle \Psi^{(0)} |O_\alpha^\dagger O_\alpha |\Psi^{(0)}\rangle - |\langle \Psi^{(0)} |O_\alpha |\Psi^{(0)}\rangle|^2, \qquad (22)$$

and use it as a practical measure of importance of the O_α's. The selection of O_α's according to $|V_\alpha|$ and S_α plays an essential role in the actual calculations.

5.2   Restoration of Symmetries

We now come to the restoration of symmetries. We implement explicitly kinematic symmetries such as angular momentum and isospin into the QMCD method, since the restoration of such symmetries proceeds only very slowly for wave functions generated stochastically. In the previous section we have presented the M-projection method to restore the magnetic quantum number, which improves the convergence drastically.

Since a nucleus has rotational symmetry, the restoration of the total angular momentum, denoted as J, is quite crucial. In the QMCD method, we diagonalize the Hamiltonian in the laboratory frame by using QMCD bases. If the QMCD bases contain all components (*i.e.*, Slater determinants) required for the coupling to a good angular momentum, the diagonalization restores the rotational symmetry. We accelerate this restoration process, by considering rotated states $\exp(-i\theta J_y) \, | \, \Phi(\sigma) \rangle$ as candidates of new basis states. We have found that the restoration of the angular momentum is remarkably improved by taking only several values of the angle θ. We refer to this method as J-drive. In addition to this y-axis rotation, the z-axis rotation is made for the M-projection.

We next discuss isospin. The isospin projection is possible in the same way as the J-projection. However, if we restrict ourselves to consider $N=Z$ nuclei, we can keep good isospin in an alternative way. In the decomposition process, eq.(2), all one-body operators can be chosen so as to carry a definite isospin $T = 0$ or 1 for the isoscalar Hamiltonian. Since the isoscalar fields are dominant over the isovector ones particularly for $T=0$ states, we start the QMCD basis generation process with the isoscalar fields. Thus, since the initial HF state has $T = 0$, the isospin is conserved at least until the isovector fields are activated. It appears that, in $N = Z$ nuclei, one obtains sufficiently good results by keeping only the isoscaler fields.

For $T \neq 0$ states, the isospin projection is definitely needed. In general, three dimensional integral over the Euler angles is required to carry out the isospin projection. However, practically we can perform the projection only by one dimensional integral. The decomposition (2) can be performed so that all one-body operators O_α conserve T_z. Therefore we can take basis states in the form of Slater determinants with definite proton and neutron numbers. In other words, all basis states can be the eigenstates of T_z. In evaluating the matrix elements of an operator O, we have to calculate, in general, the following quantity:

$$\langle \Phi(\sigma_f)| \, O \, P^T_{T_f T_i} \, | \Phi(\sigma_i) \rangle = \frac{2T + 1}{8\pi^2} \int_0^{2\pi} d\gamma \int_0^{\pi} \sin\beta \, d\beta \int_0^{2\pi} d\alpha$$
$$e^{iT_f \gamma} d^T_{T_f T_i}(\beta) e^{iT_i \alpha} \langle \Phi(\sigma_f)| \, Oe^{-iT_z \gamma} e^{-iT_y \beta} e^{-iT_z \alpha} \, | \Phi(\sigma_i) \rangle, \quad (23)$$

where $P^T_{T_f T_i}$ denotes the isospin projection operator and $d^T_{T_f T_i}(\beta)$ stands for the d-function. If the operator O conserves the z-component of isospin, the α and the γ integrals become trivial, and we have only to evaluate the remaining one dimensional β integral. Thus, one ends up with

$$\langle \Phi(\sigma_f)| \, OP^T_{T_f T_i} \, | \Phi(\sigma_i) \rangle = \frac{2T + 1}{2} \int_0^{\pi} \sin\beta \, d\beta \, d^T_{T_f T_i}(\beta) \langle \Phi(\sigma_f)| \, Oe^{-iT_y \beta} \, | \Phi(\sigma_i) \rangle,$$
$$(24)$$

where $T_i = T_f = (N - Z)/2$. The results obtained with this procedure will be presented later.

6   Application to the Realistic Shell Model

6.1 ^{24}Mg — Detailed Comparison with Exact Results —

In order to confirm the applicability of the QMCD method for realistic shell model calculations, we first consider ^{24}Mg with the USD interaction. Figure 6 shows energies and expectation values of $J \cdot J$, for 6 lowest-lying states as a function of the QMCD basis dimension compared with the exact values. In this case we start with 5 significant fields, and eventually all 144 $T = 0$ one-body operators are activated. In the process of J-drive, 3 values for θ are employed. For 800 QMCD bases, the ground state energy becomes -86.91 MeV, while the exact value is -87.08 MeV. The dimension of the m-scheme shell-model basis for the ground state is 28,503. Thus the number of bases is reduced by a factor of 1/35 with a loss of accuracy of only 0.17 MeV in the total energy. The error due to the truncation of the Hilbert space (systematic error) does not exceed 200keV in the ground-state energy in the present calculations.

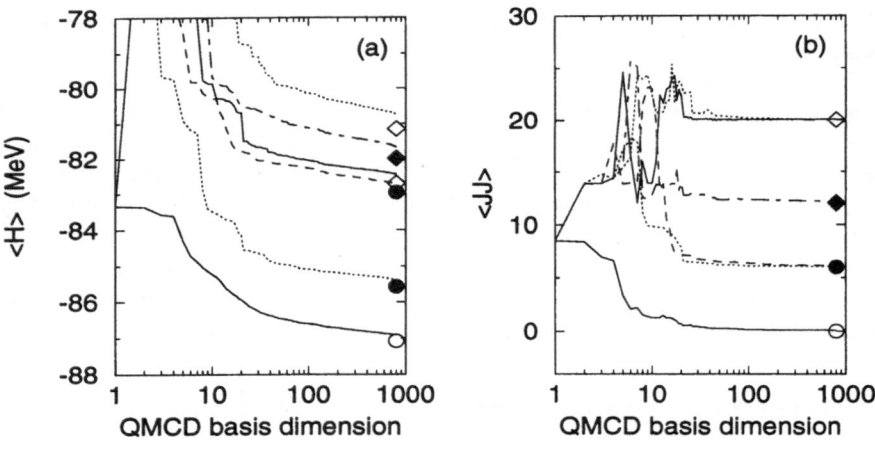

Fig. 6. (a) Energies and (b) expectation values of $J \cdot J$ of the lowest 6 states of ^{24}Mg plotted as a function of the QMCD basis dimension, with $N_t = 20$ and $\Delta\beta = 0.07$. The exact values are shown by symbols. Different symbols indicate different angular momenta.

Table 2 shows the lowest 3 energy levels, where the QMCD results for 100, 200, 400 and 800 basis dimensions are listed together with the exact results. One finds a remarkable agreement between the QMCD and exact values. Note that

Table 2. Comparison between the QMCD and the exact results for excitation energies (MeV), $B(E2)$ ($e^2\text{fm}^4$) and quadrupole moments ($e\text{fm}^2$). The effective charges $e_p + e_n = 1.78e$ are used Brown and Wildenthal 1988.

observable		QMCD dimension				exact
		100	200	400	800	
E_x	2^+_1	1.50	1.52	1.54	1.53	1.51
	2^+_2	4.33	4.25	4.23	4.18	4.12
	4^+_1	4.54	4.50	4.50	4.46	4.37
$B(E2)$	$2^+_1 \rightarrow 0^+_1$	74.1	74.2	73.2	74.2	76.1
	$2^+_2 \rightarrow 0^+_1$	7.1	7.2	7.2	7.1	6.8
	$2^+_2 \rightarrow 2^+_1$	12.1	15.3	16.8	16.2	16.6
	$4^+_1 \rightarrow 2^+_1$	103.8	104.0	102.6	102.0	101.1
	$4^+_1 \rightarrow 2^+_2$	1.8	0.6	0.4	0.5	0.5
Q	2^+_1	-18.7	-18.2	-18.4	-17.9	-17.1
	2^+_2	18.5	18.4	18.4	18.1	17.3
	4^+_1	-21.1	-21.4	-21.5	-21.2	-20.8

the accuracy of these excitation energies is better than that of absolute energies. In fact the deviations are less than 0.15MeV with only 400 basis states. In the same table, several E2 transition matrix elements and quadrupole moments are compared with the exact values. It can be found that several in-band transition $B(E2)$ values are reproduced well with only 100 basis states, and other matrix elements are also obtained with 400 basis states. Thus the QMCD method turns out to be useful especially for the study of low-lying collective states.

6.2 ^{48}Cr — Comparison with Several Approaches —

We now proceed on to full pf shell calculations. The largest calculation (Caurier et al. 1994) which has been carried out by conventional shell model diagonalization is for ^{48}Cr with the KB3 interaction (Poves 1981). Figure 7 shows the energies of several low-lying states obtained by the QMCD method, conventional shell model (Poves 1995), and the SMMC (Langanke et al. 1995). In the results of conventional shell model, different dimensions mean the different truncation schemes. The maximum number of particles allowed to jump from the $f_{7/2}$ orbit to the remaining ones, denoted t, is given in each truncation differently. In this figure results for t=0,1,2,3,5 and the complete t=8 are shown. The SMMC result corresponds to the finite temperature T=0.5MeV, and is plotted near the exact results since we cannot define the dimension for the SMMC calculations. We can

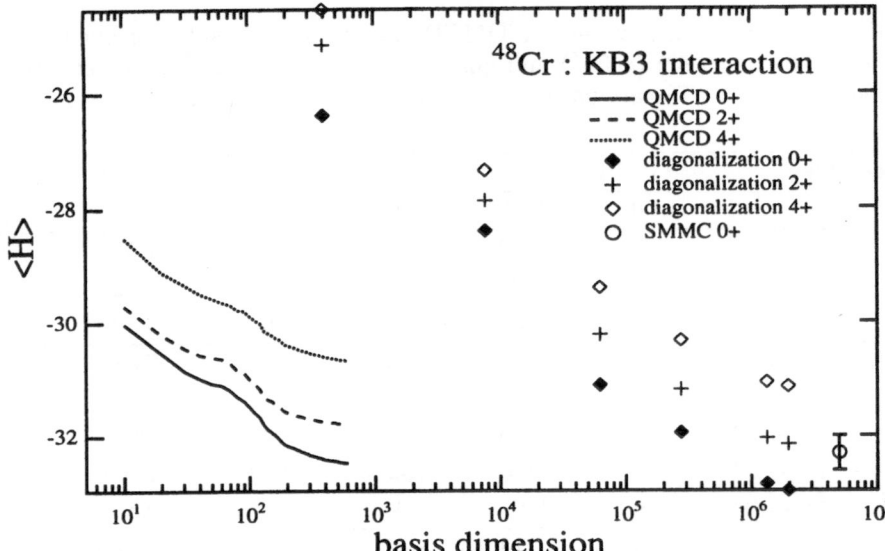

Fig. 7. Energies of lowest three states of ^{48}Cr plotted as a function of the basis dimension. The results of the QMCD method are plotted by lines, and those of the conventional shell model diagonalization are shown by symbols. The energy at a finite temperature T=0.5MeV obtained by the SMMC is shown by open circle with an error bar.

see that the QMCD method gives energies with rather good quality by taking only 600 basis states.

6.3 ^{64}Ge — First Shell Model Results of Low-lying Levels —

Next we discuss ^{64}Ge (Honma et al. 1996). The m-scheme dimension of the M=0 space is 1,087,455, 228, which is the second largest one for the $N = Z$ even-even pf shell nuclei. It is larger than the dimension for ^{48}Cr by a factor of about 550, and the exact diagonalization is hopeless in the near future. This nucleus is one of the proton-rich $N = Z$ unstable nuclei, and experimental data (Ennis et al. 1991) suggest that it is γ-soft. Thus, it is quite interesting to investigate whether we can reproduce such a structure by using a realistic interaction, the validity of which has been examined at least for the lower part of the pf shell. We adopt the FPD6 interaction (Richter et al. 1991). This interaction is derived by fitting experimental data in the mass range 41 to 49, and is suggested to be suitable for describing nuclei in the upper pf shell (Ormand and Brown 1995).

In Fig.8, calculated low-lying spectra are compared with experimental data. It is remarkable that the calculated levels show a rather good agreement with experiment without any adjustment. The γ-soft nature is evident also in the calculation. The calculated ratio of excitation energies of 2^+_2 to 2^+_1 is 1.9 and that of 4^+_1 to 2^+_1 is 2.6. Experimentally these ratios are 1.75 and 2.27, respectively. The relative magnitudes of $B(E2)$ values are shown in Fig.8. With $e_p = 1.33e$ and $e_n = 0.64e$, $B(E2;2^+_1 \to 0^+_1)=5 \times 10^2$ $(e^2\text{fm}^4)$ is obtained, which corresponds to $\beta_2 \sim 0.28$. The $B(E2)$ values of the $4^+_1 \to 2^+_1$ and $2^+_2 \to 2^+_1$ transitions are about 1.3 times larger than that of $B(E2;2^+_1 \to 0^+_1)$, suggesting γ-softness. We obtain $B(E2;2^+_2 \to 0^+_1)/B(E2;2^+_2 \to 2^+_1) \sim 2\times10^{-3}$, which is quite small similarly to the experimental value, suggesting $\gamma \sim 30°$ in triaxial deformation models (Ennis et al. 1991). Calculated quadrupole moments appear to be small (typically $|Q| < 10e\text{fm}^2$), consistently with γ-softness.

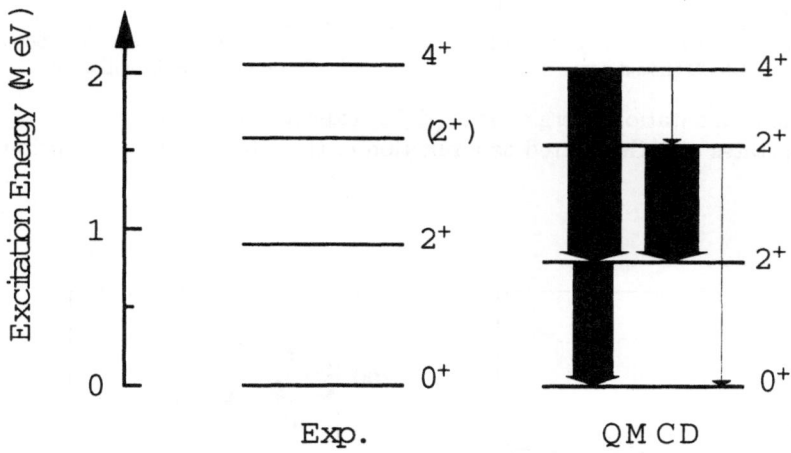

Fig. 8. Experimental and calculated energy levels of ^{64}Ge. The QMCD parameters are $N_t = 40$ and $\Delta\beta = 0.06$. The arrows designate E2 transitions with $B(E2)$'s indicated by their widths.

In Fig.9 the excitation energies and the expectation values of $J \cdot J$ are shown as a function of the QMCD basis dimension. We can see the convergence of these values. The convergence of E2 transition matrix elements can be seen in Fig.10.

The convergence of the results in Fig.8 has been examined by several calculations with different sets of random numbers. The typical deviation among different calculations (statistical error) is about 100 keV for the 2^+_2 energy level, for instance. The discrepancy between theoretical and experimental results comes partly from the systematic and statistical errors in the present method, and partly from the interaction. The former one is being reduced by improving the method.

Typical occupation numbers of $f_{7/2}$, $p_{3/2}$, $p_{1/2}$ and $f_{5/2}$ orbits are 15.1, 2.6, 0.8 and 5.5, respectively, for low-lying states. We can see that more than 6 nucleons are excited from the $(f_{7/2})^{16}$ $(p_{3/2})^8$ configuration, and that even $f_{7/2}$

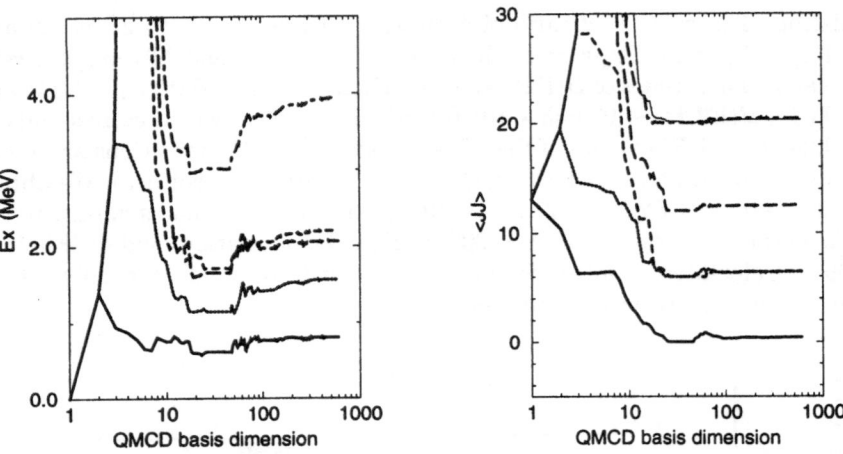

Fig. 9. (left) Excitation energies and (right) expectation values of $J \cdot J$ of the low-lying states of ^{64}Ge plotted as a function of the QMCD basis dimension.

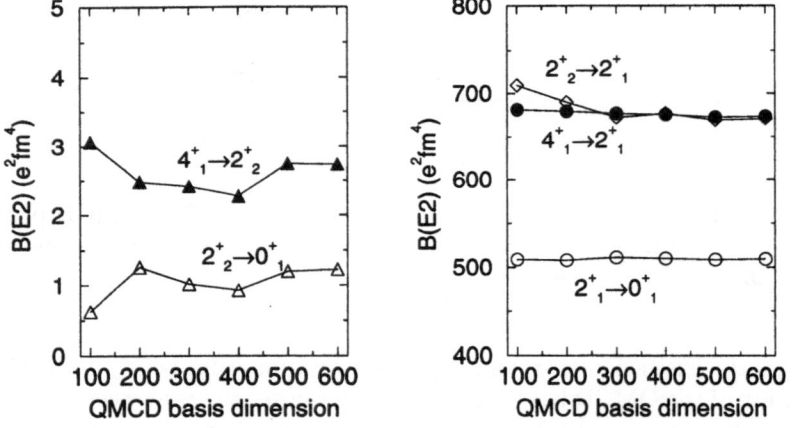

Fig. 10. Several $B(\mathrm{E}2)$ values between the low-lying states of ^{64}Ge plotted as a function of the QMCD basis dimension.

is active. One sees that all these four orbits are mixed. Because of the huge basis dimension mentioned before, the conventional shell model diagonalization is impossible.

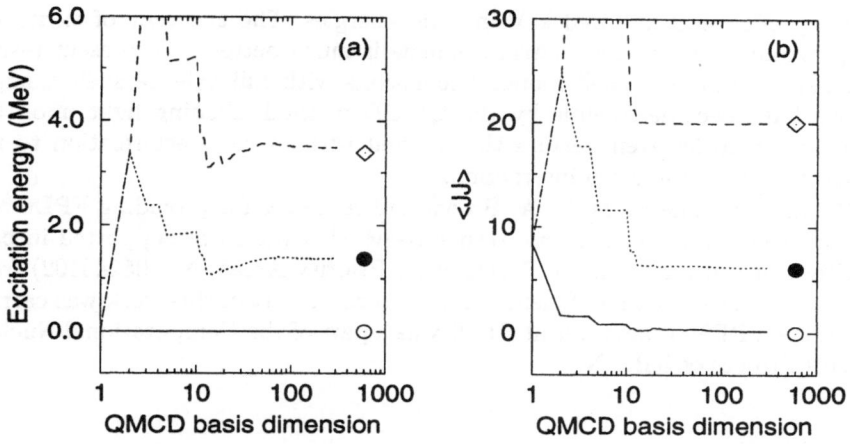

Fig. 11. (a) Excitation energies and (b) expectation values of $J \cdot J$ of the lowest three states of ^{22}Ne plotted as a function of the QMCD basis dimension. The exact values are shown by symbols. Different symbols indicate different angular momenta.

6.4 ^{22}Ne — Description of $N \neq Z$ Nuclei with Isospin Projection —

As an example of $N \neq Z$ cases, we consider ^{22}Ne with the USD interaction. In this case we have carried out exact isospin projection. In Fig.11 the excitation energies and the expectation values of $J \cdot J$ are plotted for the lowest three states as a function of the QMCD basis dimension. We can see that the QMCD method works well with isospin projection.

7   Summary

In summary, we have presented a new method for solving quantum many-body problems where particles (fermions or bosons) interact through a two-body interaction. Basis states can be generated by using the auxiliary field Monte Carlo technique. We can calculate both the ground and low-lying excited states for any interaction. As the system becomes large the QMCD basis dimension increases only gradually. Since the wave function is obtained explicitly, the transition matrix elements, including those between excited states, can be calculated directly.

It has been shown that large-scale realistic shell model calculations can be carried out by the QMCD method. The QMCD method has been improved considerably with respect to (1) the use of the local energy minima for the sampling, (2) the selection of dominant fields, and (3) the explicit implementation of kinematic symmetry requirements. Several low-lying states of large systems have been described in terms of small numbers of QMCD basis states with the accuracy of several hundred keV in total energies. The accuracy of excitation energies and E2 transition matrix elements is much better. The present results demonstrate that the shell model calculations with full valence shell configurations have become feasible by the QMCD method, shading light upon the structure of nuclei even beyond the pf shell with more direct relation to the (effective) nucleon-nucleon interaction.

We acknowledge Profs. B. A. Brown and A. Poves for providing FPD6 and KB3 two-body matrix elements, respectively. This work was supported in part by Grant-in-Aid for Scientific Research on Priority Areas (No. 05243102) from the Ministry of Education, Science and Culture. A part of this work was carried out on the VPP500 computer at RIKEN as a part of the Computational Nuclear Physics Project of RIKEN.

References

Caurier E, Zuker A. P., Poves A., Martinez-Pinedo G. (1994): Phys. Rev. **C50**, 225–236

Johnson C. W., Koonin S. E., Lang G. H., Ormand W. E. (1992): Phys. Rev. Lett. **69**, 3157–3160

Ormand W. E., Dean D. J., Johnson C. W., Lang G. H., Koonin S. E. (1994): Phys. Rev. **C49**, 1422–1427

Varga K., Liotta R. J. (1994): Phys. Rev. **C50**, R1292–R1295

Horoi M., Brown B. A., Zelevinsky V. (1994): Phys. Rev. **C50**, R2274–R2277

Koonin S. E., Dean D. J., Langanke K. (1996): to be published in Phys. Repts.

Langanke K., Dean D. J., Radha P. B., Alhassid Y., Koonin S. E (1995): Phys. Rev. **C52**, 718–725

Dean D. J., Koonin S. E., Langanke K., Radha P. B., Alhassid Y. (1995): Phys. Rev. Lett. **74**, 2909–2912

Honma M., Mizusaki T., Otsuka T. (1995): Phys. Rev. Lett. **75**, 1284–1287

Iachello F., Arima A. (1987): *The interacting boson model* (Cambridge U.P., Cambridge)

Mizusaki T., Honma M., Otsuka T. (1996): Phys. Rev. **C53**, 2786–2793

Hubbard J. (1959): Phys. Rev. Lett. **3**, 77–80

Stratonovich R. L. (1957): Dokl. Akad. Nauk. SSSR **115**, 1097–1100 [transl: Soviet Phys. Dokl. **2**, 416–419 (1957)]

De Raedt H., Frick M. (1993): Phys. Rep. **231**, 107–149

Ginocchio J. N., Kirson M. W. (1980): Nucl. Phys. **A350**, 31–60

Otsuka T., Sugita M. (1987): Phys. Rev. Lett. **59**, 1541–1544

Sugita M., Otsuka T., Gelberg A. (1989): Nucl. Phys. **A493**, 350–364

Brown B. A., Wildenthal B. H. (1988): Ann. Rev. Nucl. Part. Sci. **38**, 29–66

Elliott J. P., (1958): Proc. Roy. Soc. **A245** 128–145 and 562–581

Harvey M. (1968): Ad. Nucl. Phys. **1**, 67–182

Poves A. Zuker A. (1981): Phys. Rep. **70**, 235–314

Poves A. (1995): private communication

Honma M., Mizusaki T., Otsuka T. (1996): submitted to Phys. Rev. Lett.

Ennis P. J. *et al.* (1991): Nucl. Phys. **A535**, 392–424

Richter W. A., van der Merwe M. G., Julies R. E., Brown B. A. (1991): Nucl. Phys. **A523**, 325–353

Ormand W. E., Brown B. A. (1995): Phys. Rev. **C52**, 2455–2460

Shell-Model Applications in Nuclear Astrophysics

K. Langanke

W. K. Kellogg Radiation Laboratory, 106-38, California Institute of Technology, Pasadena, California 91125

1 Introduction and motivation

In recent years astrophysical applications have been the driving force for much research in nuclear physics. Very often these problems involve rather evolved many-body physics. In many cases mean-field approaches are the first method of choice. But often more elaborate models which take proper account of nucleonic correlations beyond the mean-field level are called for. The interacting shell model (Haxel *et al.* 1949) is generally considered to be the most fundamental theory of the nucleus short of an explicit solution of the A-body problem. This paper will discuss several astrophysical scenarios in which recently shell model calculations have significantly contributed to a better theoretical understanding. But it will also focus on important astrophysical problems and recent developments in which the nuclear many-body problem is yet treated on the mean-field level or within the many-body theory of small-amplitude vibrations, the random phase approximation. These applications can be viewed as a motivation for potential future shell model calculations.

In particular the paper discusses three different astrophysical scenarios which are all currently drawing strong research interest. At first we will review the solar neutrino problem, with special emphasis on the fate of the ^{7}Be neutrinos and the attempts of their earthbound observations. In the second part we discuss some topics important during the presupernova collapse. Particular progress has been achieved recently in the calculation of electron capture rates for the relevant nuclei in the iron-mass region and in the description of nuclei at finite temperature. The final topic deals with neutrino-nucleus interactions and their role in various aspects of the supernova, including a possible detection scheme for μ- and τ-neutrinos in water Cerenkov detectors like Superkamiokande.

2 Missing ^{7}Be neutrinos: clue to the solar neutrino puzzle?

There are currently four solar neutrino detectors operating. The ^{37}Cl detector at Homestake Mine and the two germanium detectors GALLEX and SAGE are sensitive to ν_e-induced charged-current reactions, while Kamiokande is a water

Cerenkov detector which is triggered by neutrino scattering on electrons. The latter detector is direction-sensitive and has proven that the observed neutrinos are indeed emitted by the sun. Caused by different thresholds the various detectors are sensitive to different parts of the solar neutrino spectrum. Kamiokande ($E_{thr} \approx 7$ MeV) can only detect the high-energy ^{8}B neutrinos, while the ^{37}Cl detector ($E_{thr} = 0.814$ MeV) has also additional sensitivity to ^{7}Be neutrinos (predicted to be about 24%). A new era in solar neutrino physics has been initiated by the two germanium detectors which, due to their low thresholds ($E_{thr} = 0.233$ MeV), are also sensitive to the neutrinos generated by the p+p reaction, which is the main energy source of the sun.

Provided the published results of all four detectors are correct, there now appear to be several *solar neutrino problems* (Raghavan 1995):

- All 4 detectors observe less neutrinos than predicted by the standard solar model. This puzzle, first formulated for the results of the classical ^{37}Cl detector, was originally known as the solar neutrino problem.
- As Kamiokande *only* observes ^{8}B neutrinos, their flux can be considered calibrated by this detector (about 50% of the prediction of the standard solar model). Using this calibration, the ^{37}Cl detector should observed about 3.1 SNU of ^{8}B neutrinos (1 SNU corresponds to 10^{-36} captures/target atom/s). This number is a lower limit of the total flux as it has to be incremented by the contribution from ^{7}Be neutrinos. (Note that solar hydrogen burning has to pass through ^{7}Be to reach ^{8}B.) However, the ^{37}Cl detector observes only 2.32 ± 0.26 SNU, in contradiction to the prediction based on the Kamiokande calibration.
- The neutrino flux from the p+p reaction is well constrained by our knowledge of the solar luminosity. It is about 70 SNU for the germanium detectors. Adding the ^{8}B neutrino flux, as calibrated by Kamiokande, one already has at least 80 SNU. Noting that GALLEX and SAGE combined observe 78 ± 10 SNU, these detectors do not let room for additional ^{7}Be neutrinos, whose flux in the standard solar model is predicted as 36 SNU.

The key question to the solar neutrino problems obviously is: *Where are the missing ^{7}Be neutrinos?* More precisely, it has been shown (Raghavan 1995) that the observed ratio of ^{7}Be-to-^{8}B neutrinos is in contradiction to the standard solar model as well as to frequently discussed non-standard models (Bahcall 1990). The observed ratio implies ^{8}B synthesis without a path via ^{7}Be in disagreement with the accepted solar pp-chains. It should be noted that the observed neutrino fluxes in all 4 detectors can be understood in terms of neutrino physics beyond the standard model of weak interaction (see for example Hata 1995), e.g. implying the resonant transformation of ν_e-neutrinos into ν_μ-neutrinos (MSW effect) which cannot be detected by the neutrino detectors (except for a small contribution induced by neutral current scattering events in Kamiokande). The MSW effect will be tested in future solar neutrino detectors like Superkamiokande and SNO.

To search specifically for ^{7}Be neutrinos two detectors have been proposed. BOREXINO (Raghavan 1991) is an ultrapure liquid scintillator detector which will observe neutrinos by ν-e scattering, similar to Kamiokande. This detector is particularly sensitive to ^{7}Be neutrinos. In 1988, Haxton suggested an iodine detector (Haxton 1988) which would observe solar neutrinos radiochemically via ^{127}I(ν_e,e)^{127}Xe. With a neutrino threshold of $E_{thr} = 664$ keV, the iodine detector will also mainly observe ^{7}Be and ^{8}B neutrinos, however, with a probably different ratio than the ^{37}Cl detector. Given this ratio the combined iodine and chlorine detector results might allow one to disentangle the individual ^{7}Be and ^{8}B neutrino fluxes. It is therefore very important to determine the response of the iodine detector to solar neutrinos. This problem has been attacked in recent years both experimentally and theoretically.

The ^{7}Be neutrinos will exclusively initiate a transition from the ^{127}I ground state ($J^\pi = 5/2^+$) to the first excited $3/2^+$ state in ^{127}Xe. This Gamow-Teller (GT) transition might be measurable in (p,n) experiments (Haxton 1988). Within shell model approaches of various truncation Engel, Pittel and Vogel assign a value to the Gamow-Teller strength; however, with an unsatisfying uncertainty of about a factor of two reflecting the differences among the models (Engel *et al.* 1994).

It is expected that the iodine detector response to solar ^{8}B neutrinos is dominated by Gamow-Teller transitions to states below the neutron threshold in ^{127}Xe at 7.23 MeV. Engel *et al.* have calculated the Gamow-Teller strength distribution in ^{127}Xe within the quasiparticle Tamm-Dancoff approximation (Engel *et al.* 1991). Recently the same authors have improved their study by including three-quasinucleon configurations (Engel *et al.* 1994), with the satisfying result that the Gamow-Teller strength below the neutron threshold is only slightly effected by the improvement. Furthermore the calculated relative GT strength distribution appears to be in agreement with the results of a (p,n) experiment (Sugarbaker). To reduce the uncertainty in the overall normalization of the (p,n) data and in the theoretical spectrum (truncated shell model calculations do not recover the full quenching of the GT strength, see below) an experimental group at LAMPF has attempted to measure the total cross section for absorbing ν_e neutrinos on ^{127}I (Cleveland 1993). The experiment utilized neutrinos from muon-decay at rest containing neutrinos with energies ($E_\nu \leq 53$ MeV) significantly higher than those of solar ^{8}B neutrinos ($E_\nu \leq 15$ MeV). In fact, the high-energy neutrinos, present in the LAMPF beam, will give rise to higher-multipole contributions ($L > 0$) to the cross section than Gamow-Teller transitions ($L = 0$). If these higher multipoles are considered, the truncated shell model calculation of Engel *et al.*, assuming the usual in-medium modification of the axial vector coupling strength ($g_A = 1$), predicts a cross section of $2.0 \cdot 10^{-40}$ cm^2 (Engel *et al.* 1994), more than a factor of 2 smaller than the experimental value ($(6.2 \pm 2.5) \cdot 10^{-40}$ cm^2 (Cleveland 1993). The data, however, appear to be in agreement with a result obtained in local density approximation (Kosmas and Oset 1996).

Obviously additional theoretical (and experimental) work is needed to calibrate the iodine detector for the observation of solar neutrinos. Possible benefits from an iodine solar-neutrino detector are discussed in (Engel *et al.* 1995).

3 Shell model Monte Carlo and presupernova physics

The core collapse of a massive star can no longer be stopped by electron degeneracy pressure if the core mass exceeds the appropriate Chandrasekhar mass. This will eventually happen in massive stars, as the continuing silicon burning adds material to the iron core. The description of the presupernova core collapse is a very evolved problem, invoking many subfields of physics (Bethe 1990). Important progress in the description of nuclear physics input has been achieved recently, being made possible by the development of the shell model Monte Carlo (SMMC) method (Johnson et al. 1992, Lang et al. 1993).

The SMMC method describes the nucleus by a canonical ensemble at temperature $T = \beta^{-1}$ and employs a Hubbard-Stratonovich linearization (Hubbard 1959) of the imaginary-time many-body propagator, $e^{-\beta H}$, to express observables as path integrals of one-body propagators in fluctuating auxiliary fields (Johnson et al. 1992, Lang et al. 1993). Since Monte Carlo techniques avoid an explicit enumeration of the many-body states, they can be used in model spaces far larger than those accessible to conventional methods. The Monte Carlo results are in principle exact and are in practice subject only to controllable sampling and discretization errors. The notorious "sign problem" encountered in the Monte Carlo shell model calculations with realistic interactions (Alhassid 1994) can be circumvented by a procedure suggested in Ref. (Dean et al. 1995) which is based on an extrapolation from a family of Hamiltonians that is free of the sign problem to the physical Hamiltonian. A review on the SMMC method and its applications can be found in Ref. (Koonin et al. 1996).

3.1 Gamow-Teller strength distributions in pf shell nuclei

If the core mass exceeds the Chadrasekhar mass limit, electrons are captured by nuclei. For many of the nuclei that determine the electron capture rate in this early stage of the presupernova (Aufderheide 1994), Gamow-Teller (GT$_+$) transitions contribute significantly to the electron capture rate. Due to insufficient experimental information, the GT$_+$ transition rates have so far been treated only qualitatively in presupernova collapse simulations, assuming the GT$_+$ strength to reside in a single resonance, whose energy relative to the daughter ground state has been parametrized phenomenologically (Fuller et al. 1980); the total GT$_+$ strength has been taken from the single particle model. Recent (n, p) experiments (Williams et al. 1995-Rönnquist 1993), however, show that the GT$_+$ strength is fragmented over many states, while the total strength is significantly quenched compared to the single particle model. The observed quenching for the astrophysically important mid-pf-shell nuclei has, for the first time, consistently been reproduced within SMMC studies (Langanke et al. 1995). These calculations consider all correlation within the full pf shell and invoke the standard renormalization factor of the Gamow-Teller operator (Langanke *et al.* 1995). A recent update of the GT_+ rates for use in supernova simulations assumed a constant quenching factor of 2 (Aufderheide 1994).

In a series of truncated shell model calculations, Aufderheide and collaborators have demonstrated that a strong phase space dependence makes the Gamow-Teller contributions to the presupernova electron capture rates more sensitive to the strength *distribution* in the daughter nucleus than to the total strength (Aufderheide 1993). In this work it also became apparent that complete $0\hbar\omega$ studies of the GT_+ strength distribution are desirable. Such studies are now possible using the SMMC approach, which allows to directly calculate the response function of the Gamow-Teller operator, $R_{GT}(\tau)$. The desired strength function $S_{GT}(E)$ is then obtained by an inverse Laplace transform of $R_{GT}(\tau)$, using the Maximum Entropy technique.

As first examples (Radha *et al.*) we have studied several nuclei (^{51}V, ^{54,56}Fe, ^{58,60,62}Ni, and ^{59}Co), for which the Gamow-Teller strength distribution in the daughter nucleus is known from (n,p) experiments (Williams *et al.* 1995) and (Rönnquist 1993). Note that the electron capture by these nuclei, however, plays only a minor role in the presupernova collapse. As SMMC calculates the strength function within the parent nucleus, the results have been shifted using the experimental Q-value and an appropriate Coulomb correction. For all nuclei, the SMMC approach calculates the centroid and width of the strength distribution in good agreement with data. Fig. 1 shows the response function $R_{GT}(\tau)$ for ^{54}Fe and ^{59}Co and compares the extracted GT_+ strength distribution with data. (The Gamow-Teller operator has been renormalized by the universal quenching factor 0.77). The centroid of the GT_+ strength distributions is found to be nearly independent on temperature (for odd nuclei calculations could only be performed in the temperature range $T = 0.8 - 1.2$ MeV), while its width increases with temperature.

Following the formalism as described in Refs. (Aufderheide 1994, Fuller *et al.* 1980), the Gamow-Teller contributions to the electron capture rates under typical presupernova conditions have been calculated assuming that the electrons obey a Fermi-Dirac distribution with a chemical potential adopted from the stellar trajectory at the electron-to-nucleon ratio corresponding to the respective nucleus (Aufderheide 1994). The calculation have been performed for both the SMMC and experimental GT_+ strength distributions (Williams *et al.* 1995-Rönnquist 1993). The electron capture rates so obtained agree within a factor of two for temperatures $T = (3 - 5) \times 10^9$ Kelvin, which is the relevant temperature regime in the presupernova collapse (Aufderheide 1994). Thus, it is for the first time possible to calculate with a reasonable accuracy the electron capture rate for those nuclei like ^{55}Co or ^{56}Ni which dominate the electron capture process in the early presupernova collapse (Aufderheide 1994). SMMC calculations of the electron capture rates for other pf-shell nuclei important in the early presupernova collapse are now in progress (Radha *et al.*).

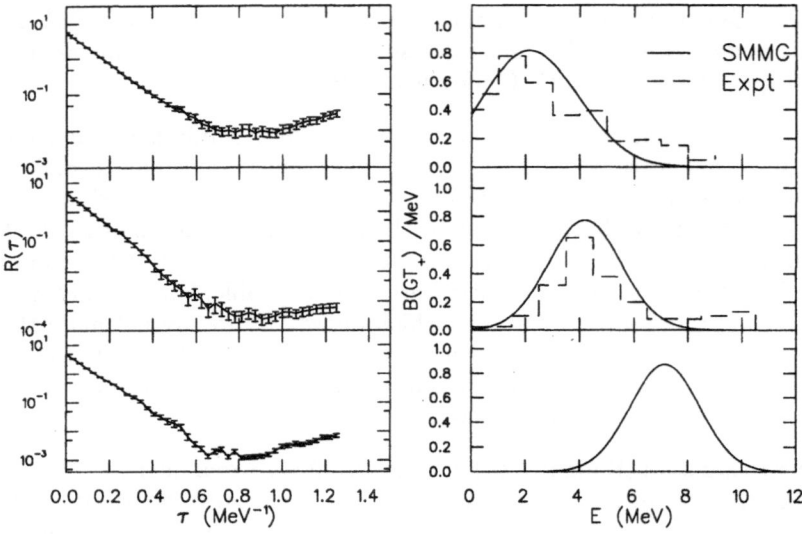

Fig. 1. SMMC GT$_+$ response functions (left side) and GT$_+$ strength distributions (right side) for ^{54}Fe (upper panel), ^{59}Co (middle panel), and ^{55}Co (lower panel). The energies refer to the daughter nuclei. The dashed histograms show the experimental strength distribution as extracted from (n,p) data (Alford *et al.* 1990).

3.2   Thermal properties of *pf* shell nuclei

The gravitational contraction during the collapse increases the temperature and density in the core and nuclear properties at finite temperature become important. Theoretical studies of nuclei at finite temperature have been based mainly on mean-field approaches, and thus only consider the temperature dependence of the most probable configuration in a given system. These approaches have been criticized due to their neglect of quantum and statistical fluctuations (Dukelsky *et al.* 1991). The SMMC method does not suffer this defect and allows the consideration of model spaces large enough to account for the relevant nucleon-nucleon correlations at low and moderate temperatures.

We have performed SMMC calculations of the thermal properties of several even-even nuclei in the mass region $A = 50 - 60$ (Dean *et al.* 1995, Langanke *et al.*, Langanke *et al.*). As a typical example, we discuss in the following our SMMC results for the nucleus ^{54}Fe, which is very abundant in the presupernova core of a massive star. Our calculations include the complete set of $1p_{3/2,1/2}0f_{7/2,5/2}$ states interacting through the realistic Brown-Richter Hamiltonian (Richter *et al.* 1991). (SMMC calculations using the modified KB3 interaction give essentially the same results.)

The calculated temperature dependence of various observables is shown in Fig. 2. In accord with general thermodynamic principles, the internal energy U steadily increases with increasing temperature (Dean et $al.$ 1995). It shows an inflection point around $T \approx 1.1$ MeV, leading to a peak in the heat capacity, $C \equiv dU/dT$, whose physical origin we will discuss below. The decrease in C for $T \gtrsim 1.4$ MeV is due to our finite model space (the Schottky effect (Civitarese et $al.$ 1989); we estimate that limitation of the model space to only the pf-shell renders our calculations of ^{54}Fe quantitatively unreliable for temperatures above this value (internal energies $U \gtrsim 15$ MeV). The same behavior is apparent in the level density parameter, $a \equiv C/2T$. The empirical value for a is $A/8$ MeV $=$ 6.8 MeV^{-1} which is in good agreement with our results for $T \approx 1.1$–1.5 MeV.

We also show in Fig. 2 the expectation values of the BCS-like proton-proton and neutron-neutron pairing fields, $\langle \hat{\Delta}^{\dagger} \hat{\Delta} \rangle$. At low temperatures, the pairing fields are significantly larger than those calculated for a non-interacting Fermi gas, indicating a strong coherence in the ground state. With increasing temperature, the pairing fields decrease, and both approach the Fermi gas values for $T \approx 1.5$ MeV and follow it closely for even higher temperatures. Associated with the breaking of pairs is a dramatic increase in the moment of inertia, $I \equiv \langle J^2 \rangle / 3T$, for $T = 1.0$–1.5 MeV; this is analogous to the rapid increase in magnetic susceptibility in a superconductor. The phase transition might be experimentally observable in the level density if one is able to perform full spectroscopy. Experiments aimed at this goal are in progress at Los Alamos.

Important for astrophysical applications, the Gamow-Teller GT$_+$ strength is nearly constant for temperatures up to 2 MeV. The $M1$ strength unquenches rapidly with heating near the transition temperature. However, for $T = 1.3$–2 MeV, $B(M1)$ remains significantly lower than the single particle estimate ($41\mu_N^2$), suggesting a persistent quenching at temperatures above the like-nucleon depairing. This finding has been confirmed in (Langanke et $al.$) which presents a detailed study of pairing correlations in pf shell nuclei. We note that an examination of the occupation numbers of the various orbitals show no unusual variation as the pairing vanishes.

Although the results discussed above are typical for even-even nuclei in this mass region (including the $N = Z$ nucleus ^{52}Fe), they are not for odd-odd $N = Z$ nuclei. $N = Z$ nuclei in the pf shell are close to the proton dripline and their properties are quite important for the $rapid$ $proton$ $capture$ $process$ (Champagne 1992). In contrast to even-even nuclei, the total Gamow-Teller strength is not constant at low temperatures, but increases by about 50% between $T = 0.4$ MeV and 1 MeV. The $B(M1)$ strength decreases significantly in the same temperature interval, while, for even-even nuclei, it increases steadily. A closer inspection of the isovector $J = 0$ and isoscalar $J = 1$ pairing correlations holds the key to the understanding of these differences. Note that at the level of the non-interacting Fermi gas, proton-proton, neutron-neutron and proton-neutron $J = 0$ correlations are identical for $N = Z$ nuclei. However, the residual inter-

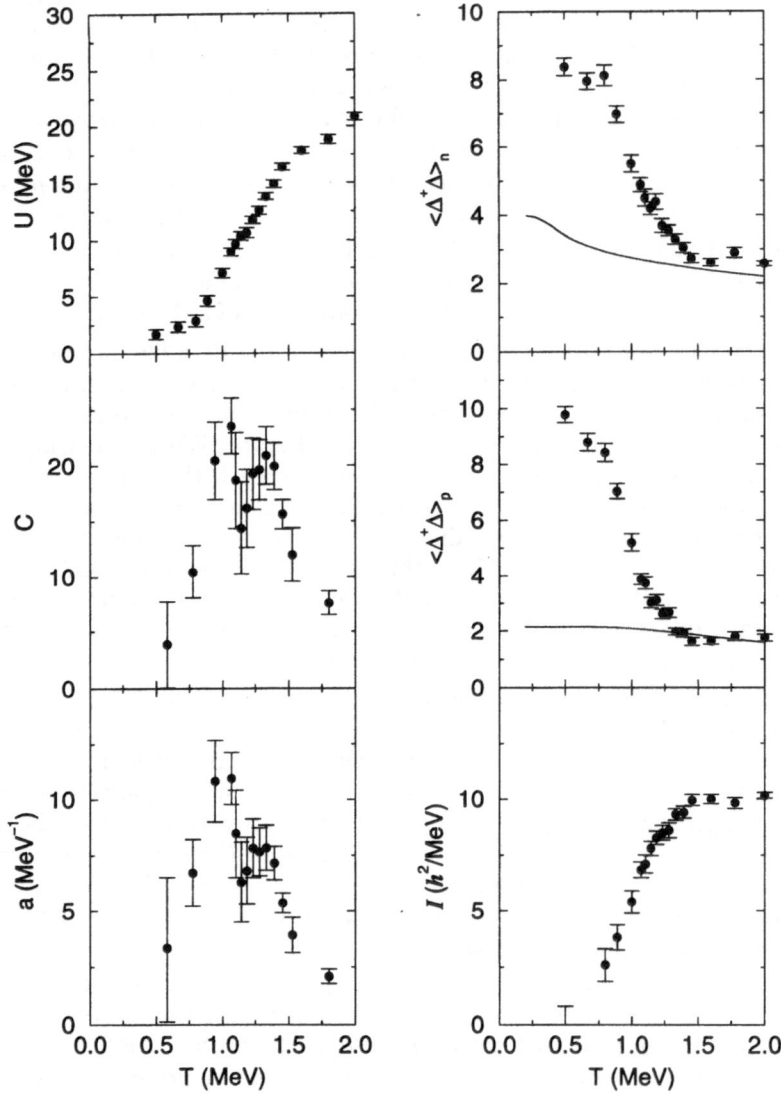

Fig. 2. Temperature dependence of various observables in ^{54}Fe. Monte Carlo points with statistical errors are shown at each temperature T. In the left-hand column, the internal energy, U, is calculated as $\langle \hat{H} \rangle - E_0$, where $\hat{H}$ is the many-body Hamiltonian and E_0 the groun d state energy. The heat capacity C is calculated by a finite-difference approximation to dU/dT, after $U(T)$ has been subjected to a three-point smoothing, and the level density parameter is $a \equiv C/2T$. In the right-hand column, we show the expectation values of the squares of the proton and neutron BCS pairing fields. For comparison, the pairing fields calculated in an uncorrelated Fermi gas are shown by the solid curve. The moment of inertia is obtained from the expectation values of the square of the total angular momentum by $I = \beta \langle \hat{J}^2 \rangle / 3$. (from Dean *et al.* 1995).

action breaks the symmetry between like-pair correlations and proton-neutron correlations in odd-odd $N = Z$ nuclei. At low temperatures isovector proton-neutron pairing dominates in ^{50}Mn, while pairing among like nucleons shows only a small excess over the Fermi gas values, in strong contrast to even-even nuclei. With increasing temperature, the isovector proton-neutron correlations decrease strongly and have essentially vanished at $T = 1$ MeV, while the isoscalar pairing strength remains about constant in this temperature region (as it does in even-even nuclei) and greatly exceeds the Fermi gas values. We also note that the pairing between like nucleons is roughly constant at $T < 1$ MeV. The change of importance between isovector and isoscalar proton-neutron correlations with temperature is nicely reflected in the isospin expectation value, which decreases from $\hat{T} = 1$ at temperatures around 0.5 MeV, corresponding to the dominance of isovector correlations, to $\hat{T} = 0$ at temperature $T = 1$ MeV, when isoscalar proton-neutron correlations are most important. The low-temperature behavior of the Gamow-Teller and $B(M1)$ strength is related to the fading of the isoscalar proton-neutron correlations.

4   Supernova neutrinos

Neutrinos play a decisive role in our current understanding of a supernova and the nucleosynthesis associated with this process. Some examples are:

1. In the now favored *delayed model* of the supernova explosion mechanism neutrinos revive the shock, which is not energetical enough to traverse the envelope of the collapsing core and stalls about a couple of hundred kilometers from the core center (Bethe 1990).
2. Within the last few years the neutrino-driven wind model has been widely discussed as the possible site of r-process nucleosynthesis (Woosley *et al.* 1994, Witti *et al.* 1994).
3. Neutrinos leaving the exploding star will pass through the outer stellar burning shells. Neutrino-induced neutral-current reactions can knock out nucleons from nuclei abundant in these shells and thus contribute to the synthesis of elements in what is coined *neutrino nucleosynthesis* (Woosley 1990).

Neutrino-reactions on nuclei are obviously important in items 2) and 3). Before reviewing some of the recent developments we have to make some remarks about the respective neutrino spectra. Most of the supernova neutrinos are generated by the cooling of the nascent neutron star, which releases its gravitational binding energy by neutrino-pair production (Cooperstein 1988). Although pairs of all three flavors are generated with equal luminosity (Qian and Fuller 1994), due to their smaller opacities ν_μ and ν_τ neutrinos and their antiparticles decouple at smaller radii, and thus higher temperatures in the core, than ν_e and $\bar{\nu}_e$ neutrinos. As the neutrinos decouple in neutron-rich matter, which is less transparent for ν_e than for $\bar{\nu}_e$, it is expected on general grounds that the neutrino spectra after decoupling obey the temperature hierarchy (Qian and Fuller 1994), $T_{\nu_x} > T_{\bar{\nu}_e} >$

T_{ν_e}, where ν_x stands for ν_μ, ν_τ and their antiparticles, which are assumed to have identical spectra. The neutrino spectra can be approximately described by Fermi-Dirac (FD) distributions with zero chemical potential and $T_{\nu_x} = 8$ MeV, $T_{\bar{\nu}_e} = 5$ MeV, and $T_{\nu_e} = 3.5$ MeV, corresponding to average neutrino energies of $\langle E_{\nu_x} \rangle = 25$ MeV, $\langle E_{\bar{\nu}_e} \rangle = 16$ MeV, and $\langle E_{\nu_e} \rangle = 11$ MeV. More elaborate investigations of neutrino production in supernovae indicate that the high-energy tail of the neutrino spectra is better described by a Fermi-Dirac distribution with a finite chemical potential (Janka and Hillebrandt 1989, Qian 1996a).

4.1   Neutrino-driven wind model for the r-process

A successful r-process within the neutrino-driven wind model depends mainly on four parameters (Qian 1996b): the entropy per baryon S, the dynamical timescale, the mass loss rate and the electron-to-baryon ratio Y_e. All parameters depend on the neutrino luminosity and are determined mostly by ν_e and $\bar{\nu}_e$ absorption on free nucleons.

Matter expelled from the neutron star surface is experiencing decreasing temperatures T on its way out, as $T \sim 1/R$ with R the distance from the core center. As long as $T_9 \geq 5$ (T_9 measures the temperature in 10^9 K), matter is in nuclear statistical equilibrium and consists mainly of free nucleons and α-particles. At lower temperatures ($T_9 \leq 5$) the α-process assembles some of this material into heavier nuclei which then becomes the seed for the subsequent r-process. For a successful r-process to produce the abundance peaks at masses $A \sim 130$ and 195, the neutron-to-seed ratio should be large. In (Woosley et al. 1994) this has been the case as the 3-body reactions assumed to start the α-process ($3\alpha \rightarrow {}^{12}$C, $\alpha + \alpha + n \rightarrow {}^9$Be) are sufficiently slow so that a high fraction of α-particles survive the α-process. As pointed out by Meyer (Meyer 1995), the slow 3-body reactions can be bypassed by the neutrino-induced spallation of ^{4}He, then followed by $t + \alpha \rightarrow {}^7$Li and ^{7}Li$+\alpha \rightarrow {}^{11}$B. Depending on the neutrino luminosity, this bypass can significantly help to process α-particles into heavy seed material, thus effectively lowering the neutron-to-seed ratio and making successful r-process nucleosynthesis upto the $A \sim 195$ peak more difficult.

Due to the high particle thresholds the ^{4}He(ν, ν') reaction is mediated by the more energetic ν_x neutrinos. Haxton has calculated the cross section for this reaction within the full sets of $2\hbar\omega$ and $4\hbar\omega$ shell model states using the Sussex residual interaction (Haxton 1988). His flavor-averaged cross section ($8.6 \cdot 10^{-43}$ cm^2 for a neutrino distribution with $T = 8$ MeV and zero-chemical potential) agrees very nicely with the results obtained within the continuum random phase approximation (CRPA) using a G-matrix residual interaction derived from the Bonn potential ($8.2 \cdot 10^{-43}$ cm^2, (Langanke and Kolbe 1993)). The cross section is dominated by collective excitations of the giant dipole resonances with $J^\pi = 1^-$ and 2^-. The good agreement between the two rather different approaches suggest that the nuclear physics involved in calculating the ^{4}He(ν, ν') cross section is well under control.

Neutrino-induced reactions can be also important during and even after the r-process. In the conventional picture (Cowan *et al.* 1991) the nuclei are in $(n, \gamma)/(\gamma, n)$ equilibrium during the r-process. The r-process path is mainly determined by neutron separation energies and the timescale is essentially set by the β-decays of the waiting-point nuclei at the magic neutron numbers $N = 50, 82$ and 126. However, in the presence of a strong neutrino flux, ν_e-induced charged-current reactions on the waiting-point nuclei might actually compete with β-decays and speed-up the passage through the bottle-necks at the magic neutron numbers. In Ref. (Qian 1996c) the relevant cross sections have been estimated assuming the dominance of Gamow-Teller transitions and placing its strength S_{GT_-} as taken from the Ikeda sum rule ($S_{GT_-} \approx 3(N - Z)$ for very neutron-rich nuclei) into one single resonance whose position has been taken from empirical parametrizations. It is then found (Qian 1996c) that, for typical neutrino luminosities and spectra, ν_e-capture rates are of order 5 s^{-1} and thus can be faster than competing β-decays for the slowest waiting-point nuclei. Of course, quantitative conclusions can only be drawn from detailed numerical simulations of the r-process. In these studies it will also be desirable to calculate the ν_e-capture rates considering the expected fragmentation of the Gamow-Teller strength in the daughter nucleus. Here more realistic shell model calculations are certainly welcome!

It is usually assumed that the r-process drops out of $(n, \gamma)/(\gamma, n)$ equilibrium in a sharp freeze-out. The very neutron-rich matter, assembled during the r-process, then decays back to the valley of stability by a sequence of β-decays. However, in the neutrino-driven wind scenario the r-process matter will still be exposed to rather strong neutrino fluxes, even after freeze-out. By both, ν_e-induced charged-current and ν_x-induced neutral-current reactions, neutrinos can inelastically interact with r-process nuclei. In these processes the final nucleus will be in an excited state and most likely decay by the emission of one or several neutrons. Thus, this *post-processing* of r-process matter after freeze-out might effect the final r-process abundance. In a first quantitative attempt Qian *et al.* have shown that post-processing might shift the $A = 195$ abundance peak to smaller masses by a few units (Qian 1996c). The other r-process abundance peaks at $A = 80$ and $A = 130$ are less effected as both the neutrino-nucleus cross section and the average number of neutrons emitted are smaller. Post-processing has been excluded (Qian 1996c, Thielemann 1996) to fill the pronounced troughs at $A \approx 115$ and 180 which are found in r-process simulations with empirical mass formulae and without consideration of neutrino processes. The quenching of shell gaps in very neutron-rich nuclei is currently discussed as a possible solution to remove these troughs (Dobaczewski *et al.* 1995, Chen *et al.* 1995).

Neutral-current cross sections have been calculated within the CRPA for selected nuclei at the various neutron magic numbers (^{78}Ni, ^{130}Cd, ^{194}Er). For a FD neutrino distribution with zero chemical potential and $T = 8$ MeV, the calculated cross sections are roughly approximated by $A \cdot 10^{-42}$ cm^2, in agreement with previous shell model results on lighter nuclei. Gamow-Teller transitions dominate. However, significant fractions of the GT strength is below the neutron thresholds, making the result sensitive to the chosen single-particle energies. Clearly shell model test calculations for some of the involved nuclei are desirable.

4.2   Neutrino nucleosynthesis

When the flux of neutrinos generated by the cooling of the neutron star passes through the overlying shells of heavy elements substantial nuclear transmutations are induced, despite the small neutrino-nucleus cross sections. Specific nuclei appear to be almost entirely (e.g. ^{11}B, ^{19}F) or in a large fraction (e.g. ^{10}B, ^{15}N) made by this neutrino nucleosynthesis (Woosley et al. 1994). Within the ν-process these nuclei appear as the product of reaction sequences induced by neutral current (ν, ν') reactions on very abundant nuclei like ^{12}C, ^{16}O or ^{20}Ne. If the inelastic excitation of these nuclei is to particle-unbound levels, these will decay by emission of protons or neutrons, in this way contributing to nucleosynthesis. (The decay by α-particle emission is energetically favored in these nuclei. However, this decay mode is strongly suppressed by isospin selection as neutrinos mainly excite $T = 1$ levels in $T = 0$ nuclei.) As the nucleon thresholds in these nuclei are relatively high, effectively only μ and τ neutrinos (and their antiparticles) with their higher average energies contribute to the ν-process.

Kolbe et al. (Kolbe 1992) has calculated the nucleon knock-out cross sections for several key nuclei consistently within the CRPA. Table 1 compares these results with those obtained by Haxton within shell model calculations with different level of truncations (Woosley 1990). Both results agree typically within a factor of 2. Uncertainties in the neutrino distribution affect the cross sections by the same order of magnitude. For example, the ^{16}O(ν, ν') cross sections changes from $5.9 \cdot 10^{-42}$ cm^2 to $3.2 \cdot 10^{-42}$ cm^2, if the FD neutrino distribution with $T = 8$ MeV and zero chemical potential is replaced by a neutrino spectrum with $T = 6.3$ MeV and $\mu = 3T$. Note that both neutrino distributions have the same average energies.

4.3   Observation of ν_μ and ν_τ neutrinos in Superkamiokande

In what is considered the birth of neutrino astrophysics, neutrinos from supernova SN1987A have been detected by the Kamiokande (Hirata et al. 1987) and IMB (Bionta et al. 1987) water Čerenkov detectors (11 and 8 events, respectively). It is generally assumed that these events originated from the $\bar{\nu}_e + p \rightarrow n + e^+$ reaction in water. The detection of ν_e and ν_x neutrinos via the $\nu + e \rightarrow \nu' + e'$ scattering or the ^{16}O$(\nu_e, e^-)^{16}$F reaction was strongly suppressed by the small effective cross sections of these processes, although the ν_e induced signal can in principle be separated by its angular distribution (Haxton 1987). The observability of supernova neutrinos has significantly improved since the Superkamiokande (SK) detector is operational (Totsuka 1992). This detector, with about 15 times the fiducial volume for supernova neutrinos of Kamiokande and a lower threshold of $E_{thr} = 5$ MeV, will be capable to detect also the recoil

Table 1. Comparison of neutrino-induced neutral current cross sections (per nucleon and averaged over flavors in units of 10^{-42}cm^2 and for a FD neutrino spectrum with $T = 8$ MeV and zero chemical potential) as calculated within different nuclear models. Exponents are given in parentheses.

reaction	CRPA (Kolbe 1992)	shell model (Woosley 1990)
^{12}C$(\nu,\nu'\,$p$)^{11}$B	2.1 (-1)	3.3 (-1)
^{12}C$(\nu,\nu'\,$n$)^{11}$C	9.5 (-2)	9.2 (-2)
^{16}O$(\nu,\nu'\,$p$)^{15}$N	1.8 (-1)	1.7 (-1)
^{16}O$(\nu,\nu'\,$n$)^{15}$O	1.0 (-1)	4.6 (-2)
^{20}Ne$(\nu,\nu'\,$p$)^{19}$F	2.4 (-1)	3.7 (-1)
^{20}Ne$(\nu,\nu'\,$n$)^{19}$Ne	1.1 (-1)	5.3 (-2)
^{24}Mg$(\nu,\nu'\,$p$)^{23}$Na	2.5 (-1)	7.3 (-1)
^{24}Mg$(\nu,\nu'\,$n$)^{23}$Mg	8.9 (-2)	1.1 (-1)
^{28}Si$(\nu,\nu'\,$p$)^{27}$Al	3.0 (-1)	5.0 (-1)
^{28}Si$(\nu,\nu'\,$n$)^{27}$Si	4.9 (-2)	3.7 (-2)
^{32}S$(\nu,\nu'\,$p$)^{31}$P	4.6 (-1)	1.0 (0)
^{32}S$(\nu,\nu'\,$n$)^{31}$S	3.6 (-2)	2.1 (-2)
^{40}Ca$(\nu,\nu'\,$p$)^{39}$K	3.5 (-1)	2.4 (-1)
^{40}Ca$(\nu,\nu'\,$n$)^{39}$Ca	1.9 (-2)	1.4 (-2)

electrons from $\nu + e \rightarrow \nu' + e'$. In principle, ν_x induced neutrino-electron scattering events can be separated from everything else in SK using their angular distributions and energy spectra (Totsuka 1992). However, only about one third of the $\nu_x + e$ scattering events will have energies distinctly larger than the recoil electrons from $\nu_e + e$ and $\bar{\nu}_e + e$ scattering. Moreover, these higher energy electron recoils have to be separated by their direction from the much more numerous positrons from $\bar{\nu}_e + p \rightarrow n + e^+$ with the same energy.

Based on the fact that SK can observe photons with energies larger than 5 MeV (Koshiba 1992), recently another signal in water Čerenkov detectors has been proposed (Langanke *et al.* 1996) which allows one to unambiguously identify ν_x induced events. Schematically the detection scheme works as follows. Supernova ν_x neutrinos, with average energies of ≈ 25 MeV, will predominantly excite 1^- and 2^- giant resonances in ^{16}O via the ^{16}O$(\nu_x,\nu'_x)^{16}$O* neutral current reaction (Kolbe 1992). These resonances are above the particle thresholds and will mainly decay by proton and neutron emission. Although these decays will be mainly to the ground states of ^{15}N and ^{15}O, respectively, some of them will go to excited states in these nuclei. If these excited states are below the particle thresholds in ^{15}N ($E^\star < 10.2$ MeV) or ^{15}O ($E^\star < 7.3$ MeV), they will decay by γ emission. As the first excited states in both nuclei ($E^\star = 5.27$ MeV in ^{15}N and $E^\star = 5.18$ MeV in ^{15}O) are at energies larger than the SK detection threshold, all of the excited states in ^{15}N and ^{15}O below the respective particle thresholds will emit photons which can be observed in SK.

The effective $^{16}O(\nu_x, \nu'_x p\gamma)$ and $^{16}O(\nu_x, \nu'_x n\gamma)$ cross sections for SK have been calculated within a two-step process combining the continuum random phase approximation with the statistical model (Langanke et al. 1996). Assuming a FD neutrino spectrum with $T = 8$ MeV the calculation yields $3.2 \cdot 10^{-42}$ cm^2 for the total γ producing cross section for each flavor of ν_x plus $\bar{\nu}_x$ in the neutral current reactions on ^{16}O. This means that Superkamiokande is expected to observe about 360 γ events in the energy window $E = 5 - 10$ MeV for a supernova going off at 10 kpc ($\approx 3 \cdot 10^4$ lightyears or the distance to the galactic center). This is to be compared with a smooth background of about 270 positron events from the $\bar{\nu}_e + p \rightarrow n + e^+$ reaction in the same energy window. The number of events produced by supernova ν_x neutrinos via the scheme proposed here is larger than the total number of events expected from ν_x-electron scattering (about 80 events (Totsuka 1992)). More importantly, the γ signal can be unambiguously identified from the observed spectrum in the SK detector, in contrast to the more difficult identification from ν_x-electron scattering.

Due to its potential importance for the conclusion to be drawn from the eventual observation of supernova ν_x neutrinos by Superkamiokande, the relevant $^{16}O(\nu, \nu'p)$ and $^{16}O(\nu, \nu'n)$ cross sections should be calculated by means of other nuclear models. An obvious choice is a multi-shell calculation.

5   Conclusions and outlook

Many astrophysical models involve complicated many-nucleon problems. Very often the shell model appears to be the obvious method of choice. However, numerical restrictions have often limited the use of the shell model. These limitations have been partly overcome in recent years by both the development of new shell model techniques and the availability of better computer hardware. It has been the intention of this manuscript to summarize some of the progress achieved recently, but also to list some of the opportunities for future shell model applications in important astrophysical scenarios.

In recent years the shell model Monte Carlo has been well established as a very useful tool for nuclear structure calculations with relevance in astrophysics. This manuscript has highlighted some of the recent applications and gives a flavor of the kind of studies that can be performed with the Monte Carlo shell model. One application has concentrated on the systematics of even-even and $N = Z$ nuclei in the fp-shell. Besides achieving satisfactory agreement with data for masses and total $M1$ and $B(E2)$ strengths for these nuclei, one is, for the first time, able to systematically reproduce the Gamow-Teller strength distribution for the nuclei in the iron-nickel region, which are essential for the early supernova stage. In another application the thermal properties of nuclei with $A = 50 - 60$ have been studied. Important links betwen nuclear properties and the temperature dependence of particular pairing correlations beyond the mean field have been established. For even-even nuclei in this mass region, the

$J = 0$ proton and neutron pairs break at around $T = 1$ MeV. This phase transition is signalled by a sharp rise of the moment of inertia and might be observable in the level density. For $N = Z$ nuclei, which are relevant in the rp-process, isovector proton-neutron correlations dominate at low temperatures. In all nuclei studied so far, isoscalar proton-neutron correlations persist to higher temperatures than isovector pairing correlations.

Neutrino-nucleus interactions play an essential role in many astrophysical scenarios. They are the detection scheme to observe solar and supernova neutrinos, thus refining our understanding of stellar astrophysics, and they are often the laboratory to search for physics beyond the standard model of weak interaction. (It has been outside the scope of this manuscript to review the nuclear physics related to the KARMEN (Drexlin *et al.* 1991) and LSND (Albert *et al.* 1995) experiments. Some relevant references might be found in Ref. (Kolbe *et al.* 1995). Only in a few exceptional cases, neutrino-nucleus interactions have been calculated using the shell model. Some studies have been performed within the framework of the (continuum) random phase approximation. But rather often the relevant cross sections have been estimated on the basis of empirical parametrizations or within crude approximation schemes. Obviously neutrino-nucleus interactions offer a wide field for future shell model calculations.

6   Acknowledgments

The presented SMMC results have been obtained in collaboration with Y. Alhassid, D.J. Dean, S.E. Koonin and P.B. Radha. The neutrino-related studies have been performed together with W.C. Haxton, E. Kolbe, Y.-Z. Qian, F.-K. Thielemann and P. Vogel. It is a pleasure to thank all my collaborators. This work was supported by NSF grants PHY94-12818 and PHY94-20470.

References

Haxel O., Jensen, J.H.D., and Suess H.E. (1949): Phys. Rev. **75**, 1766

Raghavan R.S., (1995): Science **267**, 45

Bahcall J.N. (1990):*Neutrino Astrophysics*, Cambridge University Press, Cambridge

Hata N. (1995): in *Solar Modeling*, eds. Balantekin A.B., Bahcall J.N. (World Scientific, Singapore, 1995) p.63

Raghavan R.S. (1991): in Proc. 25th ICHEP (Singapore), ed. Phua K.K., Yamaguchi, Y. (World Scientific, Singapore, 1991) p. 482

Haxton W.C. (1988): Phys. Rev. Lett. **60**, 768

Engel J. ,Pittel S. and Vogel P. (1994): Phys. Rev. **50**, 1702

Engel J.,Pittel S., Vogel P. (1991): Phys. Rev. Lett. **67**, 426

Sugarbaker E. as quoted in (Engel *et al.* 1994)

Cleveland B.T., Daily T., Distel J., Lande K., Lee, C.K., *at al.* (1993): in *Proceedings of the 23rd International Cosmic Ray Conference* eds. D.A. Leahy, R.B. Hicks and D. Venkatesan (World Scientific, Singapore, 1993) Vol. 3 p. 865

Kosmas T.S., Oset E. (1996): Phys. Rev. **C53**, 1409

Engel J., Krastev P.I., Lande, K. (1995): Phys. Rev. **C51**, 2837

Bethe H.A. (1990): Rev. Mod. Phys. **62**, 801.

Johnson C.W., Koonin, S.E., Lang, G.H., Ormand,W.E. (1992): Phys. Rev. Lett. **69**, 3157

Lang G.H., Johnson C.W., Koonin,S.E., Ormand, W.E. (1993): Phys. Rev. **C48**, 1518

Hubbard J.(1959): Phys. Rev. Lett. **3**, 77 ; Stratonovich R.L., Dokl. Akad. Nauk. SSSR **115**, 1097 (1957)

Alhassid Y., Dean D. J., Koonin S.E., Lang, G., Ormand W. E. (1994): Phys. Rev. Lett. **72**, 613

Dean D. J., Radha P. B., Langanke K., Alhassid Y., Koonin S.E. (1995): Phys. Rev. Lett. **74**, 2909

Koonin S.E., Dean D.J., Langanke K. Physics Reports, in press

Aufderheide M. B., Fushiki I., Woosley S. E., Hartmann D. H. (1994): ApJS **91**, 389

Fuller G.M., Fowler W.A., Newman M.J. (1980): ApJS **42** 447; **48** (1982) 279; ApJ **252** (1982) 715; **293** (1985) 1

Aufderheide M.B., Bloom S.D., Resler D.A and Mathews G.J. (1993): Phys. Rev. **C47** ,2961; Phys. Rev. **C48** (1993) 1677

Williams A.L., et al., Phys. Rev. **C51** ,1144

Alford W.P. et al. (1990): Nucl. Phys. **A514** ,49

Vetterli M.C. et al. (1989): Phys. Rev. **C40** ,559

El-Kateb S. et al. (1994): Phys. Rev. **C49** , 3129

Rönnquist T., et al. (1993): Nucl. Phys. **A563** , 225

Langanke K., Dean D.J., Radha P. B., Alhassid Y., Koonin S.E. (1995): Phys. Rev. **C52**, 718

Radha P.B., Koonin S.E., Langanke K., Dean D.J. to be published

Dukelsky J. , Poves A., Retamosa,J. (1991): Phys. Rev. **C44**, 2872

Langanke K., Dean D. J., Radha P. B., Koonin S. E. Nucl. Phys., in press

Langanke K., Dean D.J., Koonin S.E., Radha P.B. to be published

Richter W. A., Vandermerwe M. G., Julies R. E., Brown B. A. (1991): Nucl. Phys. **A523** ,325

Civitarese O., Dussel G. G., Zuker A. P. (1989): Phys. Rev. **C40** ,2900

Champagne A.E., Wiescher, M. (1992): Ann. R. Nucl. **42**,39

Woosley S.E. et al. (1994): Ap. J. **433**, 229

Witti J., Janka H.-Th., Takahashi K. (1994): Astron. Astrophys. **286**, 841 and 857

Woosley S.E., Hartmann D.H., Hoffman R.D., Haxton,W.C. (1990): Ap. J. **356**, 272 .

Cooperstein J. (1988): Phys. Rep. **163**, 95 .

Qian Y.-Z. , Fuller G.M. (1994): Phys. Rev. **D49**, 1762

Janka H.T., Hillebrandt W. (1989): Astron. Astrophys. **224**, 49; Astron. Astrophys. Suppl. **78**, 375 (1989).

Qian Y.-Z. private communication.

Qian Y.-Z. (1996): in *Proceedings of Nuclei in the Cosmos IV* ed. Wiescher M., Nucl. Phys., to be published; Qian Y.-Z., Woosley S.E. (1996): ApJ., in press

Meyer B.S. (1995): ApJ. **449**, L55

Haxton W.C. (1988): Phys. Rev. Lett. **60**, 1999

Langanke K., Kolbe E. (1993): in *Nuclear Physics in the Universe* eds. M.W. Guidry and M.R. Strayer (Institute of Physics Publishing, Bristol, 1993) p. 237

Cowan J.J., Thielemann F.-K. , Truran J.W. (1991): Phys. Rep. **208**, 267

Qian Y.-Z., Haxton W.C., Langanke K., Vogel,P., to be published

Thielemann F.-K. private communication

Dobaczewski J., Nazarewicz W., Werner T.R. (1995): Phys. Scr. **T56**, 15

Chen B., Dobaczewski J., Kratz K.-L., Langanke K. Pfeiffer B., Thielemann F.-K., Vogel,P. (1995): Phys. Lett. **B355**, 37

Kolbe E., Langanke K., Krewald S.,Thielemann, F.-K. (1992): Nucl. Phys. **A540**, 599

Hirata K.S., *et al.* (1987): Phys. Rev. Lett. **58**, 1490

Bionta R.M., *et al.* (1987): Phys. Rev. Lett. **58**, 1494

Haxton W. C. (1987): Phys. Rev. **D36**, 2283.

Totsuka Y. (1992): Rep. Progr. Phys. **55**, 377

Koshiba M. (1992): Phys. Rep. **220**, 229

Langanke K., Vogel P.,Kolbe E. Phys. Rev. Lett. **76**, 2629

Drexlin G.,*et al.* (1991): Phys. Lett. **B267**, 321 ; Zeitnitz B., Prog. Part. Nucl. Phys. **32**, 351 (1994)

Albert M.,*et al.* (1995): Phys. Rev. **C51**, 1065

Kolbe E., Langanke K., Thielemann F.-K., Vogel P., Phys. Rev. **C52**, 3437 (1995); Engel J., Kolbe E., Langanke K., Vogel P., submitted to Phys. Rev. C

Large Scale Continuum Shell Model Calculations for Photonuclear Reactions with Δ Isobars and Exchange Currents

Stephen R. Cotanch[1] and Thomas B. Bright[2]

[1] Department of Physics, North Carolina State University,
Raleigh, North Carolina 27695, USA
[2] Department of Radiation Sciences, Uppsala University, S-751 21 Uppsala, Sweden

Abstract. A microscopic, large scale continuum shell model calculation is described that includes many-body structure effects, nucleon correlations, Δ isobar formation and pion exchange currents. The calculation entails solving a large number, of order 100, non-local, coupled integro-differential equations obtained by projecting the A baryon Schrödinger equation onto a truncated Hilbert space for the $(A-1)$ system. A realistic, finite-ranged effective interaction is utilized containing spin, isospin, tensor and Δ excitation components and antisymmetrization is rigorously enforced. Photonuclear reactions for ^{12}C and ^{16}O are analyzed at low and medium energies in which both one-body nucleon and two-body pion exchange currents are included. Within this comprehensive framework, effects from several of the above mentioned elements were consistently assessed. Significant findings are: 1) cross sections are relatively unaffected by both Δ isobars and pion exchange currents except at high momentum transfer; 2) polarization and analyzing power observables are more sensitive to Δ effects, especially at high energies, and, to a much lesser extent, pion exchange currents; 3) a conventional continuum shell model calculation, without Δ isobars and pion exchange currents, is sufficient to obtain the correct magnitude and qualitative description for both (γ, p) and (γ, n) reactions over a wide range of energies and angles.

1 Introduction

Perhaps the quintessential, if not most successful, many-body description of the atomic nucleus is the shell model. Since its inception in 1949 by Mayer and Jensen, for which they were awarded the Nobel prize, the shell model has experienced dramatic success, especially for the quantitative understanding of spherical nuclei with magic numbers. Even today, almost 50 years later, this approach provides one of the most effective means to theoretically examine both bound and unbound many-baryon systems. In particular, one of the most powerful microscopic techniques has been the application of the shell model to nuclear reactions, an approach referred to as the continuum shell model (CSM). The CSM is most appropriate for situations in which one, and only one, baryon is in the continuum (unbounded). While this is a restriction, the CSM is ideal for investigating nucleon elastic, inelastic, charge exchange and radiative capture processes. The latter is the thrust of this chapter which, explains elements of the

CSM including large scale, supercomputer based calculations for medium energy photonuclear reactions.

To appreciate the evolution of the CSM, it is appropriate to provide a brief historic backdrop. For developments prior to 1970 see (Bloch 1966) and (Mahaux and Weidenmuller 1969). The first numerical applications of the CSM to photonuclear reactions was the pioneering work of (Buck and Hill 1967) who performed a coupled channels analysis of light nuclei in the giant dipole region. In the subsequent decade several similar analyses were reported: (Raynal *et al.* 1967, Marangoni and Saruis 1969, Birkholz 1972, Barz *et al.* 1977, Hohn *et al.* 1981, Ramavataram *et al.* 1981) and (Cavinato *et al.* 1982). Necessitated by numerical practicality, these applications generally involved restricted model spaces, zero-ranged interactions and omission of non-localities due to the Pauli principle (antisymmetrization). Agreement between theory and experiment was only qualitative. With the advent of dedicated work stations in the early 80's, Ludeking and Cotanch (1984) advanced the treatment by using an extended model space, a realistic finite-ranged interaction with tensor components and rigorously imposing antisymmetrization by retaining all non-localities. Improved but not complete detailed agreement was obtained for the giant dipole resonance excited in $^{11}B(p,\gamma)^{12}C$. Subsequently, (Ludeking and Cotanch 1986) approximately included Δ isobar degrees of freedom and applied their approach successfully to predict medium energy Bates measurements of $^{16}O(\gamma,p)^{15}N$. Finally, using supercomputer based technology, (Bright and Cotanch 1993) performed a large-scale calculation including all Δ wavefunction components to obtain a unified description of $^{16}O(\gamma,N)$ reactions and, for the first time, the correct magnitude of the cross section for $^{16}O(\gamma,n)^{15}O$. (Bright *et al.* 1996a) then included pion exchange current contributions, which they found to be small, in an analysis of $^{11}B(p,\gamma)^{12}C$.

In the following sections the CSM formalism is developed and applied to the photonuclear reactions $^{11}B(p,\gamma)^{12}C$, $^{12}C(p,\gamma)^{13}N$ and $^{16}O(\gamma,N)$ at low and medium energies. In section 2 the Hamiltonian and effective two-body interaction are defined and the basic equations for the CSM are derived. Section 3 contains a description of the electromagnetic interaction responsible for the photonuclear process and a detailed discussion of the one- and two-body (exchange) currents. Numerous applications, some very recent and unpublished, are presented in section 4 with a summary and outlook in section 5.

2 Continuum Shell Model Formulation

2.1 General Method

In the CSM approach, the mass A many-body scattering wavefunction is represented by products of mass $(A - 1)$ bound shell model eigenstates $|\beta>$ and single-particle states $|\alpha>$. Here $\beta \equiv (E_\beta, \pi_\beta, J_\beta, M_\beta, T_\beta, T_{3\beta})$ is the complete set of quantum numbers that uniquely specifies a state of the $(A - 1)$ physical system: energy E_β, parity π_β, total angular momentum J_β with projection M_β

and the total isospin T_β with projection $T_{3\beta}$, respectively. Similarly, the label $\alpha \equiv (E_\alpha, \pi_\alpha, j_\alpha, m_\alpha, l_\alpha, s_\alpha, \tau_\alpha, \tau_{3\alpha})$ is the complete set of quantum numbers that uniquely specifies a single-particle bound state: energy E_α, parity π_α, total angular momentum j_α with projection m_α, orbital l_α and spin s_α angular momenta and the isospin τ_α with projection $\tau_{3\alpha}$, respectively. The $\vec{l} + \vec{s}$ coupling scheme is chosen and the convention $\tau_3 = +\frac{1}{2}$ for a proton state is used.

The state $|\alpha>$ is a solution of the self-consistent Hartree-Fock equation

$$E_\alpha |\alpha> = (T_\alpha + V_{HF}) |\alpha>. \tag{1}$$

In this equation, T_α is the single-particle kinetic energy operator and V_{HF} is the Hartree-Fock potential energy operator. This equation has the familiar form in configuration space

$$E_\alpha \, \phi_\alpha(\vec{r}) = (T_\alpha + V_{HF}) \, \phi_\alpha(\vec{r}), \tag{2}$$

where $\phi_\alpha(\vec{r}) \; = <\vec{r}|\alpha>$ is the coordinate representation of the state $|\alpha>$. The Hartree-Fock potential V_{HF} can be written explicitly in terms of the 2-body nucleon-nucleon potential $V(\vec{r}, \vec{r}')$ as

$$V_{HF} \, \phi_\alpha(\vec{r}) \; = \; V_0(\vec{r})\phi_\alpha(\vec{r}) \; - \int d\vec{r}' U(\vec{r}, \vec{r}')\phi_\alpha(\vec{r}'), \tag{3}$$

where

$$V_0(\vec{r}) \; = \; \sum_{\alpha=1}^{A} \int d\vec{r}' \phi_\alpha^*(\vec{r}')V(\vec{r}, \vec{r}')\phi_\alpha(\vec{r}'), \tag{4}$$

and

$$U(\vec{r}, \vec{r}') \; = \; \sum_{\alpha=1}^{A} \phi_\alpha^*(\vec{r}')V(\vec{r}, \vec{r}')\phi_\alpha(\vec{r}). \tag{5}$$

The summation index α includes all nucleons within the nucleus.

Once the set of occupied bound states of the nucleus and the potential V_{HF} are determined, this same potential is used to generate the remaining unoccupied bound and continuum states. The continuum states are assumed to be finite and to have continuous derivatives. The full set of states ϕ_α is complete, orthonormal and provides a complete single-particle basis for the present model. This can be expressed as

$$\sum_{E_\alpha} \phi_\alpha^*(\vec{r})\phi_\alpha(\vec{r}') = S_\alpha^*(\hat{r})S_\alpha(\hat{r}')\frac{\delta(r - r')}{r^2}, \tag{6}$$

where the function $S(\hat{r})$ is given by

$$S(\hat{r}) = \sum_{\lambda\sigma} C(lsj, \lambda\sigma m)i^l Y_l^\lambda(\hat{r})\chi_s^\sigma\chi_\tau^{\tau_3}. \tag{7}$$

The function $S(\vec{r})$ contains all the geometrical information concerning the angular momentum, spin and isospin parts of the single-particle wavefunctions $\phi_\alpha(\vec{r})$. The quantity C is the Clebsch-Gordon coefficient coupling the orbital $\vec{l}$ and spin

$\vec{s}$ angular momenta to the total angular momentum $\vec{j}$. The phase convention in this study is one that is commonly used, for example in (Messiah 1965). The functions χ^σ and χ^τ are the single-particle spin and isospin eigenfunctions. In terms of the function $S(\vec{r})$, the single-particle wavefunctions can be written as

$$\phi(\vec{r}) = \frac{v_{jl}(E, r)}{r} S(\hat{r}). \tag{8}$$

In the CSM formalism with only one nucleon in the continuum, a many-body scattering state having total center of mass kinetic energy E, parity π, total angular momentum J with projection M and isospin T with projection T_3, may be written as a linear superposition of products of mass $(A - 1)$ and single-particle states,

$$|J^\pi> \equiv |J\pi EMTT_3 >= \sum_{cE_p} B_{cc_0}^{J^\pi}(E_p) < c|JT > a_p^\dagger|\beta>, \tag{9}$$

where $a_p^\dagger$ creates a nucleon in a single-particle state $|p>$. The many-body states $|\beta>$ are determined by diagonalizing the many-body Hamiltonian in a truncated model space that is spanned by products of bound single-particle states $|q>$, a subset of the complete set of states $|\alpha>$. In principle, $|\beta>$ can be represented by complicated antisymmetric sums of $(A - 1)$ creation operators $a_{q_i}^\dagger$ acting on the true vacuum $|0 >$,

$$|\beta> = \sum_j C(j) \prod_{i=1}^{A-1} a_{q_{j_i}}^\dagger(i) |0 > . \tag{10}$$

The set of states $|p>$ is defined to be the complement set to $|q>$; that is, the total space spanned by the set $|\alpha>$ is spanned by the sum of the two sets $|q>$ and $|p>$. This set of states $|p>$ involve all continuum wavefunctions as well as the remaining higher energy bound state configurations not included in the model space of $|\beta>$. With this specification for $|p>$ and $|q>$, it necessarily follows that

$$a_p|\beta> = 0. \tag{11}$$

The set of coefficients $C(j)$ is determined by the nuclear structure assumptions for the mass $(A - 1)$ nucleus.

The coefficients $B_{cc_0}^{J^\pi}(E_p)$ do not depend on the magnetic quantum numbers contained in the symbolic geometric coupling coefficient

$$< c|JT > = C(J_\beta j_p J, M_\beta m_p M) \times C(T_\beta t_p T, T_{3_\beta} \tau_{3_p} T_3). \tag{12}$$

The label c indicates a composite channel, defined by the collective quantum numbers for the states $|\beta>$ and $|p>$,

$$c \equiv \{E_p, \pi_p, j_p, m_p, l_p, s_p, \tau_p, \tau_{3_p}; E_\beta, \pi_\beta, J_\beta, M_\beta, T_\beta, T_{3_\beta}\}. \tag{13}$$

The label c_0 indicates the incident channel; that is, the configuration actually realized in the experiment. For example, for the reaction $^{16}O(\gamma, p)^{15}N$, c_0 represents a proton of given total angular momentum incident on the ground state of ^{15}N, or, symbolically, $(p + {}^{15}N)$. The label c can represent any possible state, such as the elastic channel $(p + {}^{15}N)$, as well as the reaction channels $(p + {}^{15}N^*)$, $(n + {}^{15}O)$, $(\Delta^+ + {}^{15}N)$, etc.

In second quantized notation, the total Hamiltonian H can be expressed in terms of the one-body kinetic energy operator T and the two-body potential energy operator V, describing the nucleon-nucleon interaction, as

$$H = \sum_{ij} < i|T|j > a_i^\dagger a_j + \frac{1}{2} \sum_{ijkl} < ij|V|kl > a_j^\dagger a_i^\dagger a_k a_l, \tag{14}$$

where i, j, k, l can be any of the possible states $|\alpha>$. This Hamiltonian is diagonalized by projecting onto the CSM basis states $a_p^\dagger |\beta>$,

$$< \beta|a_p(H - E)|J^\pi > = 0. \tag{15}$$

Inserting equation (9) and (14) into this expression gives

$$0 = \sum_{c' E_{p'}} B_{c' c_0}^{J^\pi}(E_{p'}) < c'|JT > < \beta|a_p(H - E)a_{p'}^\dagger|\beta'> \tag{16}$$

$$= \sum_{c' E_{p'}} B_{c' c_0}^{J^\pi}(E_{p'}) < c'|JT > \left(\sum_{ij} < i|T|j > < \beta|a_p a_i^\dagger a_j a_{p'}^\dagger|\beta' > \right.$$

$$\left. + \frac{1}{2} \sum_{ijkl} < ij|V|kl > < \beta|a_p a_j^\dagger a_i^\dagger a_k a_l a_{p'}^\dagger|\beta' > -E\delta_{\beta\beta'}\delta_{pp'} \right) \tag{17}$$

The expression $< \beta|a_p H a_{p'}^\dagger|\beta' >$ can be simplified using the anticommutation relations

$$\{a_i^\dagger, a_j\} = \delta_{ij}, \quad \{a_i^\dagger, a_j^\dagger\} = 0 = \{a_i, a_j\}. \tag{18}$$

Using equations (18),

$$< \beta|a_p a_i^\dagger a_j a_{p'}^\dagger|\beta' > = \delta_{\beta'\beta}\delta_{p'j}\delta_{pi} + < \beta|a_i^\dagger a_j|\beta'\overset{q}{>}\delta_{p'p}. \tag{19}$$

In this last expression, the symbol q in $< \beta|a_i^\dagger a_j|\beta'\overset{q}{>}$ indicates that the states i and j are restricted to the subspace of states $|\alpha>$ that span $|\beta>$ due to equation (11); that is, they are restricted to the set of states $|q>$. Similarly,

$$< \beta|a_p a_j^\dagger a_i^\dagger a_k a_l a_{p'}^\dagger|\beta' > = < \beta|a_i^\dagger a_k|\beta'\overset{q}{>}\delta_{p'l}\delta_{pj} + < \beta|a_j^\dagger a_l|\beta'\overset{q}{>}\delta_{p'k}\delta_{pi}$$

$$- < \beta|a_j^\dagger a_k|\beta'\overset{q}{>}\delta_{p'l}\delta_{pi} - < \beta|a_i^\dagger a_l|\beta'\overset{q}{>}\delta_{p'k}\delta_{pj} + < \beta|a_j^\dagger a_l|\beta'\overset{q}{>}\delta_{p'p}\delta_{ik}$$

$$- \sum_{\beta''} < \beta|a_j^\dagger a_k|\beta''\overset{q}{>} < \beta''|a_i^\dagger a_l|\beta'\overset{q}{>}\delta_{p'p}. \tag{20}$$

Equations (19) and (20) permit a simplification of $< \beta p \,|H|\, p'\beta' >$.

$$< \beta p \,|H|\, p'\beta' > = < p|T|p' > \delta_{\beta\beta'} + \sum_{ij}^{q} < ip|V|\widetilde{jp'} > \; < \beta|a_i^\dagger a_j|\beta' > \qquad (21)$$

$$+ \delta_{p'p}\left(\sum_{ij}^{q} < i|T|j > \; < \beta|a_i^\dagger a_j|\beta' > + \frac{1}{2}\sum_{ij}^{q}\sum_{k}^{q} < ik|V|jk > \; < \beta|a_i^\dagger a_j|\beta' > \right.$$

$$\left. - \frac{1}{2}\sum_{ij}^{q}\sum_{k}^{q}\sum_{\beta''} < ij|V|kl > \; < \beta|a_i^\dagger a_l|\beta'' > < \beta''|a_j^\dagger a_k|\beta' > \right),$$

where $|\widetilde{ab}> \equiv |ab> - |ba>$ and the q over the summation sign indicates a sum over the bound states $|q>$. Further simplification is obtained by considering the equation of motion for the mass $(A-1)$ system,

$$(H - E_\beta)\,|\beta> = 0. \qquad (22)$$

Inserting equation (14) into this expression and projecting onto the state $|\beta>$ gives

$$E_\beta\delta_{\beta\beta'} = \sum_{ij} < i|T|j > \; < \beta|a_i^\dagger a_j|\beta' > + \frac{1}{2}\sum_{ijkl} < ij|V|kl > \; < \beta|a_i^\dagger a_j^\dagger a_l a_k|\beta' >$$

$$= \sum_{ij}^{q} < i|T|j > \; < \beta|a_i^\dagger a_j|\beta' > + \frac{1}{2}\sum_{ij}^{q}\sum_{k}^{q} < ik|V|jk > \; < \beta|a_i^\dagger a_j|\beta' >$$

$$- \frac{1}{2}\sum_{ij}^{q}\sum_{kl}^{q}\sum_{\beta''} < ij|V|kl > \; < \beta|a_i^\dagger a_l|\beta'' > < \beta''|a_j^\dagger a_k|\beta' > . \qquad (23)$$

A similar procedure can be used to obtain an expression for $< p|T|p' >$. Consider the Hartree-Fock equation for the single-particle state $|p>$. In second quantized notation, the Hartree-Fock Hamiltonian operator can be written (Bohr and Mottelson 1969) as

$$H_{HF} = \sum_{ij} < i|T|j > a_i^\dagger a_j + \sum_{ij}^{q}\sum_{k}^{q} < ik|V|\widetilde{jk} > a_i^\dagger a_j. \qquad (24)$$

The matrix element of this operator between states $|p>$ is the single-particle energy. Thus

$$< p|H_{HF}|p' > = E_p\delta_{p'p} = < p|T|p' > + \sum_{k}^{q} < pk|V|\widetilde{p'k} > . \qquad (25)$$

With the aid of equations (23) and (25), equation (21) becomes

$$< \beta p \,|H|\, p'\beta' > = (E_p + E_\beta)\,\delta_{\beta'\beta}\delta_{p'p} - < p|V_{HF}|p' > \delta_{\beta'\beta}$$
$$+ \sum_{ij}^{q} < ip|V|\widetilde{jp'} > < \beta|a_i^\dagger a_j|\beta' > . \qquad (26)$$

The second and third terms can be combined by rewriting $< pk|V|\widetilde{p'k} >$ as

$$\sum_{k}^{q} < pk|V|\widetilde{p'k} > = \sum_{ij}^{q} < pi|V|\widetilde{p'j} > < \gamma|a_i^\dagger a_j|\gamma >, \qquad (27)$$

where $|\gamma >$ is the shell model many-body wavefunction for the ground state of the mass A system and is spanned by the single-particle basis states $|q>$. Then

$$< \beta p \,|H|\, p'\beta' > = (E_p + E_\beta)\,\delta_{\beta'\beta}\delta_{p'p} \qquad (28)$$
$$+ \sum_{q'q} < qp|V|\widetilde{q'p'} > [< \beta|a_q^\dagger a_{q'}|\beta' > - < \gamma|a_q^\dagger a_{q'}|\gamma > \delta_{\beta'\beta}]$$

In this last expression and hereafter, the sum over ij with its restriction to the states $|q>$ is replaced with the more obvious sum over $q'q$. The quantity $< \gamma|a_q^\dagger a_{q'}|\gamma >$ is a measure of the occupation probability of a particle having quantum numbers q in that configuration. The quantity $< \beta|a_q^\dagger a_{q'}|\beta' >$ is the one-body transition density for the $(A-1)$ nucleus. The quantities $< \beta|a_q^\dagger a_{q'}|\beta' >$ and $< \gamma|a_q^\dagger a_{q'}|\gamma >$ incorporate all assumptions about the nuclear structure. For brevity, define the quantity $\rho(\beta q, \beta'q')$, which has been called the density matrix in the literature (Ramavataram $et\ al.$ 1981), as

$$\rho(\beta q, \beta'q') \equiv < \beta|a_q^\dagger a_{q'}|\beta' > - \delta_{\beta'\beta} < \gamma|a_q^\dagger a_{q'}|\gamma > . \qquad (29)$$

Then

$$< \beta p \,|H|\, p'\beta' > = (E_p + E_\beta)\,\delta_{\beta'\beta}\delta_{p'p} + \sum_{q'q} \rho(\beta q, \beta'q') < pq|V|\widetilde{p'q'} > . \qquad (30)$$

This last expression considerably simplifies the coupled channels equations derived from diagonalizing the full nuclear Hamiltonian. Equation (17) becomes

$$(E - E_p - E_\beta)B_{cc0}^{J^\pi}(E_p) < c|JT >$$
$$= \sum_{c'E_{p'}} B_{c'c0}^{J^\pi}(E_{p'}) < c'|JT > \sum_{q'q} < pq|V|\widetilde{p'q'} > \rho(\beta q, \beta'q'). \qquad (31)$$

This expression represents an infinite set of coupled algebraic equations for the coefficients $B_{cc0}^{J^\pi}(E_p)$. However, these coefficients are not explicitly needed since the infinite sum over E_p' can be performed by transforming from the energy to the coordinate representation through the introduction of the physically meaningful scattering wavefunction $\Psi_{cc_0}^{J^\pi}$, defined by

$$\Psi_{cc_0}^{J^\pi}(\vec{r}) = \sum_{E_p} B_{cc_0}^{J^\pi}(E_p) < c|JT > \phi_p(\vec{r})$$

$$= f_{cc_0}^{J^\pi}(r) < c|JT > \frac{1}{r} S_p(\hat{r}),$$ \hfill (32)

where

$$f_{cc_0}^{J^\pi}(r) \equiv \sum_{E_p} B_{cc_0}^{J^\pi}(E_p) v(r, E_p),$$ \hfill (33)

and the subscripts for $v(r, E_p)$ have been suppressed, to be reintroduced only when required. This transformation is achieved by multiplying equation (31) by $\phi_p(\vec{r}_0)$ and performing the sum over E_p. Then

$$\sum_{E_p} B_{cc_0}^{J^\pi}(E_p) v(r_0, E_p) < c|JT > \frac{S_p(\hat{r}_0)}{r_0} (E - E_p - E_\beta)$$ \hfill (34)

$$= \sum_{\tilde{c}'} \sum_{E_{p'}} \sum_{E_p} B_{c'c_0}^{J^\pi}(E_{p'}) v(r_0, E_p) < c'|JT > \frac{S_p(\hat{r}_0)}{r_0} \sum_{q'q} < pq|V|\widetilde{p'q'} > \rho(\beta q, \beta' q'),$$

where explicit reference to $E_{p'}$ is removed from the label c', defining $\tilde{c}'$. To sum the left side of this equation, E_p is replaced by again using the Hartree-Fock equation (2) with $\alpha = p$. Performing the sum over $E_{p'}$ on the right side and the summation over E_p on the left side yields

$$(T + V_{HF} + E_\beta - E) < c|JT > \frac{S_p(\hat{r}_0)}{r_0} f_{cc_0}^{J^\pi}(r_0)$$

$$= \sum_{\tilde{c}'} \sum_{E_p} v(r_0, E_p) < c'|JT > \frac{S_p(\hat{r}_0)}{r_0} \sum_{q'q} \rho(\beta q, \beta' q')$$

$$\times \left\{ < \phi_p(\vec{r}_1)\phi_q(\vec{r}_2) | V(\vec{r}_1, \vec{r}_2) | f_{c'c_0}^{J^\pi}(r_1) \frac{S_{p'}(\hat{r}_1)}{r_1} \phi_{q'}(\vec{r}_2) > \right.$$

$$\left. - < \phi_p(\vec{r}_1)\phi_q(\vec{r}_2) | V(\vec{r}_1, \vec{r}_2) | \phi_{q'}(\vec{r}_1) f_{c'c_0}^{J^\pi}(r_2) \frac{S_{p'}(\hat{r}_2)}{r_2} > \right\},$$ \hfill (35)

or

$$(T + V_{HF} + E_\beta - E) \Psi_{cc_0}^{J^\pi}(\vec{r}_0) = \sum_{\tilde{c}'} \sum_{E_p} \phi_p(\vec{r}_0) \sum_{q'q} \rho(\beta q, \beta' q')$$

$$\times < \phi_p(\vec{r}_1)\phi_q(\vec{r}_2)|V(\vec{r}_1, \vec{r}_2) \left\{ |\Psi_{c'c_0}^{J^\pi}(\vec{r}_1)\phi_{q'}(\vec{r}_2) > -|\phi_{q'}(\vec{r}_1)\Psi_{c'c_0}^{J^\pi}(\vec{r}_2) > \right\}.$$ \hfill (36)

The remaining sum over E_p is of the form

$$P(\vec{r}_0, \vec{r}_1) \equiv \sum_{E_p} |\phi_p(\vec{r}_0) >< \phi_p(\vec{r}_1)|. \qquad (37)$$

This sum is simply a projection onto the space spanned by the states $|p>$. Since the space spanned by $|\alpha>$ is assumed to be complete,

$$\sum_{E_\alpha} |\phi_\alpha(\vec{r}_0) >< \phi_\alpha(\vec{r}_1)| = \frac{\delta(r_0 - r_1)}{r_0^2} S_{\bar{\alpha}}(\hat{r}_0) S_{\bar{\alpha}}^*(\hat{r}_1), \qquad (38)$$

where $\bar{\alpha}$ refers to all quantum numbers for the state $|\alpha>$ with the exception of E_α. Equation (37) can be rewritten as

$$P(\vec{r}_0, \vec{r}_1) = \frac{\delta(r_0 - r_1)}{r_0^2} S_{\bar{\alpha}}(\hat{r}_0) S_{\bar{\alpha}}^*(\hat{r}_1) - \sum_{E_{q''}} |\phi_{q''}(\vec{r}_0) >< \phi_{q''}(\vec{r}_1)|. \qquad (39)$$

Except for the energy, the quantum numbers of the state $|q'' >$ are the same as those of the state $|p>$. Therefore the infinite sum over E_p has been replaced by a finite sum over $E_{q''}$.

The model restrictions thus far are three: only one particle is allowed in the space spanned by the states $|p>$; truncation to a finite set of $(A - 1)$ states $|\beta>$ that are spanned by the the states $|q>$; the nucleon-nucleon interaction is, at most, two-body in nature. With these restrictions, the nuclear scattering problem is reduced to solving a finite set of coupled integro-differential equations for the coupled channels scattering wavefunctions,

$$(T + V_{HF} + E_\beta - E)\, \Psi_{cc_0}^{J^\pi}(\vec{r}_0)$$

$$= \sum_{c'} \sum_{q'q} \rho(\beta q, \beta' q') \int \int d\vec{r}_1 d\vec{r}_2 \left[\frac{\delta(r_0 - r_1)}{r_0^2} S_{\bar{p}}(\hat{r}_0) S_{\bar{p}}^*(\hat{r}_1) - \sum_{E_{q''}} \phi_{q''}(\vec{r}_0)\phi_{q''}^*(\vec{r}_1) \right]$$

$$\times \left(\phi_q^*(\vec{r}_2)\, V(\vec{r}_1, \vec{r}_2)\, \left\{ \Psi_{c'c_0}^{J^\pi}(\vec{r}_1)\phi_{q'}(\vec{r}_2) - \phi_{q'}(\vec{r}_1)\Psi_{c'c_0}^{J^\pi}(\vec{r}_2) \right\} \right). \qquad (40)$$

Therefore, the infinite set of coupled algebraic equations, equation (17), having as solutions the expansion coefficients of equation (9), has been replaced by a finite set of coupled integro-differential equations, equation (40), with attending physical boundary conditions. The solution to the coupled channels equations directly yields the physical scattering and reaction amplitudes for nucleons incident on the mass $(A - 1)$ states $|\beta>$. The complete wavefunction, equation (9), is then a finite sum of products of the above solutions and the mass $(A - 1)$ states $|\beta>$,

$$|J^\pi > = \sum_c \Psi_{cc_0}^{J^\pi}(\vec{r}_0) < \vec{r}|\beta>. \qquad (41)$$

2.2 Angular Integrations

The coupled channels equations, equation (40), can be further reduced by integration of the angular coordinates to yield a set of coupled channels radial equations. To this end, the two-body interaction $V(\vec{r}_1, \vec{r}_2)$ is expanded in spherical harmonics. The details of this expansion depend on the form of the interaction, which, in this study, has both central and tensor components,

$$
V(\vec{r}_1, \vec{r}_2) = V_0(r) \{ a_o + a_\sigma (\vec{\sigma}_1 \cdot \vec{\sigma}_2) + a_\tau (\vec{\tau}_1 \cdot \vec{\tau}_2) + a_{\sigma\tau} (\vec{\sigma}_1 \cdot \vec{\sigma}_2)(\vec{\tau}_1 \cdot \vec{\tau}_2) \}
$$
$$
+ V_t(r) S_{12} (\vec{\tau}_1 \cdot \vec{\tau}_2), \tag{42}
$$

with $r = |\vec{r}_1 - \vec{r}_2|$ the relative separation of particles 1 and 2, and $S_{12} = S^2 \cdot Y^2(\hat{r})$, where S^2 and Y^2 are second rank tensors in spin and coordinate space, respectively, and are given by the tensor products $S^2 \equiv (\sigma_1 \otimes \sigma_2)^2$ and $Y^2 \equiv (\hat{r}_1 \otimes \hat{r}_2)^2$. The central interaction, $V_0(r) \equiv y_0(\mu_c r) - \frac{\Lambda}{\mu_c} y_0(\Lambda r)$, is a scalar operator with respect to the relative coordinate $\vec{r}$ and, consistent with boson exchange, has a Yukawa radial dependence with a partial wave expansion

$$
y_0(\alpha r) = \sum_{L_1 M_1} -4\pi j_{L_1}(i\alpha r_<) h_{L_1}^{(1)}(i\alpha r_>) Y_{L_1}^{M_1}(\hat{r}_1) Y_{L_1}^{M_1}{}^*(\hat{r}_2)
$$
$$
= \sum_{L_1} \tilde{V}_{L_1 L_1}(r_1, r_2) (Y^{L_1}(\hat{r}_1) \otimes Y^{L_1}(\hat{r}_2))^0 . \tag{43}
$$

The tensor interaction, $V_t(r) \equiv y_2(\mu_t r) - \frac{\Lambda}{\mu_t} y_2(\Lambda r)$, is a tensor operator of rank 2 with respect to the relative coordinate, also with a Yukawa radial dependence. The Γ component on this tensor has the expansion (Wong and Anderson 1971)

$$
y_2(\alpha r) Y_2^\Gamma(\hat{r}) = \sum_{L_1 M_1 L_2 M_2} (-1)^{L_2 + 1 + \Gamma} i^{L_1 - L_2} (4\pi)^{3/2} \sqrt{1/20}
$$
$$
\times < L_1 \| Y_2 \| L_2 > \begin{pmatrix} L_1 & L_2 & 2 \\ M_1 & M_2 & \Gamma \end{pmatrix} Y_{L_1}^{M_1}(\hat{r}_1) Y_{L_2}^{M_2}{}^*(\hat{r}_2)
$$
$$
\times \begin{Bmatrix} j_{L_1}(i\alpha r_1) h_{L_2}^{(1)}(i\alpha r_2), & r_2 > r_1 \\ j_{L_2}(i\alpha r_2) h_{L_1}^{(1)}(i\alpha r_1), & r_1 > r_2 \end{Bmatrix}
$$
$$
= \sum_{L_1} \sum_{L_2} \tilde{V}_{L_1 L_2}(r_1, r_2) (Y^{L_1}(\hat{r}_1) \otimes Y^{L_2}(\hat{r}_2))_2^\Gamma . \tag{44}
$$

In these expressions, $y_0(x) \equiv \frac{e^{(-x)}}{x}$ and $y_2(x) \equiv (1 + \frac{3}{x} + \frac{3}{x^2}) y_0(x)$. The reduced matrix elements and the Wigner-j coefficients conform to the conventions employed in (Messiah 1965).

Continuing the reduction to the radial equations, equations (8), (32) and (40) are now combined and the orthogonality of the Clebsch-Gordon coefficients is used to bring $< c|JT >$ to the right side of equation (40). In the resulting expression, c_z will refer to the magnetic substates of c. Then, both sides are multiplied by the spin-angle function $S_p^*(\hat{r}_0)$ to utilize orthogonality by integrating over the spin, isospin and angular coordinates yielding

$$\frac{1}{r_0}\left\{\frac{\hbar^2}{2M}\left[\frac{l(l+1)}{r_0^2}-\frac{d^2}{dr_0^2}\right]+V_{HF}-(E-E_\beta)\right\}f_{cc_0}^{J^\pi}(r_0) \tag{45}$$

$$=\sum_{c_z}<c|JT>\sum_{q'q}\rho(\beta q,\beta'q')\sum_{\bar{c}'}<c'|JT>\sum_{L_1 L_2}\sum_{I_v}$$

$$\left\{\int_0^\infty\int_0^\infty dr_1 dr_2 r_1^2 r_2^2\frac{\delta(r_0-r_1)}{r_0^2}\frac{v_q(r_2,E_q)}{r_2}\right.$$

$$\int d^2\Omega_1 d^2\Omega_2 S_{\bar{p}}^*(\hat{r}_1)S_q^*(\hat{r}_2)\tilde{V}_{L_1 L_2}^{I_v}(r_1,r_2)S^{I_v}\cdot(Y^{L_1}(\hat{r}_1)\otimes Y^{L_2}(\hat{r}_2))^{I_v}$$

$$\times\left.\left[S_{c'}(\hat{r}_1)S_{q'}(\hat{r}_2)\frac{f_{c'c_0}^{J^\pi}(r_1)}{r_1}\frac{v_{q'}(r_2,E_{q'})}{r_2}-S_{q'}(\hat{r}_1)S_{c'}(\hat{r}_2)\frac{v_{q'}(r_1,E_{q'})}{r_1}\frac{f_{c'c_0}^{J^\pi}(r_2)}{r_2}\right]\right\}$$

$$-\sum_{c_z}<c|JT>\sum_{q'q}\rho(\beta q,\beta'q')\sum_{\bar{c}'}<c'|JT>\sum_{L_1 L_2}\sum_{I_v}$$

$$\left\{\int_0^\infty\int_0^\infty dr_1 dr_2 r_1^2 r_2^2\sum_{E_{q''}}\delta_{\widetilde{q''p}}\frac{v_{q''}(r_0,E_{q''})}{r_0}\frac{v_{q''}(r_1,E_{q''})}{r_1}\frac{v_q(r_2,E_q)}{r_2}\right.$$

$$\int d^2\Omega_1 d^2\Omega_2 S_{q''}^*(\hat{r}_1)S_q^*(\hat{r}_2)\tilde{V}_{L_1 L_2}^{I_v}(r_1,r_2)S^{I_v}\cdot(Y^{L_1}(\hat{r}_1)\otimes Y^{L_2}(\hat{r}_2))^{I_v}$$

$$\times\left.\left[S_{c'}(\hat{r}_1)S_{q'}(\hat{r}_2)\frac{f_{c'c_0}^{J^\pi}(r_1)}{r_1}\frac{v_{q'}(r_2,E_{q'})}{r_2}-S_{q'}(\hat{r}_1)S_{c'}(\hat{r}_2)\frac{v_{q'}(r_1,E_{q'})}{r_1}\frac{f_{c'c_0}^{J^\pi}(r_2)}{r_2}\right]\right\}$$

where I_v is summed over all interaction potentials, four central ($I_v = 0$) potentials and one tensor ($I_v = 2$) potential, denoted by $\tilde{V}_{L_1 L_2}^{I_v}$, with the attending spherical harmonic tensor products $(Y^{L_1}(\hat{r}_1)\otimes Y^{L_2}(\hat{r}_2))^{I_v}$. The symbol $\mathcal{S}^{I_v}$ represents the spin and isospin operators of equation (42). The integrals over $d^2\Omega_1, d^2\Omega_2$ are also understood to include the integrals over spin and isospin. The symbol $\delta_{\widetilde{q''p}}$ is shorthand for the product of δ functions, $\delta_{s_{q''}s_p}\,\delta_{l_{q''}l_p}\,\delta_{j_{q''}j_p}$ $\delta_{\tau_{q''}\tau_p}$ arising from the orthogonality of the functions $S_p(r_0)$ and $S_{q''}(r_0)$. Simplifying equation (45) by performing all integrations yields

$$0 = H_c(r)f_{cc_0}^{J^\pi}(r)+\sum_{c'}V_{cc'}(r)f_{c'c_0}^{J^\pi}(r)$$

$$+\sum_{c'}\int_0^\infty U_{cc'}(r',r)f_{c'c_0}^{J^\pi}(r')dr'+\sum_{E_q}I_{c,q}v_q(r,E_q), \tag{46}$$

where

$$H_c(r)\equiv\frac{\hbar^2}{2M}\left[\frac{l(l+1)}{r^2}-\frac{d^2}{dr^2}\right]+V_{HF}-E_c, \tag{47}$$

$$V_{c'c}(r) \equiv \sum_{c_z} \sum_{I_v} <c|JT> \sum_{q'q} \rho(\beta q, \beta' q') \sum_{\tilde{c}'} <c'|JT> \sum_{L_1 L_2}$$

$$\times \left\{ \int \int d^2\Omega d^2\Omega' S_{\tilde{p}}^*(\hat{r}) S_q^*(\hat{r}') S^{I_v} \cdot (Y^{L_1}(\hat{r_1}) \otimes Y^{L_2}(\hat{r_2}))^{I_v} S_{c'}(\hat{r}) S_{q'}(\hat{r}') \right\}$$

$$\times \int_0^\infty dr' v_q(r', E_{q'}) \tilde{V}_{L_1 L_2}^{I_v}(r, r') v_{q'}(r', E_{q'}), \tag{48}$$

$$U_{c'c}(r', r) \equiv \sum_{c_z} \sum_{I_v} <c|JT> \sum_{q'q} \rho(\beta q, \beta' q') \sum_{\tilde{c}'} <c'|JT> \sum_{L_1 L_2}$$

$$\times \left\{ \int \int d^2\Omega d^2\Omega' S_{\tilde{p}}^*(\hat{r}) S_q^*(\hat{r}') S^{I_v} \cdot (Y^{L_1}(\hat{r_1}) \otimes Y^{L_2}(\hat{r_2}))^{I_v} S_{q'}(\hat{r}) S_{c'}(\hat{r}') \right\}$$

$$\times v_q(r', E_{q'}) \tilde{V}_{L_1 L_2}^{I_v}(r, r') v_{q'}(r, E_{q'}), \tag{49}$$

and

$$I_{c,q''} \equiv - \sum_{c_z} <c|JT> \sum_{q'q} \rho(\beta q, \beta' q') \sum_{\tilde{c}'} <c'|JT> \sum_{L_1 L_2}$$

$$\times \left\{ \int \int d^2\Omega_1 d^2\Omega_2 S_{q''}^*(\hat{r_1}) S_q^*(\hat{r_2}) S^{I_v} \cdot (Y^{L_1}(\hat{r_1}) \otimes Y^{L_2}(\hat{r_2}))^{I_v} S_{c'}(\hat{r_1}) S_{q'}(\hat{r_2}) \right\}$$

$$\times \int_0^\infty \int_0^\infty dr_1 dr_2 v_{q''}(r_1, E_{q''}) v_q(r_2, E_q) \tilde{V}_{L_1 L_2}^{I_v}(r_1, r_2) f_{c'c_0}^{J^*}(r_1) v_{q'}(r_2, E_{q'})$$

$$+ \sum_{c_z} <c|JT> \sum_{q'q} \rho(\beta q, \beta' q') \sum_{\tilde{c}'} <c'|JT> \sum_{L_1 L_2}$$

$$\times \left\{ \int \int d^2\Omega_1 d^2\Omega_2 S_{q''}^*(\hat{r_1}) S_q^*(\hat{r_2}) S^{I_v} \cdot (Y^{L_1}(\hat{r_1}) \otimes Y^{L_2}(\hat{r_2}))^{I_v} S_{q'}(\hat{r_1}) S_{c'}(\hat{r_2}) \right\}$$

$$\times \int_0^\infty \int_0^\infty dr_1 dr_2 v_{q''}(r_1, E_{q''}) v_q(r_2, E_q) \tilde{V}_{L_1 L_2}^{I_v}(r_1, r_2) v_{q'}(r_1, E_{q'}) f_{c'c_0}^{J^*}(r_2). \tag{50}$$

Hence the angular and spin coordinates have been successfully separated leaving only radial equations that are dependent on the energy.

2.3 Radial Solutions

Equation (46) represents a set of integro-differential equations with an inhomogeneous source term for which there is a complete set of solutions for each set of boundary conditions c_0.

Following previous investigators, the Hartree-Fock potential V_{HF} is replaced by a real, channel dependent Wood-Saxon potential V_{WS} containing a spin-orbit component. This replacement is made both in the solution of the above radial equations and also in the original determination of the bound state wavefunctions $\phi_q(\vec{r})$.

The label $c = 1...N$ enumerates the N distinct physical channels contributing to the scattering process. The first two terms of equation (46),

$$\sum_{c'} [H_c(r)\delta_{c'c} + V_{c'c}(r)]\, h^i_{c'}(r) = 0, \tag{51}$$

represent N second order homogeneous equations having $2N$, N component solutions, of which N solutions will be regular near the origin. Each solution, i, involves a different boundary condition outside the nucleus and corresponds to N different physical scattering situations - N different choices for the elastic channel c_0. To obtain an accurate numerical solution, it is necessary to generate the most general solution by solving the equations N times with N different starting conditions. The obvious choice for the starting condition is to require zero incident flux for all channels but c_0.

Ignoring the non-local term for the moment, the remainder of equation (46) is then a particular differential equation where the last term is the source term for the equation:

$$\sum_{c'} [H_c(r)\delta_{c'c} + V_{c'c}(r)]\, p_{c'}(r) = -\sum_{E_q} I_{c,q} v_q(r, E_q). \tag{52}$$

There is one unique N-component solution to this particular equation. The direct determination of this solution, however, requires the evaluation of the constant $I_{c,q}$ in equation (52), which is possible by iteration but computationally time consuming. Alternatively, M particular equations are introduced

$$\sum_{c'} [H_c(r)\delta_{c'c} + V_{c'c}(r)]\, p^j_{c'}(r) = v_j(r)\delta_{\widetilde{jc}}, \tag{53}$$

where M is the number of E_q bound states in the source term of equation (46). The symbol $\delta_{\widetilde{jc}}$ has been defined above. The strength of each of these M source terms, being related to the constants $I_{c,q}$, are unknown. Writing b_j for these strengths, the unique particular solution $p_{c'}(r)$ of equation (52) is given by a linear combination of the particular solutions $p^j_{c'}(r)$ of equation (53),

$$p_{c'}(r) = \sum_j b_j p^j_{c'}(r). \tag{54}$$

The solutions to equation (46), neglecting the non-local term, are then a sum of the particular solution and a linear combination of the homogeneous solutions

$$f_c(r) = \sum_{i=1}^{N} a_i h^i_c(r) + \sum_{j=1}^{M} b_j p^j_c(r). \tag{55}$$

The coefficients a_i and b_j are determined by matching this general solution to the physical boundary conditions,

$$f_c(R1) = \Phi_c(R1)\delta_{cc_0} + \sqrt{k_{c_0}/k_c}\, S_{cc_0}\Theta_c(R1),$$
$$f_c(R2) = \Phi_c(R2)\delta_{cc_0} + \sqrt{k_{c_0}/k_c}\, S_{cc_0}\Theta_c(R2). \tag{56}$$

The functions $\Phi_c(r)$ and $\Theta_c(r)$ are the incoming and outgoing asymptotic solutions. If $k_c^2 > 0$, they are Coulomb waves, if the incident particle is charged, or Hankel functions, if the incident particle has zero charge. Otherwise, if $k_c^2 < 0$, they are Whittaker functions, which decay exponentially, describing a virtual channel in which flux cannot escape because it is kinematically forbidden. The matrix S_{cc_0} is the scattering matrix or S-matrix.

Because of the introduction of the M unknowns b_j, the $2N$ constraints, equation (56), are insufficient to determine a unique solution. To enumerate, there are N coefficients $\{a_i\}$, N values of the S-matrix for a given incident channel c_0, $\{S_{cc_0}\}$, and M coefficients $\{b_j\}$, for a total of $2N + M$ unknowns. Additional constraints follow from the orthogonality of the scattering wavefunction, which exists in the space spanned by p, and the bound states, which exist in the space spanned by q. The functions $p_c^j(r)$ arose by considering the last term of equation (46) as a source term. This source generates the orthogonality of the scattering wavefunction $f_c(r)$ and the bound states $v_k(r)$, originating from the Pauli projection operator in equation (40). Thus the scattering wavefunctions are said to be Pauli blocked by the bound states as specified by orthogonality $< v_k|f_c >= 0$, where $< v_k|f_c > \equiv \int_0^\infty v_k(r)f_c(r)dr$. This provides M additional equations

$$< v_k|f_c > = \sum_{i=1}^N a_i < v_k|h_c^i > + \sum_{j=1}^M b_j < v_k|p_c^j >= 0. \qquad (57)$$

For convenience, define $H_{kc}^j \equiv < v_k|h_c^i >$ and $P_{kc}^j \equiv < v_k|p_c^j >$. There are then $2N + M$ equations to solve simultaneously for $2N + M$ unknowns,

$$\sum_{i=1}^N a_i h_c^i(R1) + \sum_{j=1}^M b_j p_c^j(R1) = \Phi_c(R1)\delta_{cc_0} + \sqrt{k_{c_0}/k_c}S_{cc_0}\Theta_c(R1)$$

$$\sum_{i=1}^N a_i h_c^i(R2) + \sum_{j=1}^M b_j p_c^j(R2) = \Phi_c(R2)\delta_{cc_0} + \sqrt{k_{c_0}/k_c}S_{cc_0}\Theta_c(R2)$$

$$\sum_{i=1}^N a_i H_{kc}^i + \sum_{j=1}^M b_j P_{kc}^j = 0, \qquad (58)$$

where $c = 1, 2, ..., N$ and $k = 1, 2, ..., M$.

Finally, the non-local interaction term, arising from the anticommutation relations, is incorporated using an iterative process. A source term for the non-local coupling can be defined,

$$\Upsilon_c^{l+1}(r) = -\sum_{c'} \int_0^\infty U_{c'c}(r',r)f_{c'}^l(r')dr', \qquad (59)$$

where $f_{c'}^0(r)$ is the set of solutions, equation (55), determined by equations (58). With this definition, the non-local term previously neglected in equation (46) can be incorporated. As before, this results in a particular equation

$$\sum_{c'} [H_c(r)\delta_{c'c} + V_{c'c}(r)] q^l_{c'}(r) = \Upsilon^l_c(r), \qquad (60)$$

where $q^l_c(r)$ is the particular solution in channel c. The index l specifies that the source term is the l-th approximate source term defined by equation (59) by using the $(l-1)$-th solution. The complete l-th solution is then a sum of the particular solutions and the homogeneous solution,

$$f_c(r) \simeq f^l_c(r) = \sum_{i=1}^{N} a_i h^i_c(r) + \sum_{j=1}^{M} b_j p^j_c(r) + q^l_c(r). \qquad (61)$$

The boundary conditions are suitably modified to include this extra particular solution by writing

$$\sum_{i=1}^{N} a_i h^i_c(R1) + \sum_{j=1}^{M} b_j p^j_c(R1) + q^l_c(R1) = \Phi_c(R1)\delta_{cc_0} + \sqrt{k_{c_0}/k_c} S_{cc_0} \Theta_c(R1)$$

$$\sum_{i=1}^{N} a_i h^i_c(R2) + \sum_{j=1}^{M} b_j p^j_c(R2) + q^l_c(R2) = \Phi_c(R2)\delta_{cc_0} + \sqrt{k_{c_0}/k_c} S_{cc_0} \Theta_c(R2)$$

$$\sum_{i=1}^{N} a_i H^i_{kc} + \sum_{j=1}^{M} b_j P^j_{kc} + <v_k|q^l_c> = 0, \qquad (62)$$

where $c = 1, 2, ..., N$ and $k = 1, 2, ..., M$. This process is repeated until convergence.

The solution of the radial equations in this section, along with the angular and spin components determined in the previous section, completely specifies the scattering nuclear wavefunction. The solution also yields the scattering matrix, which, in principle, can be used to determine cross sections for both elastic and inelastic nuclear reactions.

3   Electromagnetic Interaction

3.1   Transition Amplitude

The electromagnetic transition amplitude $\mathcal{M}$ is written, quite generally, in terms of the nuclear current, $\vec{j}(\vec{r})$ and the final (initial) nuclear wavefunctions $\Psi_{f(i)}$

$$\mathcal{M} = -\frac{1}{c} \int \Psi^\dagger_f(\vec{r}_1, \vec{r}_2, ..., \vec{r}_A) \vec{j}(\vec{r}) \cdot \vec{A}(\vec{r}) \Psi_i(\vec{r}_1, \vec{r}_2, ..., \vec{r}_A) d^3\vec{r} d^3\vec{r}_1 d^3\vec{r}_2...d^3\vec{r}_A. \qquad (63)$$

The final nuclear wavefunction is given by equation (41). The initial nuclear wavefunction, written as $|\Upsilon>$ below, is provided by the same shell model calculation that generates the wavefunctions for the mass $(A-1)$ states.

The electromagnetic potential $\vec{A}(\vec{r})$ is decomposed by the usual multipole expansion,

$$\vec{A}(\vec{r}) = -\sqrt{\frac{2\pi\hbar c}{Vk_\gamma}}\hat{\epsilon}_\lambda e^{i\vec{k}_\gamma\cdot\vec{r}}$$

$$= -\sqrt{\frac{2\pi\hbar c}{Vk_\gamma}}\sqrt{2\pi}\sum_{L\geq 1}i^L\sqrt{2L+1}\sum_\nu D_{\nu\lambda}^L(\vec{R}_\gamma)$$

$$\times\left[\lambda j_L(k_\gamma r)\bar{Y}_{LL1}^\nu(\hat{r}) + \frac{1}{k_\gamma}\nabla\times\left(j_L(k_\gamma r)\bar{Y}_{LL1}^\nu(\hat{r})\right)\right], \qquad (64)$$

using box normalization with volume V, and where k_γ and $\hat{\epsilon}_\lambda$ are the momentum and polarization of the photon, respectively. $D_{\nu\lambda}^L(\vec{R}_\gamma)$ is the rotation matrix, where $\vec{R}_\gamma$ is the angle between the direction of the quantization axis (defined to be the direction of the incident nucleon) and that of the photon. $\bar{Y}_{LJ1}^\nu(\hat{r})$ is the vector spherical harmonic defined by

$$\bar{Y}_{LJ1}^\nu(\theta, \phi) \equiv \sum_{m\lambda} C(J1L, m\lambda\nu)Y_J^m(\theta, \phi)\hat{\epsilon}_\lambda, \qquad (65)$$

and $j_L(k_\gamma r)$ are spherical bessel functions.

Since the nuclear Hamiltonian contains an isospin dependence, and therefore does not commute with the charge density, the nuclear current in principle contains many-body contributions. The form of the nuclear many-body current, of course, depends on the exact form of the nuclear potential V and the requirement of gauge invariance or, equivalently, the continuity equation $\nabla\cdot\vec{j} + \frac{i}{\hbar}[H, \rho] = 0$. The nuclear current is assumed to be a sum of one-body and two-body parts. In second quantized notation, the one- and two-body current density operators can be written as

$$\hat{j}^1(\vec{r}) = \sum_{ij} <i|\vec{j}(\vec{r})|j> a_i^\dagger a_j \qquad (66)$$

$$\hat{j}^2(\vec{r}) = \frac{1}{2}\sum_{ijkl} <ij|\vec{j}(\vec{r})|kl> a_i^\dagger a_j^\dagger a_l a_k. \qquad (67)$$

The electromagnetic transition amplitude, for an arbitrary nuclear one-body current and photon polarization λ is

$$\mathcal{M}_{1-body}^\lambda = -\frac{1}{c}\sqrt{\frac{2\pi\hbar c}{Vk_\gamma}}\sqrt{2\pi}\sum_{cq} <\beta|a_q|\Upsilon> \sum_{J\geq 1}i^J\sqrt{2J+1}\sum_\nu D_{\nu\lambda}^J(\vec{R}_\gamma)$$

$$\times\left\{i\sqrt{\frac{J+1}{2J+1}}\delta_{J,L-1} - i\sqrt{\frac{J}{2J+1}}\delta_{J,L+1} + \lambda\delta_{J,L}\right\}$$

$$\times\int d\vec{r} <c\left|\vec{j}(\vec{r})\cdot\left[j_L(kr)\bar{Y}_{JL1}^\nu(\hat{r})\right]\right|q>. \qquad (68)$$

Similarly, for an arbitrary nuclear two-body current and photon polarization, the electromagnetic transition amplitude is

$$\mathcal{M}_{2-body}^{\lambda} = +\frac{1}{c}\sqrt{\frac{2\pi\hbar c}{V k_{\gamma}}}\sqrt{2\pi}\sum_{cq} < \beta|a_q|\Upsilon > \sum_{J\geq 1} i^J\sqrt{2J+1}\sum_{\nu} D_{\nu\lambda}^J(\vec{R}_{\gamma})$$

$$\times \left\{ i\sqrt{\frac{J+1}{2J+1}}\delta_{J,L-1} - i\sqrt{\frac{J}{2J+1}}\delta_{J,L+1} + \lambda\delta_{J,L} \right\}$$

$$\times \sum_{q'} \int d\vec{r} < cq'|\vec{j}(\vec{r}) \cdot [j_L(kr)\bar{Y}_{JL1}^{\nu}(\hat{r})] | \widetilde{qq'} > . \tag{69}$$

In these expressions, the quantity $< \beta|a_q|\Upsilon >$ is the spectroscopic amplitude obtained from the same shell model calculation which provides the density matrix in the previous section.

The one-body current contains the usual convective and magnetic currents, $\vec{j}(\vec{r}) = \vec{j}_c(\vec{r}) + \vec{j}_m(\vec{r})$, with

$$\vec{j}_c(\vec{r}) = \sum_q \frac{e}{2m} [1 + \tau_3(q)]\frac{1}{2}\left[\vec{P}_q\delta(\vec{r} - \vec{r}_q) + \delta(\vec{r} - \vec{r}_q)\vec{P}_q\right] \tag{70}$$

and

$$\vec{j}_m(\vec{r}) = \frac{e\hbar}{4m}\sum_q \left[\frac{1}{2}(g_p + g_n) + \tau_3(q)(g_p - g_n)\right]\nabla \times \vec{\sigma}\delta(\vec{r} - \vec{r}_q), \tag{71}$$

where m is the mass of the proton and $g_p(g_n)$ is the g-factor of the proton(neutron). Although the form of the two-body current is in principle determined by the nucleon-nucleon potential, in the present study the two-body current is restricted to that arising from the one pion exchange potential. As this potential gives rise to the long range part of the true nucleon-nucleon potential, this restriction should be a reasonable approximation to the full meson exchange current. Non-relativistically the pion exchange current is a sum of the Kroll-Ruderman and pion pole terms, which are given by

$$\vec{j}^{KR}(\vec{r}; \vec{r}_1, \vec{r}_2) = -e\left(\frac{f^2}{4\pi}\right)(\vec{\tau}_1 \times \vec{\tau}_2)_3$$

$$\times [\vec{\sigma}_1\delta(\vec{r}_1 - \vec{r})(\vec{\sigma}_2 \cdot \hat{r}) + \vec{\sigma}_2\delta(\vec{r}_2 - \vec{r})(\vec{\sigma}_1 \cdot \hat{r})]\left(1 + \frac{1}{m_\pi r}\right)\frac{e^{-m_\pi r}}{m_\pi r} \tag{72}$$

and

$$\vec{j}^{pole}(\vec{r}; \vec{r}_1, \vec{r}_2) = -e\left(\frac{f^2}{16\pi^2}\right)(\vec{\tau}_1 \times \vec{\tau}_2)_3$$

$$\times(\nabla_1 - \nabla_2)(\vec{\sigma}_1 \cdot \nabla_1)(\vec{\sigma}_2 \cdot \nabla_2)\frac{e^{-m_\pi|\vec{r}_1 - \vec{r}|}}{m_\pi|\vec{r} - \vec{r}_1|}\frac{e^{-m_\pi|\vec{r}_2 - \vec{r}|}}{m_\pi|\vec{r} - \vec{r}_2|}, \tag{73}$$

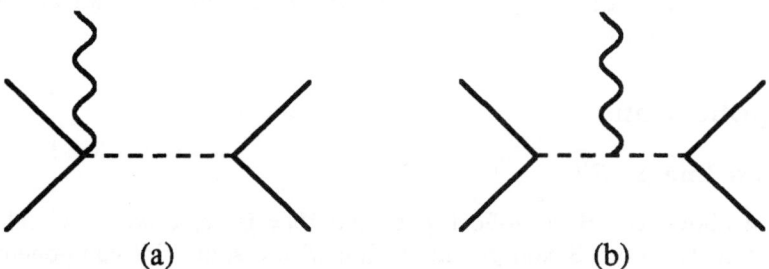

Fig. 1. Feynman diagrams for the (a) Kroll-Ruderman and (b) pion pole currents.

where $\vec{r} \equiv \vec{r}_1 - \vec{r}_2$ is the point at which the virtual pion couples to the photon. These two currents are diagrammed in figure 1.

There can exist additional many-body currents that are longitudinal in nature and therefore are not constrained by the requirement of gauge invariance. These currents are model dependent. The most important exchange current of this type is expected to be the current associated with the excitation of the $\Delta(1232)$ resonance. The excitation of the Δ resonance by the exchange of a meson followed by the interaction with the electromagnetic field is certainly two-body in nature. However, by expanding the model space in the coupled channels formalism to include Δ degrees of freedom, Δ isobars can be incorporated explicitly into the nuclear scattering wavefunction. Because the Δ has spin $\frac{3}{2}$, the number of coupled channels is increased by a factor of three. The electromagnetic interaction with the Δ is then a one-body radiative transition to a nucleon. Further, the Δ is then treated in a way identical to that of the nucleons, allowing for the possibility of multi-scattering of the Δ resonance within the nucleus. The different treatments are illustrated schematically in figure 2. The form of the

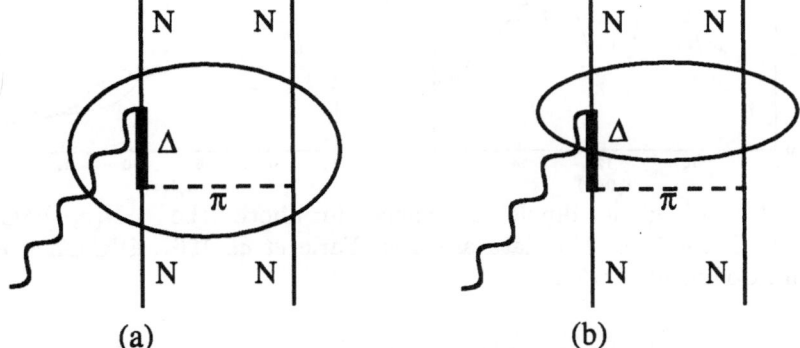

Fig. 2. Δ isobar exchange current treatment: (a) two-body elementary exchange current; (b) one-body electromagnetic current with Δ isobar components allowed in the nuclear wavefunction.

$NN \rightarrow N\Delta$ and $N\Delta \rightarrow N\Delta$ interactions are taken from (Niephaus *et al.* 1979). The details of how this is applied to the present formulation can be found in (Bright and Cotanch 1993).

4   Applications

4.1   Low Energy Fit

As noted above, the Hartree-Fock potential V_{HF} is replaced by a real, channel dependent, Woods-Saxon potential V_{WS} with a spin-orbit component. This potential then determines both the bound and continuum single-particle states. The bound state single-particle energies have been taken from (Birkholz 1972). The potential depth for the continuum single-particle states for the first few partial waves are set to be the same as that for the bound states. The potential depths for the higher partial waves decrease slowly to 40 MeV. The parameters of the effective interaction, which are given in table 1, are determined by a fit to

a_o	a_σ	a_τ	$a_{\sigma\tau}$	V_c MeV	V_t MeV	μ_c fm^{-1}	μ_t fm^{-1}	Λ fm^{-1}
0.0125	0.25	0.0425	0.10	62.0	5.1	0.73	6.0	0.714

Table 1. Residual interaction parameters.

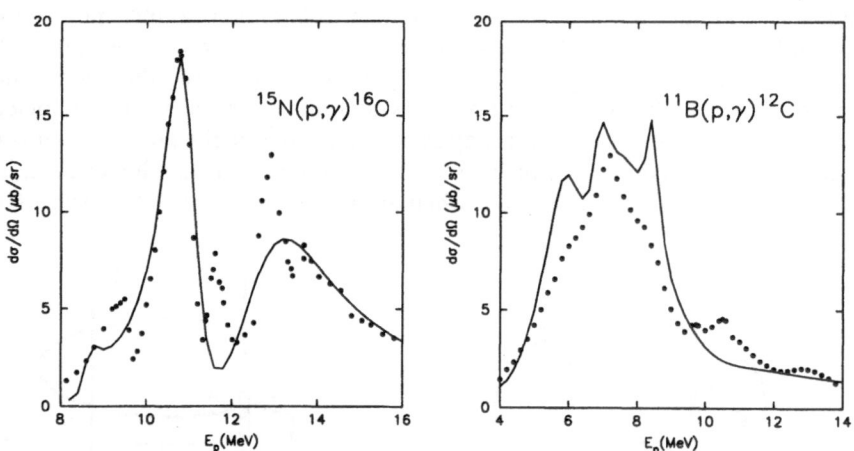

Fig. 3. Fit of giant dipole resonance for both the ^{15}N$(p, \gamma)^{16}$O and ^{11}B$(p, \gamma)^{12}$C reactions. The data are from Earle *et al.* 1967, O'Connell *et al.* 1978 and Collins *et al.* 1982.

the low energy photonuclear cross section energy spectra shown in figure 3. Our effective interaction strengths and ranges are consistent with other phenomenological forces used in this mass region. Our CSM approach provides a reasonable description of the giant dipole cross sections for both ^{16}O and ^{12}C. Having specified all model parameters, we now apply our model to medium energy reactions and compare predictions with available data.

4.2   The ^{11}B$(p, \gamma)^{12}$C Reaction

To calculate the nuclear scattering wavefunctions $\Psi_{cc_0}^{J^\pi}(\vec{r})$, we truncate the model space for the mass 11 system to include the low lying states of the physical ^{11}B and ^{11}C nuclei. In particular, we restrict the space to include only the negative parity states $J^\pi = \left(\frac{3}{2}\right)_1^- \left(\frac{1}{2}\right)_1^- \left(\frac{5}{2}\right)_1^- \left(\frac{3}{2}\right)_2^- \left(\frac{7}{2}\right)_1^-$. We include all nucleon and Δ channels that are consistent with angular momentum conservation, leading to a maximum of 144 channels for all but the first few multipole states $|J^\pi >$.

Recent measurements by (Bright et al. 1996a) have resolved transitions to discrete excited states of ^{12}C and we therefore consider not only the ground state of ^{12}C, but also the 2^+ and 0^+ states of ^{12}C at 4.4 and 7.7 MeV excitation energy, respectively. The data for the ^{11}B$(p, \gamma)^{12}$C reaction is compared to our calculation in figure 4. In this figure, the solid curve will be referred to as our baseline calculation. The baseline calculation is the complete coupled channels calculation outlined above but includes only the one-body currents and does not include Δ degrees of freedom. The dashed line is the calculation in which the pion exchange current, the two-body currents of equations 72 and 73, are added to the baseline calculation. The dotted line shows the effect of including only Δ degrees of freedom with the baseline calculation. Note that the contributions from the two-body currents are relatively small and thus most of the electromagnetic interaction takes place via the one-body currents. Indeed, our calculations demonstrate that in this reaction the pion exchange current makes a vanishingly small contribution, while the effects due to Δ exchange become appreciable only at the highest momentum transfers investigated. Δ effects are expected to increase as the energy is increased approaching the Δ threshold, as will be evident in the next section for the reactions ^{16}O$(\gamma, p)^{15}$N and ^{16}O$(\gamma, n)^{15}$O.

In general, effects from both pion and Δ exchange currents tend to increase with increasing momentum transfer, since a many-body exchange current mechanism facilitates additional momentum sharing between the nucleons. However, our CSM calculation, which includes nuclear correlation effects and exchange currents, shows that the one-body electromagnetic mechanism is the most important contribution to this reaction, in contrast with other recent calculations (Mori et al. 1995, Benenti et al. 1994). The importance of the one-body mechanism is illustrated in figure 5 where the overall magnitude and shape of the angular distributions are predominantly described through the direct knock-out reaction mechanism, with corrections arising through the two-body residual interaction. The direct knock-out reaction prediction (dashed line) shown in this figure is obtained from the baseline calculation by setting the residual interaction to zero. In accordance with the above momentum sharing argument, however, this simple direct knock-out interpretation becomes less reliable at higher momentum transfers. The direct knock-out and residual interaction reaction mechanisms are

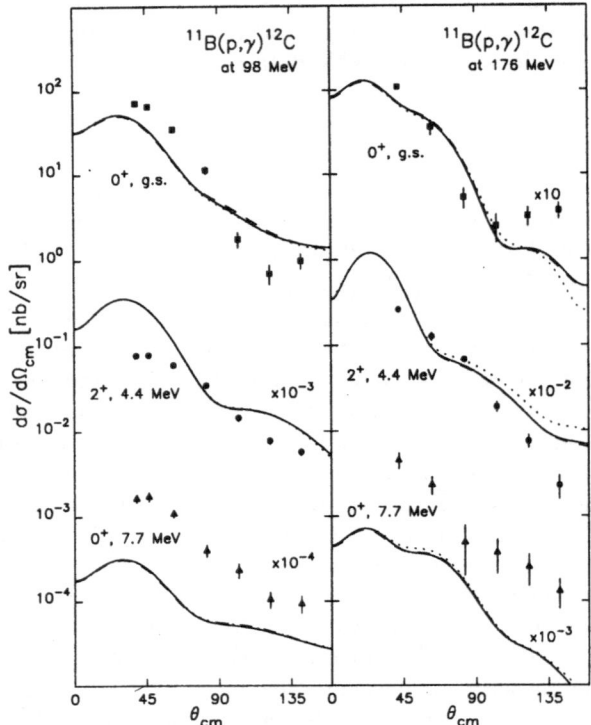

Fig. 4. Angular distribution of the differential cross section for transitions to the ground state, 4.4 MeV and 7.7 MeV state in ^{12}C. The errors shown are statistical. The curves show the results from the continuum shell model calculations. Solid line - baseline calculation. Dashed line - includes pion exchange currents. Dotted line - includes Δ degrees of freedom. The data and calculations have been scaled by the scaling factors given in the figure. The data are from Bright *et al.* 1996a and Höistad *et al.* 1992.

illustrated schematically in figures 6(a) and 6(b), where the exchanged meson represents the two-body residual interaction and the electromagnetic interaction is strictly one-body in nature. For completeness, a schematic representation of the residual interaction reaction mechanism along with a two-body electromagnetic interaction is illustrated in figure 6(c) for the pion pole current. A similar diagram exists for the Kroll-Ruderman pion current.

Our calculation is sensitive to the choice of the interaction used to describe the nuclear structure of the mass 11 and mass 12 systems and to the choice for the single-particle potentials, as figure 7 shows. In this figure, the various curves represent different sets of single-particle potentials or different interactions used in the OXBASH (Brown *et al.* 1994) shell model code which has been used to

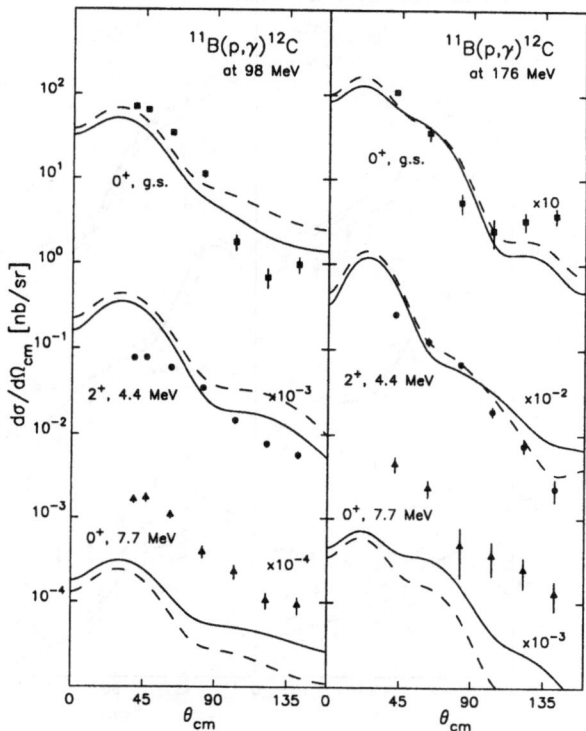

Fig. 5. Study of reaction mechanism in the ^{11}B$(p,\gamma)^{12}$C reaction. The curves show the results from the continuum shell model calculations. Solid line - baseline calculation. Dashed line - direct knock-out calculation. The data is the same as in figure 4.

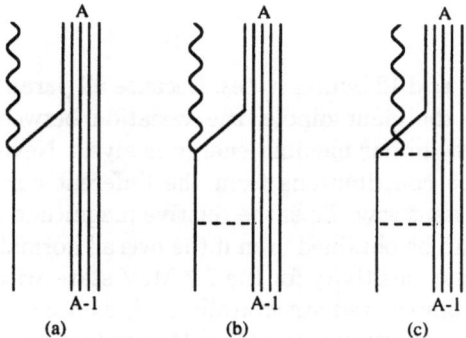

Fig. 6. (a) Direct reaction mechanism; (b) residual interaction reaction mechanism with one-body electromagnetic current; (c) residual interaction reaction mechanism with two-body pion-pole electromagnetic current.

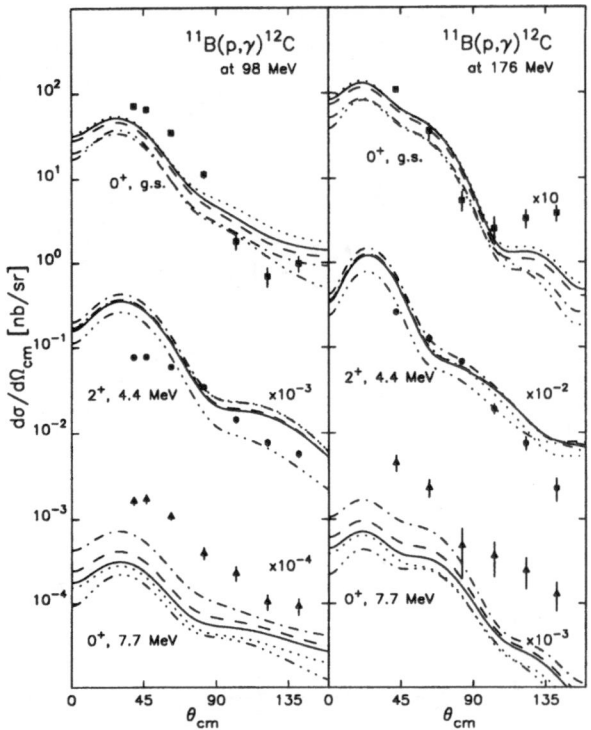

Fig. 7. Sensitivity to nuclear structure calculation and single-particle potentials. The solid line is the baseline calculation of figure 4, which uses the OXBASH interaction MP. The dashed, dotted, and dash-dot curves are obtained by using OXBASH interactions CKPOT, PEWT and SPSDVH, respectively. The remaining curve is obtained by increasing the single-particle potentials by 20%. The data is the same as in figure 4.

generate the mass 11 and 12 bound states. Because all parameterizations provide an equivalent fit to the giant dipole, the variation between curves represents the model uncertainty in our medium energy analysis. Nevertheless, for a given set of potentials, the contributions from the different current sources can be compared in a consistent way. Thus the relative magnitude of the one-body and two-body currents can be obtained even if the overall normalization is uncertain. Moreover, the extreme sensitivity for the 7.7 MeV state arises because this state is apparently very complicated structurally and, as a consequence, is not part of the data base that determines the OXBASH potentials. As a result, a much more extensive model space than used in this study would be needed to describe adequately the transition to this state. Conversely, the present data would guide in the construction of a proper wavefunction for this state.

4.3 The ^{12}C$(p, \gamma)^{13}$N Reaction

Although the dominant reaction mechanism for this reaction, a 1p-0h process, is simpler than the corresponding mechanism in the previous reaction, a 1p-1h process, the description of ^{12}C$(p, \gamma)^{13}$N is complicated by the uncertainty in the nuclear structure.

In our analysis we restrict the mass 12 states to the ground, 4.4 and 7.7 MeV states of ^{12}C. This choice precludes any effects due to exchange currents so that there is only the conventional calculation. The final states in ^{13}N considered are the $\frac{1}{2}^-$ ground state, the $\frac{1}{2}^+$ at 2.4 MeV and the $\frac{3}{2}^-$ and $\frac{5}{2}^+$ states at 3.5 MeV that are not resolved experimentally. Our calculation is compared to recent data from (Bright *et al.* 1996b) in figure 8. The solid line uses the OXBASH

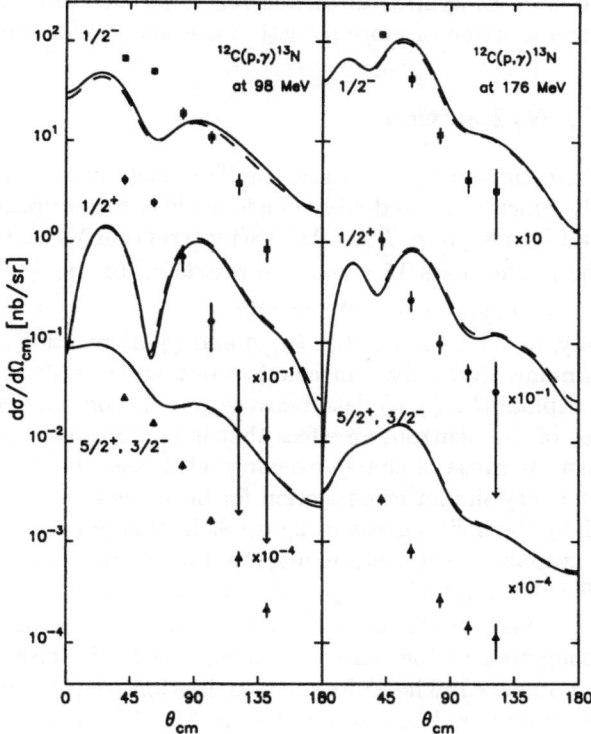

Fig. 8. Angular distribution of the differential cross section for transitions to the ground state, 2.4 MeV and 3.5 MeV state in ^{13}N. The errors shown are statistical. The curves show the results from the continuum shell model calculations. Solid line - baseline calculation. Dashed line - direct knockout calculation. The data are from Bright *et al.* 1996b.

interaction SPSDVH including nucleons in the sd shell. The dashed line is a direct knockout calculation, obtained from the full calculation by setting the two-body residual interaction strength to zero. Not unexpectedly, this reaction is dominated by a direct knockout process. Our reaction description to the ground state is reasonable and similar in quality to the description of $^{11}B(p, \gamma)^{12}C$. The reaction to the excited states is not described well, however, perhaps because the shell model description of the positive parity states in terms of the sd shell or the single-particle representation of the sd shell energy levels is inadequate. Although a shell model description of the $\frac{3}{2}^-$ final state does not require sd shell components, it does entail a more complex 2p-1h reaction mechanism and is therefore strongly suppressed compared to the 1p-0h leading reaction mechanism for the reaction to the other excited states.

The ground state of ^{13}N, as well as the states discussed in the previous section, except as noted therein, are well described by the shell model using only the 1p shell. The CSM as applied to carbon reactions therefore provides useful insight into the construction of more realistic wavefunction for the sd shell.

4.4   The $^{16}O(\gamma, N)$ Reaction

Of all reactions investigated in this study, the ^{16}O reactions are the best understood structurally. Since ^{16}O is a doubly magic nucleus a description in terms of the 1p shell should be adequate. The OXBASH interaction MP is therefore used for this case. The model mass 15 states are restricted to the $\frac{1}{2}^-$ ground state and the $\frac{3}{2}^-$ first excited states of ^{15}N and ^{15}O.

Experimentally, it is found that the (γ, p) and (γ, n) reaction cross sections are remarkably similar. Certainly, a mechanism restricted to direct knock-out is not sufficient to explain the (γ, n) data because the photon only couples to the magnetic moment of the neutron, an effect that is far too weak. In the coupled channels formalism, because of charge-exchange channels, the conventional calculation predicts a very similar cross section for both the (γ, n) and (γ, p) data, as demonstrated by the solid curves in figure 9. In this figure the angular distribution of the available (γ, n) data is displayed along with the corresponding (γ, p) data. As for the mass 12 system, effects due to pion exchange currents (dashed curve), although larger, are still very small. However, Δ isobar effects have increased compared to the mass 12 system due to the larger momentum and higher energy involved. The Δ isobar contribution (dotted curve) makes a correction to the conventional calculation that agrees with the overall magnitude and shape of the measured angular distribution. The dash-dot curve represents the full calculation.

Assuming that the same reaction mechanism should explain both (γ, p) and (γ, n) reactions, we find that the process represented in figure 6(b), the residual interaction reaction mechanism with a one-body electromagnetic current, describes both (γ, n) and (γ, p) data reasonably well. For the (γ, p) reaction, the electromagnetic amplitude is governed predominantly by the one-body convec-

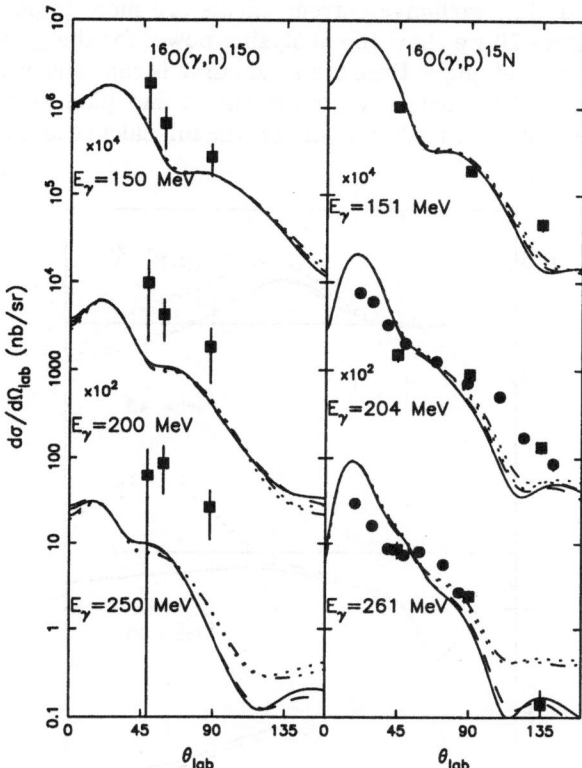

Fig. 9. Angular distributions for the reactions $^{16}O(\gamma, n)^{15}O$ and $^{16}O(\gamma, p)^{15}N$ at E_γ=150, 200 and 250 MeV. The curves show the results from the continuum shell model calculations. Solid line - baseline calculation. Dotted line - Δ + baseline calculation. Dashed line - exchange + conventional calculation. Dash-dot - full calculation. The data are from Beise *et al.* 1989, Leitch *et al.* 1985 and Adams *et al.* 1977. The data and calculations have been scaled by the scaling factors given in the figure.

tive current. This is also true for the (γ, n) reaction where the one-body convective current now enters through the exchange of a charged pion between the ejected nucleon and the residual nucleus. This charged meson exchange, of course, also determines the two-body current. Nevertheless, we find that by performing a consistent microscopic calculation and including all the non-localities that arise from antisymmetrization, we are able to describe quantitatively the photonuclear cross sections for both neutron and proton by means of one-body electromagnetic currents alone and that the effects from two-body exchange currents are small in this reaction.

We have found that exchange current effects are more important for spin observables. In figure 10 we show the analyzing power for the (p, γ) reaction as a function of energy and angle. Here the solid curve is the conventional baseline result, the dotted and dashed curves represent Δ and pion exchange current effects, respectively, and the dash-dot curve is the full calculation. For increasing

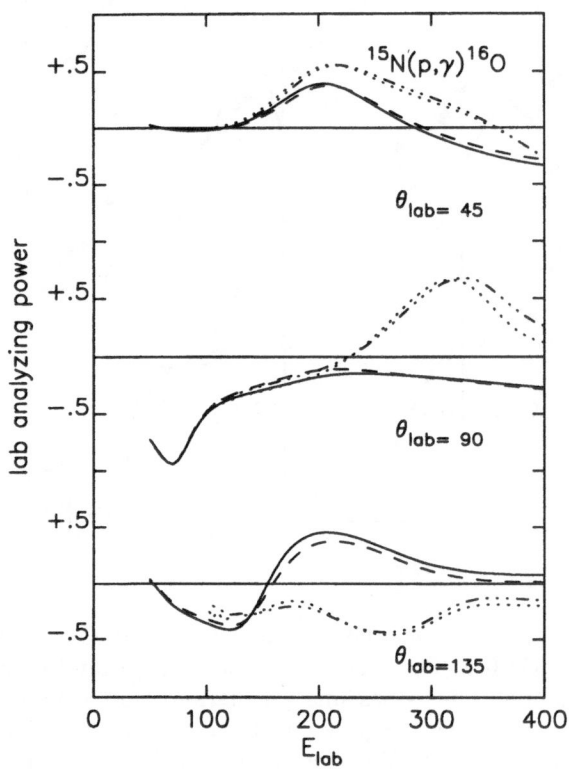

Fig. 10. Analyzing power for the reaction $^{15}N(p, \gamma)^{16}O$ as a function of energy and angle. Solid line - conventional calculation. Dashed line - conventional + pion exchange calculation. Dotted - conventional calculation + Δ. Dash-dot - full calculation.

momentum transfer, there is a sizable effect from the Δ isobars, and, to a lesser extent, pion exchange currents. We conclude that spin observable measurements offer an ideal method for investigating exchange currents.

5   Summary and Outlook

In this chapter we have reviewed the key elements of the CSM and documented its utility for a few selected applications. Although we have only focused on photonuclear reactions for light nuclei, we have nevertheless conducted a comprehensive, large-scale calculation to assess consistently the important theoretical ingredients. Perhaps surprising to some, we find that conventional physics is most significant: a reasonably large model space, realistic interactions, antisymmetrization, nucleon correlations, neutron and proton channels and gauge invariance but only at the one-body current level. With the exception of spin observables, effects from Δ isobars and two-body pion exchange currents are in general modest except at high momentum transfer.

While our CSM application provides a qualitative, semi-quantitative description of the data, detailed agreement is still lacking. We contend this deficiency stems primarily from the treatment of the $(A - 1)$ wavefunctions and the effective interaction, especially at small distances. The treatment could be improved by increasing the number of $(A - 1)$ excited basis states as well as by utilizing more sophisticated, multi-particle, multi-hole wavefunctions. Further, an improved effective interaction having a soft-core that is consistent with the basis wavefunctions should be implemented. Although increasing the number of basis states increases the number of coupled channels, these improvements could be computationally achieved by treating Δ isobar excitation as an exchange current, because the effect is not large. This would reduce the number of channels by 2/3 because there are twice as many Δ as nucleon channels.

Other future developments should include extending the CSM to $(e, e'N)$ and (N, e^+e^-) relevant to future measurements at CEBAF and present experiments at Uppsala, respectively. These reactions, involving space and time-like virtual photons, would provide important, complementary structure information through the longitudinal current matrix elements. Also to be investigated are polarization and analyzing power measurements, especially to complete the assessment of Δ and meson exchange current effects. Related, though more challenging for the CSM, studies of (γ, NN) and $(e, e'NN)$ should be initiated. Exchange currents effects are expected to be larger in these reactions because the two nucleons can more effectively share the momentum transfer which, due to the mathematical structure of the exchange current, enhances exchange current matrix elements.

In summary, the CSM approach is an attractive, powerful framework for investigating a wide variety of nucleon-nucleus phenomena. It is particularly well suited for photonuclear reactions involving single-nucleon capture or knockout and it permits controllable approximations that are amendable to systematic improvement from future computational advances. This is indeed fortunate since it appears that solving the conventional many-body problem as accurately as possible still remains a fundamental goal– just as it was 50 years ago.

References

Adams, G. S., Kinney, E. R., Matthews, J. L., Sapp, W. W., Soos, T., Owens, R. O., Turley, R. S. and Pignault, G. (1977): Nucl. Phys. **A279**, 385

Barz, H. W., Rotter, J. and Hohn, J. (1977): Nucl. Phys. **A275**, 111

Beise, E. J., Dodson, G., Garcon, M., Hoibraten, S., Maher, C., Pham, L.D., Redwine, R. P., Sapp, W., Wilson, K.E., Wood, S. A. and Deady, M. (1989): Phys. Rev. Lett. **62**, 2593

Benenti, G., Giusti, C. and Pacati, F. D. (1994): Nucl. Phys. **A574**, 716

Birkholz, J. (1972): Nucl. Phys. **A189**, 385

Bloch, C. ed., *Many-Body Description of Nuclear Structure and Reactions*, (1966), Academic Press

Bohr, DaAage and Mottelson, Ben. R, *Nuclear Structure*, Vol. I, (1969), W.A. Benjamin, Inc., appendix 3B

Bright, T. B. and Cotanch, S. R. (1993): Phys. Rev. Lett. **17**, 2563

Bright, T. B., Höistad B., Johansson, R., Traneus, E. and Cotanch, S. R. (1996): Nucl. Phys. **A**, in press

Bright, T. B., Höistad B., Johansson, R., Traneus, E. and Cotanch, S. R. (1996): Nucl. Phys. **A**, in preparation

Brown, B. A., Etchegoyen, A., Rae, W. D. M. and Godwin, N. S. (1984): MSUCL Report No. 524

Buck B. and Hill, A. D. (1967): Nucl. Phys. **A95**, 271

Cavinato, M., Marangoni, M., Ottaviani, P. L. and Saruis, A. M. (1982): Nucl. Phys. **A373**, 445

Collins, M. T. *et al.* (1982): Phys. Rev. **C26**, 332

Earle, E. D. and Tanner, N. W. (1967): Nucl. Phys. **A95** 241

Hohn, J., Kayser, J., Pilz, W., Schmidt, D. and Seeliger, D. (1981): J. Physics **G7**, 803

Höistad B., Nilsson, E., Thun, J., Dahlgren, S., Isaksson, S., Adams, G. S., Landberg, C., Bright, T. B. and Cotanch, S. R. (1992): Phys. Lett. **B276**, 294

Leitch, M. J., Matthews, J. L., Sapp, W. W., Sargent, C. P., Wood, S. A., Findlay, D. J. S. and Owens, R. O. (1985): Phys. Rev. **C31**, 1633

Ludeking, L. D. and Cotanch, S. R. (1986): A.I.P. Conference Proceedings **150**, 542

Mahaux, Claude and Weidenmuller, Hans A., *Shell-Model Approach to Nuclear Reactions*, (1969), North-Holland Publishing Company

Marangoni, M. and Saruis, A. M. (1969): Nucl. Phys. **A132**, 649

Messiah, Albert(1965): *Quantum Mechanics*, Vol. II, (1965), North-Holland Publishing Company

Mori, K., Harty, P. D., Fujii, Y., Konno,O., Maeda,K., Nomura, I., O'Keefe, G. J., Ryckebusch, J., Suda, T., Terasawa, T., Thompson, M. N. and Torizuka, Y. (1995): Phys. Rev. **C51**, 2611

Niephaus, G.-H., Gari, M. and Sommer, B. (1979): Phys. Rev. **C20**, 1096

O'Connell, W. J. and Hanna, S. S. (1978): Phys. Rev. **C17**, 892

Ramavataram, S., Ramavataram, K. and Do Dang, G. (1981): Prog. Theor. Phys. **65**, 1928

Raynal, J., Melkanoff, M. A. and Sawada, T. (1967): Nucl. Phys. **A101**, 369

Wong, C. and Anderson, J. D. (1971): Phys. Rev. **C3**, 1904

Kerman–Klein Method for Nuclear Structure: Accomplishments and Opportunities

Abraham Klein and Pavlos Protopapas

Physics Dept., University of Pennsylvania,
Philadelphia, PA 19104-6396

Abstract. The Kerman–Klein method was invented in 1962 to restore the broken symmetry of mean field solutions of the nuclear many body problem. It was seen to contain a natural mechanism, within a shell model framework, for decoupling collective modes from non–collective ones and for studying the interplay between collective and individual degrees of freedom. We present a general formulation of the theory in terms of single particle coefficients of fractional parentage. Some results of current applications of a semi–microscopic version of this theory are described. We contrast several fully microscopic formulations of the theory and summarize early applications of these to the study of nuclear shape changes and to the properties of semi–magic nuclei. Other applications such as an ongoing study of a nuclear version of the Born-Oppenheimer approximation and to quantum field theories with soliton solutions are discussed briefly.

1 Introduction

If by the shell model one means the diagonalization of large Hamiltonian matrices in a suitably chosen basis, then the present chapter does not belong in this volume. If one is willing to broaden one's definition to include methods that elucidate (mainly) low energy nuclear structure within the framework of the shell model, we may proceed without misgivings. The importance of these methods is that they can be applied successfully to problems not yet accessible by standard diagonalization procedures and that they provide the shell–model basis of successful phenomenological models.

One of the salient aspects of nuclear structure is the existence of collective modes, which, at low energy especially, are to a good approximation decoupled from the single particle motion. Historically there have been two principal methods for understanding these phenomena. In the first of these, it is assumed that the shell model Hamiltonian possesses a continuous symmetry higher than rotational invariance, in which the latter is embedded. (The father of such approaches is the SU(3) symmetry postulated by Elliott 1958a, Elliott 1958b.) It implies that to lowest approximation the Hamiltonian can be written in terms of generators of the associated algebra and that the eigenstates will belong to a single irreducible representation. In practice, however, the symmetry will be broken and representations will be mixed. Where the mixing is not great, there still results a technical simplification and an increase in understanding. In addition to the SU(3) scheme applicable to the first half of the s-d shell, other well-known schemes are the generalized seniority (broken pair) approximation applicable to semi-magic nuclei such as Ni and Sn (Allaart *et al* 1988)*, the pseudo-SU(3) scheme devised for heavy deformed nuclei (Draayer 1992), and the Fermion Dynamical Symmetry model (FDSM) (Wu *et al* 1994), which claims a wider range of applicability than the other models. The connection of FDSM with basic elements of the shell model is less apparent than for the other symmetries, but it has the intriguing property that it can be viewed as a Fermion generalization of the phenomenological Interacting Boson Model.

The second approach starts with the construction or assumption of a mean field as the average field that determines the local single particle structure and has associated with it a shape, usually spherical or axially deformed. Collective excitations of this substratum have been studied with the help of the time–dependent Hartree–Fock or Hartree–Fock–Bogoliubov equations **. A special case of these equations is the Random Phase approximation for vibrational motion, which is generally recognized as the basic method for the identification of (vibrational) collective modes. As opposed to the method described in the previous paragraph, these results emerge without any special assumptions about higher symmetries of the shell model Hamiltonian. Instead, it has been emphasized that the nature of the results, i. e., the special spin-isospin structure of a given state, can be tied to the existence of a large overlap of the effective nuclear interaction with a related separable interaction (quadrupole-quadrupole, etc.). The validity of the representation of a realistic shell model interaction in terms of a small number of separable interactions has been reaffirmed most recently in the work of Zucker and associates reported in this volume.

There are two serious deficiencies of the methods alluded to in the previous paragraph. One is the violation of fundamental symmetry properties of the nuclear Hamiltonian. The spherical mean field violates translational invariance, the axially deformed field rotational invariance, and the mean field treatment of pairing number conservation. Two methods have been used in nuclear physics to obtain approximate solutions that satisfy the required symmetries. Most widely used has been the method of generator coordinates. (The most recent applications of this method have been discussed by Heenen *et al* (1992) and Berger *et al* (1992).) The second method used is that which forms the subject matter of this chapter. In connection with this problem, one often speaks of the restoration of broken symmetries. This characterization is relevant to the method of generator coordinates that explicitly acts on symmetry violating solutions and by suitable linear combination turns then into symmetry preserving solutions. However this characterization is not relevant to our method, which is constructed *ab initio* as a symmetry preserving formulation. Afterwards one notes what approximations will lead back to the standard mean field solutions.

The second deficiency is that the RPA, because of its rudimentary character, allows two different interpretations. One is that the so–called vibration is a

---

* See also the article by Covello and associates, this volume.
** For a textbook account see Ring and Schuck (1980).

single approximate shell model state with "collective" properties relative to the ground state, but this is hardly a prediction of a vibrational spectrum. A second is that it is a quasi-boson approximation, and a completely harmonic spectrum is "predicted". The first interpretation arises because the RPA can be derived as a lowest approximation shell model calculation, albeit one formulated in a basis redundant by a factor of two and carried out by an equation of motion approach. The doubling of the basis results in an improved treatment of the ground state compared to the mean field approximation (ground state correlations). This interpretation has led to a higher order generalization of a redundant basis method evaluated by an equation of motion technique (Rowe 1968, Rowe 1970). In this method, there is no concept of boson, the calculations are more involved at every stage than a corresponding shell model diagonalization (doubling of the space) and in any event, the results have to be analyzed further for evidence of multiple excitation, as for the usual shell model. The second interpretation suggests that one map the shell model Hamiltonian onto a boson space whose elements are determined by the solutions of the RPA with the avowed purpose of ending with an anharmonic vibrational description (Ring and Schuck 1980, Klein and Marshalek 1991). In application to the problem of extending the RPA, our method is closer to the second point of view, but avoids its extremes by not introducing bosons.

Concerning the contents of this work, in Sec. 2 we describe our method in general terms. In Sec. 3 we present a theory of generalized single particle coefficients of fractional parentage (CFP) that provides a complete and general formulation, though, as seen later, not a unique one. A semi-microscopic approximation is applied in Sec. 4 to the study of rotational spectra of odd nuclei. In Sec. 5 we return to the fully microscopic theory and argue that there are advantages to replacing the CFP version by a density matrix version, more appropriate to the study of the properties of even nuclei. Early applications of both methods are discussed briefly. Section 6 focuses particularly on decoupling procedures for vibrational spectra with reference to an application to Ni. Finally, because of space limitations, we can only mention in Sec. 7 other applications to nuclear physics, especially to large amplitude collective motion, and to other fields, such as to field theories with soliton solutions.

2   Fundamental Ideas

We begin with a succinct exposition of the elements of the method in order to emphasize that it is a rather general technique for solving problems in quantum mechanics, either for particles or for fields. This was not fully apparent from the earliest publications (Kerman and Klein 1962, Kerman and Klein 1963, Kerman and Klein 1965, Celenza et al 1965), which were concerned with the nuclear many body problem, which remains among the most challenging of the problems studied.***

*** For reviews see Klein 1983, Klein and Walet 1992.

For applications based on the nuclear shell model, we might naturally start (but see below) with the anticommutation relations of a set of shell model Fermion annihilation and creation operators, a_α and $a_\beta^\dagger$, respectively, where

$$a_\alpha a_\beta^\dagger + a_\beta^\dagger a_\alpha = \delta_{\alpha\beta} \ . \tag{1}$$

All other anticommuatators vanish, and the Hamiltonian is a number conserving polynomial in these operators. (A specific form can be found in Sec. 4.)

The single fermion operators connect states of even nuclei to those of neighboring odd nuclei. The study of such matrix elements, the single particle coefficients of fractional parentage (CFP), is essential for some problems, and was indeed emphasized in our earliest work. It was realized later, however, that for fully microscopic applications of our techniques, there were important technical advantages to decoupling the study of the properties of even nuclei from those of their odd neighbors, as illustrated by a number of studies (Dreizler and Klein 1969, Dreiss and Klein 1969, Vallières et al 1973, Dasso et al 1973, Klein 1974, Vassanji et al 1978, Li et al 1979, Li and Klein 1979).

From the point of view of a complete version of the theory, the formulation in terms of CFP remains the most fundamental one. Nevertheless we postpone further discussion of it, since it will be derived and applied in the next two sections. Instead, for the remainder of this section, we discuss the fundamental concepts of our approach within the context of a method in which the CFP do not enter directly.

The decoupling mentioned above can be done by working only with elementary multipole and pairing operators,

$$Q_{\alpha\beta} = a_\alpha^\dagger a_\beta \ , \tag{2}$$

$$P_{\alpha\beta}^\dagger = a_\alpha^\dagger a_\beta^\dagger \ , \tag{3}$$

respectively (together with the hermitian conjugate of (3)), or with suitable linear combinations of these operators. The set of such operators $\mathbf{X} = \{X_i\}$ forms a Lie algebra under commutation,

$$[X_i, X_j] = f_{ijk} X_k \ . \tag{4}$$

We furthermore assume that the Hamiltonian has been rewritten as a polynomial, $H(\mathbf{X})$ in the generators, X_i. Then together with (4), which we refer to as the kinematic constraints, we have the dynamic constraints, also known as the equations of motion,

$$\begin{aligned}[X_i, H] &\equiv F_i(\mathbf{X}) \\ &= A_{ij} X_j + B_{ijk} X_j X_k + \cdots \ , \end{aligned} \tag{5}$$

where henceforth the summation convention is understood.

To implement these conditions, we introduce the eigenstates, $|n\rangle$, of the Hamiltonian,

$$H|n\rangle = E_n|n\rangle \ , \tag{6}$$

and study matrix elements of (4) and of (5). Since our object is to obtain equations which involve matrix elements of individual generators, we evaluate a matrix element of a product by means of the completeness relation,

$$\langle n|X_j X_k|n'\rangle = (X_j)_{nn''}(X_k)_{n''n'} \ , \tag{7}$$

where we are using the obvious notation, $\langle n|X_j|n'\rangle = (X_j)_{nn'}$. We thus obtain two sets of "sum rules",

$$f_{ijk}(X_k)_{nn'} = (X_i)_{nn''}(X_j)_{n''n'} - (X_j)_{nn''}(X_i)_{n''n'} \ , \tag{8}$$

$$(E_{n'} - E_n)(X_i)_{nn'} = A_{ij}(X_j)_{nn'} + B_{ijk}(X_j)_{nn''}(X_k)_{n''n'} + \cdots \ . \tag{9}$$

These sum rules must be satisfied by a complete set of solutions obtained by any means, such as the diagonalization of the Hamiltonian in a basis. In favorable physical cases, many of the sum rules will be exhausted by a small number of states. Conversely we may ask: Can one start with the sum rules, in which the kinematics and the dynamics are treated on an equal footing, to construct an independent scheme for solving problems in quantum mechanics? With suitable amplifications and restrictions, as will be described below, this question can be answered in the affirmative for a wide class of problems. The emergent theory is a form of Heisenberg matrix mechanics, with a modern twist. So far it has proved most successful in physical situations where sum rules converge rapidly. A vital part of the task is to recognize conditions under which this will occur.

Before going on to analyze different classes of applications, we have to delimit more carefully the composition of the equations needed in order to have a closed scheme for calculation. We must also discuss the related question of the novelty of our treatment of the kinematics compared to more traditional approaches. In the traditional approach to the study of bound states of a many particle system of fermions, such as the shell model for nuclei, one first constructs a set of basis states, taking all the pains necessary to insure that these states satisfy the Pauli principle. The subsequent dynamical treatment may, nevertheless, be of poor quality.

In our approach, the situation is quite different. Take the case of our first papers, where we worked with the algebra of individual Fermion operators. The output of the method consisted of energy differences and of matrix elements of the Fermion operators connecting different eigenstates of the system. How do we know that this set of numbers corresponds to matrix elements between states satisfying the Pauli principle? The answer is that this requirement is embodied in the anticommutation relations, which are utilized in the scheme as a set of sum rules. To the extent that these sum rules are not satisfied exactly, there is some violation of the exclusion principle. This is not necessarily a matter of grave concern, however, since if we do a good job of satisfying all the sum rules, kinematic and dynamic, we can be assured that the errors involved in Pauli principle violation are no more serious than the errors arising from our approximate treatment of the dynamics.

Consider next the case that we use the Lie algebra of fermion pairs to study the properties of even nuclei. Since the realization of the algebra in terms of Fermion pairs is only a particular realization, how do we impose the Pauli principle in this case? We want to guarantee to the extent possible that we are working in the antisymmetric representation of the appropriate algebra.

One method is to recall the theorem that an irreducible representation of a Lie algebra is uniquely specified by the eigenvalues of its Casimir invariants. These relations are of the form

$$\mathcal{P}^{(n)}(\mathbf{X}) = \lambda^{(n)}, \quad , \tag{10}$$

i.e., of the form of a polynomial of degree n in the generators set equal to its eigenvalue $\lambda^{(n)}$. These conditions must be added to the commutation relations and to the equations of motion to have a complete scheme. The associated sum rules then give additional constraints necessary to have a well–determined system of equations. In practice we have used both this method and variants of it that will be described *in situ*.

As opposed to the treatment of the Pauli principle, which is unconventional in our method, the formulation of approximation schemes that conserve invariance properties of the Hamiltonian is elementary and therefore constitutes one of its greatest assets. The cases of angular momentum and number conservation will be dealt with directly in the next section.

3   Formulation in Terms of Single–Particle Coefficients of Fractional Parentage

3.1   Equations of Motion

We start with a shell-model Hamiltonian of the form

$$
\begin{aligned}
H = & \sum_{\alpha} h_a a_{\alpha}^{\dagger} a_{\alpha} \\
& + \frac{1}{4} \sum_{abcd} \sum_{LM_L} F_{acdb}(L) B_{LM_L}^{\dagger}(ac) B_{LM_L}(db) \\
& + \frac{1}{4} \sum_{abcd} \sum_{ML_M} G_{abcd}(L) A_{LM_L}^{\dagger}(ab) A_{LM_L}(cd) \ , \tag{11}
\end{aligned}
$$

general enough to deal with all the applications to be considered. Here h_a are the spherical single-particle energies referred to the nearest closed shell, α refers to the standard set of single-particle quantum numbers, including in particular the pair (j_a, m_a) and a refers to the same set with m_a omitted. $B_{LM_L}^{\dagger}$ is the particle-hole multipole operator,

$$
\begin{aligned}
B_{LM_L}^{\dagger}(ab) & \equiv \sum_{m_a m_b} s_{\beta}(j_a m_a j_b - m_b | LM_L) a_{\alpha}^{\dagger} a_{\beta} \\
& = (-1)^{j_a + j_b - M_L + 1} B_{L - M_L}(ba) \ , \tag{12}
\end{aligned}
$$

and $A^{\dagger}_{LM_L}$ is the particle-particle multipole operator,

$$A^{\dagger}_{LM_L}(ab) \equiv \sum_{m_a m_b} (j_a m_a j_b - m_b|LM_L)a^{\dagger}_{\alpha} a^{\dagger}_{\bar{\beta}} , \qquad (13)$$

where $(j_1 m_1 j_2 m_2|jm)$ is a Clebsch-Gordon (CG) coefficient, $s_{\alpha} = (-1)^{j_a - m_a}$, and a bar indicates reversal of the sign of the magnetic quantum number. The coefficients F are the particle-hole matrix elements,

$$F_{acdb}(L) \equiv \sum_{m's} s_{\gamma} s_{\beta}(j_a m_a j_c - m_c|LM_L)(j_d m_d j_b - m_b|LM_L)V_{\alpha\beta\gamma\delta} , \qquad (14)$$

and G are the particle-particle matrix elements

$$G_{abcd}(L) \equiv \sum_{m's}(j_a m_a j_b - m_b|LM_L)(j_c m_c j_d - m_d|LM_L)V_{\alpha\beta\gamma\delta} . \qquad (15)$$

Our initial task is to obtain equations for the states and energies of an odd nucleus assuming that properties of immediately neighboring even nuclei are known. The states of the odd nucleus (particle number A) are designated as $|J\mu\nu\rangle$, where ν denotes all quantum numbers besides the angular momentum J and its projection μ. The states of the neighboring even nuclei with particle numbers $(A \pm 1)$ are written, in a parallel notation, as $|IMn(A \pm 1)\rangle$. The corresponding eigenvalues are $E_{J\nu}$ and $E^{(A\pm1)}_{In}$, respectively. The procedure is to calculate the operator equations of motion for a_{α} and for $a^{\dagger}_{\alpha}$ and to then form matrix elements that determine the single-particle coefficients of fractional parentage,

$$V_{J\mu\nu}(\alpha; IMn) = \langle J\mu\nu|a_{\alpha}|IMn(A+1)\rangle , \qquad (16)$$

$$U_{J\mu\nu}(\alpha; IMn) = \langle J\mu\nu|a^{\dagger}_{\bar{\alpha}}|IMn(A-1)\rangle . \qquad (17)$$

In order to obtain the equations in the most desirable form for the application of the completeness relation, it is necessary to use the anticommutation relations so as to insure that every term that contributes to the interaction contains a product of a single particle operator and of a multipole or pairing operator in that order. This leads, in different cases, to the replacement of the single particle energy, h_{α}, by either

$$h'_a = h_a - \frac{1}{4}\sum_{Lj_c} F_{acac}(L)\frac{2L+1}{2j_a+1} , \qquad (18)$$

or by

$$h''_a = h'_a - \sum_{Lj_c} \frac{2L+1}{2j_a+1}(G_{acac} + \frac{1}{2}F_{acac}) . \qquad (19)$$

In terms of a convenient and physically meaningful set of energy differences and sets of multipole fields and pairing fields defined below, we thus obtain generalized matrix equations of the Hartree-Bogoliubov form

$$\mathcal{E}_{J\nu}V_{J\mu\nu}(\alpha;IMn) = (\epsilon' + \omega^{(A+1)} + \Gamma^{(A+1)})_{\alpha IMn,\gamma I'M'n'}V_{J\mu\nu}(\gamma;I'M'n')$$
$$+\Delta_{\alpha IMn,\gamma I'M'n'}U_{J\mu\nu}(\gamma;I'M'n') \ , \tag{20}$$

$$\mathcal{E}_{J\nu}U_{J\mu\nu}(\alpha;IMn) = (-\epsilon'' + \omega^{(A-1)} - \Gamma^{(A-1)\dagger})_{\bar\alpha IMn,\bar\gamma I'M'n'}U_{J\mu\nu}(\gamma;I'M'n')$$
$$-\Delta^{\dagger}_{\bar\alpha IMn,\bar\gamma I'M'n'}V_{J\mu\nu}(\gamma;I'M'n') \ . \tag{21}$$

Here

$$\mathcal{E}_{J\nu} = -E_{J\nu} + \frac{1}{2}(E_0^{(A+1)} + E_0^{(A-1)}) \ , \tag{22}$$

$$\epsilon'_{\alpha IMn,\gamma I'M'n'} = \delta_{\alpha\gamma}\delta_{II'}\delta_{MM'}\delta_{nn'}(h'_a - \lambda_A) \ , \tag{23}$$

$$\lambda_A = \frac{1}{2}(E_0^{(A+1)} - E_0^{(A-1)}) \ , \tag{24}$$

$$\omega^{(A\pm1)}_{\alpha INn,\gamma I'M'n'} = \delta_{\alpha\gamma}\delta_{II'}\delta_{MM'}\delta_{nn'}(E_{In}^{(A\pm1)} - E_0^{(A\pm1)}) \ , \tag{25}$$

$$\Gamma^{(A\pm1)}_{\alpha IMn,\gamma I'M'n'} = \frac{1}{2}\sum_{LM_L}\sum_{bd}s_\gamma(j_a m_a j_c - m_c|LM_L)F_{acdb}(L)$$
$$\langle I'M'n'(A\pm1)|B_{LM_L}(db)|IMn(A\pm1)\rangle \ , \tag{26}$$

$$\Delta_{\alpha IMn,\gamma I'M'n'} = \frac{1}{2}\sum_{LM_L}\sum_{bd}(j_a m_a j_c - m_c|LM_L)G_{acdb}(L)$$
$$\langle I'M'n'(A-1)|A_{LM_L}(db)|IMn(A+1)\rangle \ . \tag{27}$$

Furthermore $E_0^{(A\pm1)}$ refer to the ground state energies of the neighboring even nuclei, the matrix elements of $\Gamma^{\dagger}$ are derived from those of (26) simply by the replacement of the operator B by $B^{\dagger}$, and the matrix elements of $\Delta^{\dagger}$ are similarly derived from those of Δ by the replacement of A by $A^{\dagger}$ together with the interchange $A\pm1 \to A\mp1$. Finally ϵ''_a is obtained from ϵ'_a by the replacement of h'_a by h''_a.

In order to specify a scale for the solutions, we take a suitable matrix element of the summed anticommutator,

$$\sum_\alpha \{a_\alpha, a_\alpha^{\dagger}\} = \Omega \ , \tag{28}$$

$$\Omega = \sum_a (2j_a + 1) \equiv \sum_a \Omega_a \ . \tag{29}$$

We thus find

$$\frac{1}{\Omega}\sum_{\alpha IMn}[|U_{J\mu\nu}(\alpha;IMn)|^2 + |V_{J\mu\nu}(\alpha;IMn)|^2 = 1 \ . \tag{30}$$

All of the above equations are still exact and, in particular, rotationally invariant and number conserving. In practice, however, we shall have to impose restrictions on the number and nature of the core states included in any application, as well as on the size of the single-particle space. As we have tried to explain in the previous section, any restriction on the number of states included results in some violation of the Pauli principle, at the same time that the other symmetries continue to be satisfied.

3.2 Transition Matrix Elements; Reduced Matrix Elements of Equations of Motion

We next apply the formalism to the computation of matrix elements of single-particle tensor operators, T_{LM_L}, that we write in the form

$$T_{LM_L} = \sum_{\beta\gamma} t_{\beta\gamma} a_\beta^\dagger a_\gamma \ . \tag{31}$$

We wish to calculate the matrix element $\langle J'\mu'\nu'|T_{LM_L}|J\mu\nu\rangle$ between eigenstates of the odd system. To carry through the calculation, we substitute for the ket the formally exact expression (Do Dang 1970, Dönau and Frauendorf 1977a)

$$|J\mu\nu\rangle = \frac{1}{\Omega}\sum_{\alpha,IMK}[U_{J\mu\nu}(\alpha,IMK)a_{\bar{\alpha}}^\dagger|\underline{IMK}\rangle$$

$$+V_{J\mu\nu}(\alpha,IMK)a_\alpha|\overline{IMK}\rangle] \ , \tag{32}$$

where an underline identifies the lighter of the two cores and an overline the heavier one. By using the commutation relations and completeness, this leads to the following expression for the transition element:

$$\langle J'\mu'\nu'|T_{LM_L}|J\mu\nu\rangle = \frac{1}{\Omega}\sum_{\alpha,IMK,I'M'K'}[U_{J'\mu'\nu'}(\alpha,I'M'K')U_{J\mu\nu}(\alpha,IMK)$$

$$\times\langle\underline{I'M'K'}|T_{LM_L}|\underline{IMK}\rangle$$

$$+[V_{J'\mu'\nu'}(\alpha,I'M'K')V_{J\mu\nu}(\alpha,IMK)\langle\overline{I'M'K'}|T_{LM_L}|\overline{IMK}\rangle$$

$$+\frac{1}{\Omega}\sum_{\alpha,\gamma,IMK}t_{\alpha\gamma}[U_{J'\mu'\nu'}(\bar{\alpha},IMK)U_{J\mu\nu}(\bar{\gamma},IMK)$$

$$-V_{J\mu\nu}(\alpha,IMK)V_{J'\mu'\nu'}(\gamma,IMK)] \ . \tag{33}$$

This is simplified by use of the Wigner-Eckart theorem with the following definitions of the reduced matrix elements:

$$\langle J'\mu'\nu'|T_{LM_L}|J\mu\nu\rangle = \frac{(-1)^{J-\mu}}{\sqrt{2L+1}}(J'\mu'J-\mu|LM_L)\langle J'\nu'||T_L||J\nu\rangle \ , \tag{34}$$

$$\langle I'M'K'|T_{LM_L}|IMK\rangle = \frac{(-1)^{I-M}}{\sqrt{2L+1}}(I'M'I-M|LM_L)$$

$$\times\langle I'K'||T_L||IK\rangle \ , \tag{35}$$

$$t_{\alpha\gamma} = \frac{(-1)^{j_c-m_c}}{\sqrt{2L+1}}(j_am_aj_c-m_c|LM_L)t_{ac} \ , \tag{36}$$

$$V_{J\mu\nu}(\alpha, IMK) = \frac{(-1)^{J-\mu}}{\sqrt{2j_a+1}}(IMJ-\mu|j_a m_a)v_{J\nu}(\alpha IK) , \tag{37}$$

$$U_{J\mu\nu}(\alpha, IMK) = \frac{(-1)^{J-\mu+j_a+m_a}}{\sqrt{2j_a+1}}(IMJ-\mu|j_a m_a)u_{J\nu}(\alpha IK) . \tag{38}$$

With the help of these definitions, we obtain the formula for the reduced matrix element that is applied in practice:

$$
\begin{aligned}
\langle J'\nu'\|T_L\|J\nu\rangle = \frac{1}{\Omega} & \sum_{\alpha IKI'K'} (-1)^{j_a+J'+I+L} \begin{Bmatrix} I & I' & L \\ J' & J & j_a \end{Bmatrix} \\
& \times [u_{J\nu}(\alpha IK)u_{J'\nu'}(\alpha I'K')\langle \underline{I'K'}\|T_L\|\underline{IK}\rangle \\
& + v_{J\nu}(\alpha IK)v_{J'\nu'}(\alpha I'K')\langle \overline{I'K'}\|T_L\|\overline{IK}\rangle] \\
& + \frac{1}{\Omega} \sum_{\alpha cIK} t_{ac}[(-1)^{j_a+I+J+L} \begin{Bmatrix} j_a & j_c & L \\ J & J' & I \end{Bmatrix} u_{J'\nu'}(\alpha IK)u_{J\nu}(cIK) \\
& + (-1)^{j_a+I+J+1} \begin{Bmatrix} j_a & j_c & L \\ J' & J & I \end{Bmatrix} v_{J\nu}(\alpha IK)v_{J'\nu'}(cIK)]. \tag{39}
\end{aligned}
$$

The expression just derived can be evaluated provided we can supply matrix elements of the tensor operator in the neighboring even nuclei, which we call the collective matrix elements, and have solutions of the equations for the CFP. The expression has the characteristic form of a sum of collective and non–collective terms.

Having introduced the definitions (35), (37), (38), it is convenient at this point to also give the equations of motion in the form of reduced matrix elements. In conformity with the applications we have carried out and shall describe in the next section, we consider the case where the quantum number n refers to the K bands of axially symmetric core nuclei and limit ourselves to even values of K. We thus find the equations

$$
\begin{aligned}
\mathcal{E}_{J\nu}v_{J\nu}(\alpha IK) = & (\epsilon_a + \omega_{IK})v_{J\nu}(\alpha IK) \\
& + \sum_{cI'K'} \Gamma(\alpha IK, cI'K')v_{J\nu}(cI'K') \\
& + \sum_{cI'K'} \Delta(\alpha IK, cI'K')u_{J\nu}(cI'K') , \tag{40}
\end{aligned}
$$

$$
\begin{aligned}
\mathcal{E}_{J\nu}u_{J\nu}(\alpha IK) = & (-\epsilon_a + \omega_{IK})u_{J\nu}(\alpha IK) \\
& - \sum_{cI'K'} \Gamma(\alpha IK, cI'K')u_{J\nu}(cI'K') \\
& + \sum_{cI'K'} \Delta(\alpha IK, cI'K')v_{J\nu}(cI'K') , \tag{41}
\end{aligned}
$$

$$
\begin{aligned}
\Gamma(\alpha IK, cI'K') = & \frac{1}{2} \sum_{Lbd} F_{acdb}(L)\sqrt{2L+1} \\
& \times (-1)^{j_a+I+J} \begin{Bmatrix} j_a & j_c & L \\ I' & I & J \end{Bmatrix} \langle I'K'\|B_L(db)\|IK\rangle , \tag{42}
\end{aligned}
$$

$$\Delta(aIK, cI'K') = \frac{1}{2} \sum_{Lbd} G_{acdb}(L)\sqrt{2L+1}$$

$$\times \ (-1)^{j_a+I+J} \begin{Bmatrix} j_a & j_c & L \\ I' & I & J \end{Bmatrix} \langle I'K'\|A_L(db)\|IK\rangle \ . \qquad (43)$$

The reduced matrix elements B_L and A_L appearing in the last two equations are defined exactly as in (35). In Eqs. (40) and (41) we have set $\epsilon'_a = \epsilon''_a = \epsilon_a$.

Finally we record the reduced matrix element of the normalization condition (30),

$$\frac{1}{\Omega(2J+1)} \sum_{aIK} [|u_{J\nu}(aIK)|^2 + |v_{J\nu}(aIK)|^2] = 1 \ . \qquad (44)$$

The theory described above was originally intended as a fully microscopic generalization of mean field methods. Considerable effort was invested in this direction in the period following initial formulation, with admittedly modest success. Nevertheless, at the time of our last publications along these lines almost two decades ago (Vallières *et al* 1973, Li *et al* 1979, Li and Klein 1979), two viable procedures for application to both vibrational and rotational motion had been developed and applied to first examples. This work will be described in Secs. 5 and 6. Recently, applications of a more modest but nevertheless useful character have been carried out. We interrupt the presentation of the formal aspects of the theory in order to give an account of these applications.

4   Semi–Microscopic Core–Particle Coupling Model

The derivation of the previous theory emphasized the view that if we knew the properties of two neighboring even nuclei, we could deduce therefrom the properties of the intervening odd nucleus. Under these conditions (40), (41 and (44) define a linear eigenvalue problem.

To be specific, we shall consider below the case of axially deformed nuclei. For the moment we take for granted that we have settled on how to delimit the number of core states to be included. In order to have a linear problem, we have to supply for these a set of energy differences and a set of reduced matrix elements of the elementary multipole and pairing operators that occur in the interaction. For the latter quantities this far exceeds the possibility of empirical input or even input from any existing phenomenological model. We thus have two choices. The microscopic choice that interests us ultimately requires a calculation of these properties of the core nuclei. We shall discuss this problem beginning in the next section.

The alternative choice is that made by the semi-microscopic core-particle models. It is to simplify the theoretical description of the cores to the extent that the reduced matrix elements that enter can be taken from experiment or from existing phenomenology. For this purpose, we choose, provisionally, the standard monopole pairing plus quadrupole-quadrupole model. For the $L = 0$ pairing we have

$$-\sum_b G_{aabb} A_{00}(ab) \equiv 2\Delta_a \sqrt{2j_a + 1} \tag{45}$$

$$\cong 2\Delta\sqrt{2j_a + 1} \; , \tag{46}$$

and for the only non-vanishing multipole element,

$$F_{abcd}(2) = -\kappa_2 F_{ab} F_{cd} \; , \tag{47}$$

$$Q_M = \sum_{ab} F_{ab} B_{2M}(ab) \; , \tag{48}$$

where Q_M is the mass quadrupole moment operator.

In 1968 (Do Dang *et al* 1968a) we showed how existing core-particle coupling models were special cases of the Kerman-Klein formalism. Later Dönau and Frauendorf formulated and applied a semi-microscopic theory that generalized existing models using some of our ideas (Dönau and Frauendorf 1977a, Dönau and Frauendorf 1977b). Some additional applications were made later by Chen *et al* (1986), but, for reasons that have nothing to do with the quality of the method, it fell into disuse until reactivated and extended by the authors (Protopapas *et al* 1994, Protopapas *et al* 1996a, Protopapas *et al* 1996b). In the following we describe a few simple applications and comment on work in progress.

4.1   Application to Strongly Deformed Nuclei

The simplest of our applications is to several strongly deformed rare earth nuclei. We shall later quote results for ^{157}Gd and ^{159}Dy (Protopapas *et al* 1994, Protopapas *et al* 1996a). We begin with a qualitative account of the assumptions:

1. *Choice and treatment of core states.* The simplest assumption is to start with the ground state bands of the neighboring even nuclei. From comparison of the results of such a calculation with experiment it was apparent that some of the bands seen experimentally were associated with excited bands of the core. In the final calculations, we therefore included the observed lowest lying excited bands with observed quadrupole strength to the ground band. The bands previously fitted were affected little by this generalization, expressing the dominance of intraband transition strengths over interband strengths, that is, after all, the physical basis for our ability to construct viable approximations for deformed nuclei. For the nuclei studied, we could, with impunity, ignore number conservation, replacing the two neighboring cores by an average nucleus. We have since devised an algorithm that strictly conserves nucleon number and verified that this approximation is accurate.

2. *Single particle states.* We have included essentially all bound single particle levels. (In that sense we have a much larger configuration space than the standard shell model.) The energies were determined, in the first instance,

from an appropriate Woods-Saxon Hamiltonian. However, in order to improve the fit to band heads, we permitted a fluctuation of ±5% for the single particle energies near the Fermi surface.

3. *Other parameters.* The chemical potential λ was taken from experiment, as was the energy gap Δ. The energy differences in the even nuclei and the quadrupole transition elements (assuming a standard proportionality between mass and electric quadrupole elements) were taken from experiment when these were available. In any event we found that for the chosen nuclei these were well approximated by standard formulas of the geometric model (Bohr and Mottelson 1975)) that could be used to extrapolate for unknown values. That left only a single completely free fitting parameter, namely, the strength, κ_2, of the quadrupole-quadrupole interaction

Technically, we would appear to have, for each value of J, (since angular momentum is conserved) a straightforward diagonalization problem. Matters are not quite as trivial as they might appear, however. First we must recall, that just as in the standard (BCS) pairing theory, we have twice too many solutions. In the latter case, they divide neatly into positive energy physical solutions and negative energy unphysical solutions. In our case, primarily because of the presence of core excitations energies, this does not occur. In our early work, we followed a method suggested by Dönau (1984) in which one first turns off enough of the core-particle Hamiltonian so as to reach a limit which is equivalent to the BCS case. One then tracks the physical solutions as one "slowly" tunes back to the full Hamiltonian. Recently, we have achieved a considerable simplification of this procedure by invoking the "no-crossing theorem" for interacting states.

We would also like to sort the calculated states into K bands. Since we work in the laboratory rather than in the intrinsic system, this two requires some additional attention, that we shall not describe.

The first application that we shall consider is to cases of relatively cleanly separated rotational bands. In Fig. 1 we show results for the energy levels of the nucleus ^{159}Dy. On the left are the results when the core is approximated by the ground band alone; on the right one sees that the previously missing bands are accounted for when the beta and gamma bands are added to the description of the cores. The results are satisfactory except for the $K = (1/2)$ band where the theory seems to predict too much staggering. Similar results were found for ^{155}Gd and ^{157}Gd. We plan to examine the effect of a small residual interaction of magnetic dipole type on this behavior. It is to be emphasized that our results do not require artificial renormalization of the Coriolis coupling. In our equations the experimental Coriolis coupling and the recoil effect are automatically included, though in hidden form because the calculations are carried out in the laboratory system.

In Fig. 2 we exhibit our results for BE(2) values for the ground state band of ^{157}Gd in comparison with calculations based on the conventional particle-rotor model and the cranking model. Both of these models are limiting cases of our method. As is the custom we have given separately the transitions with $\Delta J = 2$

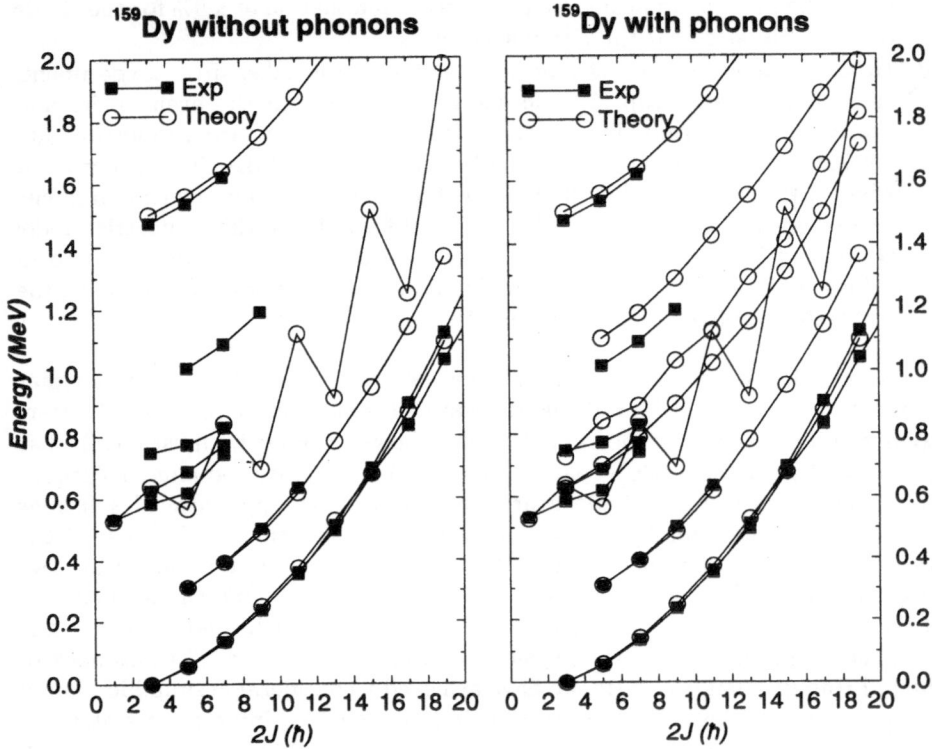

Fig. 1. Energy levels of ^{159}Dy. On the left are shown the results when only the ground band of the core is included, on the right the improved results when the beta and gamma bands are added.

and $\Delta J = 1$. It is important to remark finally that the fits are made without an effective charge parameter. This is not surprising, since the collective contribution is taken from experiment and the single particle contribution is quite small. On the other hand we have also obtained a good fit to the magnetic dipole transitions, where collective and single particle contributions are of comparable size. We find a similar quality of fit for ^{155}Gd.

4.2   Application to Backbending; other Work in Progress

We have also examined the phenomenon of backbending in odd nuclei and its relation to the same behavior in the neighboring cores. We consider the example (Protopapas *et al* 1996b) of the proton spectra of ^{165}Lu with neighbors ^{164}Yb and ^{166}Hf. Our object was to see if we could improve the fit found previously for this

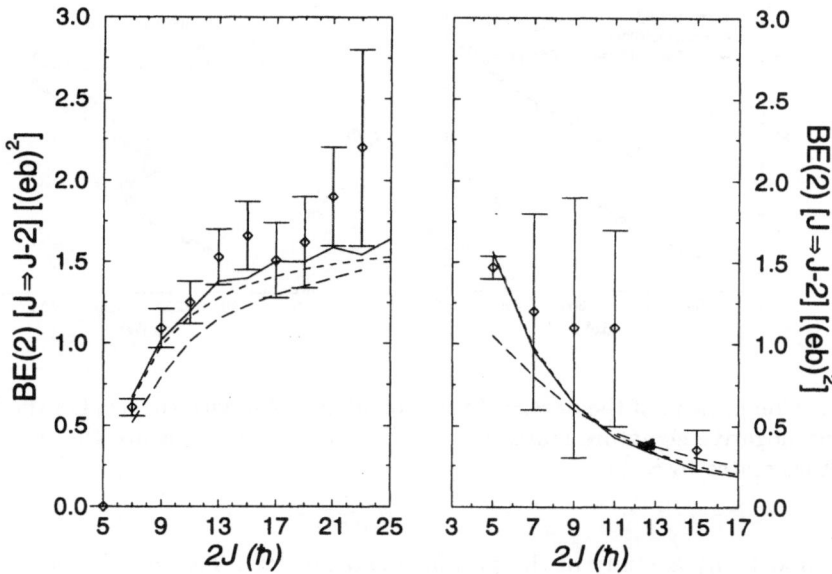

Fig. 2. Comparison of our results for electric quadrupole transitions in ^{157}Gd with other models. The particle–rotor results were calculated from the standard formula found in (Bohr and Mottelson 1975). The results for the cranking model as well as experimental values with error bars were taken from Kusakari *et al* 1992. The short dashed–line refers to the particle–rotor model, the long–dashed line to the cranking model, and the solid line to the current work.

nucleus (Chen *et al* 1986). In Fig. 3 we compare the results of our calculation of backbending for the yrast band of ^{165}Lu with experiment and with the previous calculation. A similar comparison is made for magnetic dipole transition rates in Fig. 4. The superiority of our results can be explained easily. We fitted the backbending in the adjacent even nuclei using a phenomenological two band model that fits the data with high precision. On the other hand Chen *et al* (1986) use a semimicroscopic model that takes into account the microscopic structure of the s–band. This model leads to deviations from observation that are reflected in the properties of ^{165}Lu. The relation between the semimicroscopic model for odd nuclei and a microscopic model for the neighboring even cores will be discussed briefly in the next section.

The applications quoted above are the first ones carried out with the our semimicroscopic core–particle coupling model. Two additional investigations are nearing completion: (i) The transformation of our formalism to the intrinsic system so as to exhibit more closely the connection with the conventional particle–

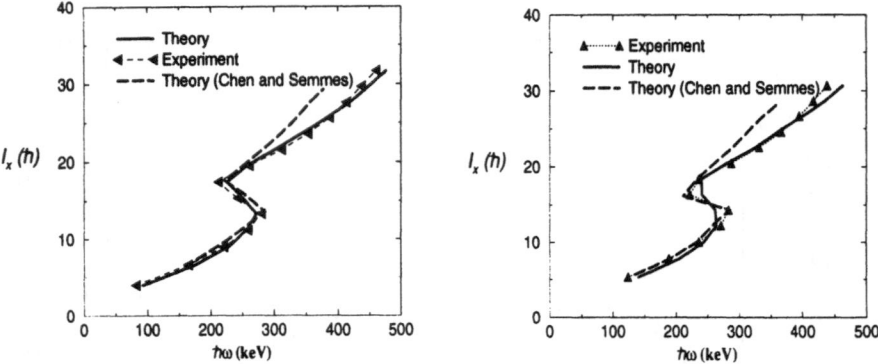

Fig. 3. Comparison of the observed yrast band of ^{165}Lu with theory. On the left are the negative signature states ($\alpha = -\frac{1}{2}$), and on the right are the positive signature states ($\alpha = \frac{1}{2}$).

rotor model. (ii) A study of the proton spectra of ^{123}I. The interest of such a choice is that in contrast to the examples chosen above, here the neighboring even cores have quite distinct properties. ^{122}Te has a vibrational spectrum and ^{124}Xe is a transitional nucleus. In this application (Protopapas 1995), therefore, we take particle number strictly into account. The reason for this particular choice of nuclei is that it has been treated previously (Dönau and Hagemann 1979) in a more rudimentary application of the same theory used by us. Other applications are planned, including a test of a completely microscopic form.

5   Further Theoretical Developments

In this section we shall discuss how the theory first developed in Sec. 3 can be utilized as a completely microscopic theory. Under these circumstances it is not necessary, though it may still be convenient, to choose an interaction as highly simplified as the one used for the semimicroscopic formulation.

5.1   Generalized Hartree–Bogoliubov Equations

We shall now see that the scheme defined by the reduced matrix elements of the equations of motion (40) and (41) and the reduced matrix elements of the normalization condition (44) can be considered the basis for a fully microscopic theory. (This was in fact the point of view advanced in our original papers.) This application requires an additional and admittedly new use of the completeness relation in which the intermediate states are those of the *odd* nucleus. Recalling

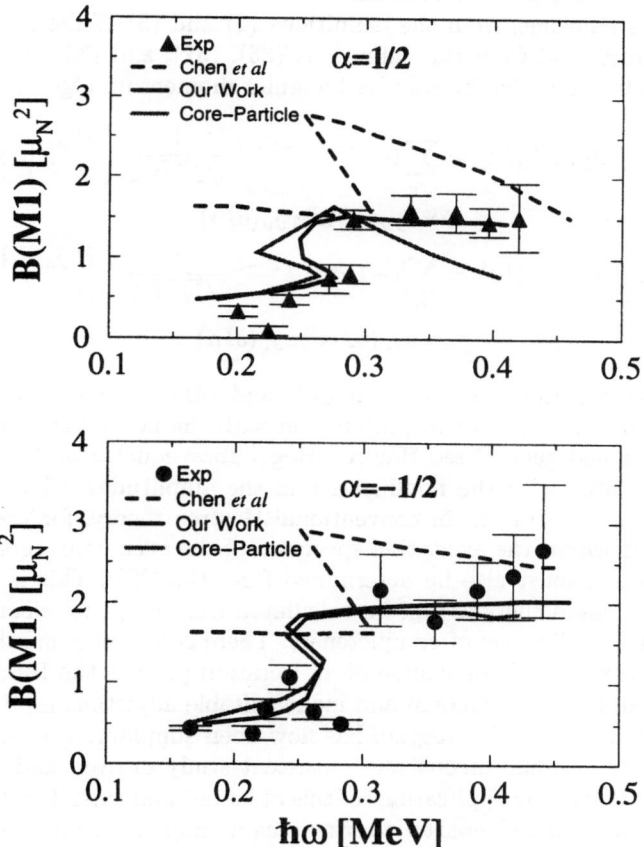

Fig. 4. $B(m1)$ transition rates in units, μ_N^2, of the square of nuclear Bohr magneton. On the left are the positive signature states and on the right the negative signature states. The experimental data (Jonsson *et al* 1984) is represented by the solid symbols, the strong coupling results are shown as the chain line, our work appears as the solid line, and the results of (Chen *et al* 1986) are given as the dashed line.

the definitions (16) and (17) of the CFP's, we have (assuming real amplitudes)

$$\langle I'M'n'|a_\alpha^\dagger a_\beta|IMn\rangle = \sum_{J\mu\nu} V_{J\mu\nu}(\alpha; I'M'n')V_{J\mu\nu}(\beta; IMn) , \qquad (49)$$

$$\langle \overline{I'M'n'}|a_\alpha^\dagger a_{\bar\beta}^\dagger|\underline{IMn}\rangle = \sum_{J\mu\nu} V_{J\mu\nu}(\alpha; I'M'n')U_{J\mu\nu}(\beta; IMn) . \qquad (50)$$

Since the present considerations can be relevant to other than deformed nuclei, we have reverted to a general notation.

From these relations, from the definitions (2) and (3) of the multipole and pairing operators, and from the definitions (35), (37), and (38) of the reduced matrix elements, we derive by standard angular momentum algebra

$$\langle I'n'\|B_L(ab)\|In\rangle = \sum_{J\nu}(-1)^{j_b+J+I'-L}\frac{1}{\sqrt{2L+1}}\left\{\begin{array}{ccc}I' & j_a & J\\ j_b & I & L\end{array}\right\}$$
$$\times v_{J\nu}(aI'n')v_{J\nu}(bIn)\ , \tag{51}$$

$$\langle I'n'\|A_L(ab)\|In\rangle = \sum_{J\nu}(-1)^{j_b+J+I'-L}\frac{1}{\sqrt{2L+1}}\left\{\begin{array}{ccc}I' & j_a & J\\ j_b & I & L\end{array}\right\}$$
$$\times v_{J\nu}(aI'n')u_{J\nu}(bIn)\ . \tag{52}$$

With these equations, we see that (40) and (41) become a set of nonlinear equations for the cfp that in conjunction with the normalization conditions are properly named generalized Hartree–Bogoliubov equations. Aside from the difficulty associated with the nonlinearity in the amplitudes, which is qualitatively of the same nature as in conventional Hartree theory, for the theory to be fully self consistent the excitation spectrum of the even cores represented by the quantities ω_{In} must also be determined from the CFP. This is feasible, in principle, since these energies can be calculated from diagonal elements of the Hamiltonian by application of completeness. There is a further subtlety in that we have to check that in the course of an iteration process the Hamiltonian of the even system remains diagonal and make suitable adjustments, if necessary.

As explained below this program has now been supplanted by another that does not require a simultaneous self–consistent study of even and odd nuclei. Nevertheless microscopic applications of this older method exist, for the standard pairing force model (single–particle terms plus monopole pairing) (Do Dang *et al* 1968b) and for isotopes of the semi–magic nuclei Ni (Wu 1973) and Sn (Dreiss *et al* 1971). The main objection to these calculations is that the configuration spaces of single-particle states and/or of core states admitted was too small to make the physics convincing. Undoubtedly one can do much better these days if one were so inclined. Below we shall outline a method that we feel has important advantages compared to that just discussed.

5.2   Generalized Density Matrix Method

Our object is to decouple the study of even nuclei from that of odd nuclei. Since the Hamiltonian is a polynomial in multipole and pairing operators and these operators are closed under commutation, the required equations of motion for these operators can be derived *ab initio*, as sketched in Sec. 2. However, they can also be derived directly from the previous single particle equations by eliminating the eigenvalues $\mathcal{E}_{J\nu}$ and making use of the relations (51) and (52) between reduced matrix elements of the multipole and pairing tensors and bilinear sums of reduced matrix elements of the CFP. It is uninformative, for the purposes of this report, to record these equations in detail. It suffices to think of them as generalized non-linear RPA equations, the energy differences of the even nuclei serving the role of eigenvalues. Indeed it is easy to show that the equations customarily identified as the RPA are a special case of these equations.

As we have emphasized in Sec. 2, the equations of motion by themselves do not define a complete scheme. We must bring in sum rules based on the operator algebra and some expression of the Pauli principle. We shall describe how the latter is incorporated, since this is point least likely to be familiar to the reader.

For illustrative purposes, we consider the case of a single j level (Vallières *et al* 1973). Define the multipole and pairing operators

$$g_{mm'} = a_m^\dagger a_{m'} - \frac{1}{2}\delta_{mm'} \equiv G_{mm'} - \frac{1}{2}\delta_{mm'} \,, \tag{53}$$

$$F_{mm'} = a_{m'} a_m = (F_{mm'}^\dagger)^\dagger \,. \tag{54}$$

The operators g and F are generators of the Lie algebra $O(4j+2)$. The restrictions of the Pauli principle can be expressed within the framework of the specific realization of the algebra, by exhibiting an identity among the generators whose derivation specifically utilizes the anticommutation relations, namely

$$\frac{1}{2}\{G_{mm'}, G_{m''m'''}\} + \frac{1}{2}\{F_{mm''}^\dagger, F_{m'''m'}\}$$
$$-\frac{1}{2}\delta_{m'm}G_{m''m'''} - \frac{1}{2}\delta_{m'''m''}G_{mm'}$$
$$= \frac{1}{2}(\delta_{mm'''}\delta_{m'm''} - \delta_{mm'}\delta_{m''m'''}) \,. \tag{55}$$

Since the right side of (55) is a constant, the quadratic function of the generators on the left hand side commutes with all the generators.

Representation dependent constraints that are independent of the commutation relations of the Lie algebra follow from this condition by projecting out independent spherical tensor components. The tensor of rank zero is the quadratic Casimir invariant. Tensors of higher rank express the essential content of higher order Casimir invariants.

Only a limited application has been made so far of the formalism described above. Vallières (Vallières *et al* 1973) has studied both the vibrational and the rotational regimes of a single j model with monopole pairing and quadrupole interactions. By varying the strength of the ratio of the two interactions he was able to describe the shape phase transition from spherical near the pairing limit to deformed near the quadrupole limit. This promising beginning has not been followed up. For applications to semi–magic nuclei, described in the next section, we employed a method similar in spirit to the present one but distinct in detail.

We recall that once the problem posed in this subsection has been solved with sufficient accuracy the problem of odd nuclei becomes linear.

It should be noted in passing that Belyaev and Zelevinsky (Belyaev and Zelevinsky 1970, Belyaev and Zelevinsky 1974) formulated a theory of even nuclei, conceptually equivalent to the method of this section. Early applications, reviewed by Zelevinsky (Zelevinsky 1983), were mainly to rotational motion, but quite recently there has even been an application to chaotic motion (Zelevinsky 1993).

6   Decoupling Procedures for Vibrational Motion

In this section, we shall describe the ideas necessary to transform the sum rule formulation of the previous sections into a practical calculus for vibrational motion. Though our main interest is in the nuclear many body problem, it is instructive to begin with a very simple problem and build from there.

6.1   Anharmonic Oscillator

We study an anharmonic oscillator described by the Hamiltonian

$$H = \frac{1}{2}(p^2 + x^2) + \frac{1}{4}\mu x^4 \ . \tag{56}$$

For further study it is slightly more convenient to use a Lagrangian formulation of this problem. This is obtained by using the equation of motion ($\hbar = 1$),

$$p = -i[x, H] \ , \tag{57}$$

to eliminate the momentum operator from the remaining equation of motion and from the commutation relation. This yields the operator equations

$$[x, [H, x]] = 1 \ , \tag{58}$$

$$[[x, H], H] = x + \mu x^3 \ , \tag{59}$$

which we wish to convert into sum rules. Here we may simplify matters by quoting a theorem that is not difficult to prove (Li et al 1975), namely that the off–diagonal matrix elements of the commutation rule are not independent of the equations of motion. It suffices therefore to consider the equations

$$\sum_{n'}(E_{n'} - E_n)|x_{nn'}|^2 = 1 \ , \tag{60}$$

$$(E_{n'} - E_n)^2 x_{nn'} = x_{nn'} + \mu x_{nn''} x_{n''n'''} x_{n'''n'} \ . \tag{61}$$

We have solved these equations in two regimes. In the first, purely quantum regime, we wish to find the energy differences and matrix elements characterizing the lowest states (Li et al 1975). In the second, semiclassical limit, applicable to large values of the quantum numbers, we can formulate a semiclassical quantization procedure (Greenberg et al 1995). Here we shall describe the purely quantum problem. Our ability to solve this problem depends on the result that even for very large values of the anharmonic coupling μ, we can order the states in sequences $n, n+1, n+3 \ldots n+2k+1 \ldots$ with two properties,

$$x_{n,n+1} >> x_{n,n+3} >> \cdots >> x_{n,n+2k+1} >> \cdots , \qquad (62)$$

$$x_{n,n+2k+1} \cong x_{n+m,n+m+2k+1} . \qquad (63)$$

The content of these two conditions can be characterized as follows: The first condition asserts that in respect to a reference state the remaining states are layered, in that the main or harmonic amplitude dominates the crossover amplitudes, which themselves are ordered according to the "distance" (in energy) between the two states involved. The second condition asserts that the order of magnitude of a given matrix element depends, approximately, only on the difference of the quantum numbers of the two states. These same conditions apply to a vibrational spectrum of any many body system, which is our reason for looking at this simple problem.

The validity of condition (62) is easily established from perturbation theory in the weak anharmonicity limit. This is not acceptable, since we wish to solve our problem for an arbitrary strength of the coupling constant μ. We are aware of three arguments that buttress this condition in the general case, but mention only one: The wave function associated with the state $|n\rangle$ has n nodes, that associated with $x|n\rangle$ has $n+1$ nodes. We therefore expect maximum overlap of two wave functions that have the same number of nodes and the same phase at the origin. Correspondingly, we anticipate that the larger the difference in the number of nodes, the greater the opportunity for cancellation of contributions and the smaller the net result.

We expect condition (63) to be satisfied as long as m is not too large. The role played by a condition of this type, as will be seen below, is to provide what we term a closure approximation for our equations. Both conditions play a role in deciding how to define a consistent approximation.

To see how to utilize the proposed scheme, we consider the simplest possible viable example, which we shall afterward generalize to a full scheme. Relative to the ground state, we expect the matrix element $x_{01} = x_{10}$ to be dominant. We therefore try an approximation in which we drop all elements $x_{0,2k+1}$, $k \geq 1$. Under these conditions, (60) and (61) take the forms,

$$(E_{n'} - E_n)|x_{01}|^2 = 1 , \qquad (64)$$

$$(E_{n'} - E_n)^2 x_{01} = x_{01} + \mu(x_{01}x_{10}x_{01} + x_{01}x_{12}x_{21}) . \qquad (65)$$

To have a consistent numerical treatment, the inclusion of the last term in (65) is required by (63). We thus see that we have two equations but three variables. We cannot cure this problem by more sum rules, since if we treat them consistently, according to the assumptions as to size that we have imposed (and as is verified *a posteriori*), we shall alway encounter a failure of closure. Therefore at some point we shall have to impose a closure approximation. In this simple case, the harmonic oscillator relation $x_{12} = \sqrt{2}\, x_{01}$ suggests itself as a good starting point.

The fact that the resulting simple nonlinear algebraic scheme gives relatively accurate results is already of some interest, but of more interest is that it suggests all the elements of a sequence of increasingly more accurate approximations. In general, we choose an odd integer $N = 2K + 1$. We may then write sum rules that, closure aside, provide equations for the set of elements $x_{n,n+1}$, with $n + 1 \leq N$. Additional accuracy of the scheme is obtained by including $x_{n,n+3}$ and still smaller elements. This requires both a more accurate evaluation of the sum rules previously recorded as well as the addition of sum rules based on the equations of motion for the smaller elements. Typically, we use closure approximations of the form

$$x_{N,N+1} \cong \sqrt{\frac{N+1}{N}} x_{N-1,N} \cong x_{N-1,N} \ , \tag{66}$$

$$x_{N,N+2k+1} \cong x_{N-2k-1,N} \ , \tag{67}$$

for N large enough. Intuition suggests and experience shows that as N increases, results for the lowest lying states become increasingly more accurate and also insensitive to the precision of the closure approximation.

To complete the present discussion requires several additional observations. For the simple one-dimensional model under discussion, there is no difficulty in treating a nonlinear scheme in which all amplitudes, large and small, are included on an equal footing. We shall find below that similar ideas can be applied to the study of collective vibrational motion for a many body system. In this case, it is increasingly inconvenient, numerically, to treat crossover amplitudes with the same respect as the main harmonic ones. It is simpler to start with a nonlinear scheme that contains only the harmonic amplitudes and the corresponding energy differences and to include the smaller amplitudes by an iterative procedure once the starting scheme has been solved. Second, it is time to notice that the calculational scheme yields matrix elements and energy differences. To obtain "absolute" energies, we need a reference energy. All one needs to do is to calculate the diagonal matrix element H_{00} by the same sum rule approach as used in the calculational scheme.

For extensive results of the model discussed above see Li *et al* (1975).

7   Low Energy Properties of Even Ni Isotopes

We have already alluded in Sec. 2 to a number of applications of the Lie algebraic method designed to study the low energy properties of systems of even numbers of nuclei. A number of these were artificially constructed (schematic) models designed to illustrate some, but never all, of the complexity associated with the effort to understand collective motion in various regimes starting from a shell model. Though highly instructive and very reassuring as to the power of our methods, space considerations prevent us from describing any of this work in detail. Instead we turn directly to the last and most ambitious project attempted of its type, a study of the Ni isotopes (Li *et al* 1979, Li and Klein 1979). We judge the outcome of this undertaking to be very promising, although the work remains incomplete in a number of respects.

There is also an important respect in which the work to be described is not a pure example of the Lie algebraic program preserving all symmetries, as outlined in the previous pages of this chapter. We thought it might be useful, in trying to "sell" our work, to present it as a nonlinear generalization of the standard approach to spherical nuclei. The standard "collective" treatment of such nuclei is based on the number nonconserving BCS approximation for the ground state, followed by the QRPA treatment of excited states. We shall describe a method that contains these results as a limiting case.

When we turn to problems of realistic nuclear physics, we encounter at least two additional problems, compared to schematic models, that are related to each other technically, and arise from the fact that a realistic microscopic model has both collective and non-collective degrees of of freedom. The first problem is to decouple these two sets of degrees of freedom to the extent possible, but also to describe how we might successively include the coupling where this is dictated by the physics. The second problem, already discussed in some detail is how to take into account the restrictions imposed by the Pauli principle. It is simplest to take up these problems as we encounter them in the detailed exposition.

The starting Hamiltonian for this study is a version of the pairing plus quadrupole model,

$$
\begin{aligned}
H &= H_{sp} + H_p + H_{Q-Q} \\
&= \sum_a h_a N_a - G \sum_{ab} \sqrt{\Omega_a \Omega_a} A^{(0)\dagger}(a) A^{(0)\dagger}(b) \\
&\quad - \frac{1}{2}\chi \sum_q (Q_q^\dagger Q_q)_{\text{modified}} ,
\end{aligned}
\tag{68}
$$

where the first two terms constitute the standard pairing Hamiltonian and

$$
Q_q^\dagger = \sum_{\alpha\beta} (\alpha|r^2 Y_q^2(\hat{r})|\beta) a_\alpha^\dagger a_\beta
\tag{69}
$$

is the quadrupole operator associated with the valence shell. The subscript *modified* means that pairing matrix elements have been removed from the operator. This was found to be necessary because of the large and unphysical effect of such matrix elements when the level degeneracies are small, as they are for the assumed model of $Ni(2p_{1/2}, 2p_{3/2}, 1f_{5/2})$. Ultimately other modifications were called for in the Hamiltonian.

We outline the method used to study the spectrum of the Hamiltonian for a nucleus with a quadrupole phonon spectrum and a "superconducting" ground state.

1. We start with a Bogoliubov-Valatin transformation to "tilde" operators (quasiparticles)

$$
a_\alpha^\dagger = u_\alpha \tilde{a}_\alpha^\dagger + s_\alpha v_\alpha \tilde{a}_\alpha ,
\tag{70}
$$

$$u_\alpha^2 + v_\alpha^2 = 1 , \tag{71}$$

where the coefficients u_α, v_α will be determined by a method which is an improvement over the usual BCS procedure. Under this transformation the pair and multipole operators $A_M^{(j)\dagger}$ and $B_M^{(J)}$ undergo linear transformations which convert them into linear combinations of the corresponding quasiparticle operators $\tilde{A}_M^{(J)\dagger}$ and $\tilde{B}_M^{(J)}$. With these changes H becomes (symbolically)

$$H = \tilde{H}(u, v, \tilde{A}, \tilde{A}^\dagger, \tilde{B}) , \tag{72}$$

where only the operators for $J = 0$ and $J = 2$ intervene. To insure number conservation, on the average, and also to damp number fluctuations, we work with an auxiliary Hamiltonian of Lipkin– Nogami type (Nogami 1964),

$$H' = H - \lambda_1 N - \lambda_2 N^2 . \tag{73}$$

2. From the resulting operator we drop all terms containing $\tilde{A}^{(0)\dagger}(a)$ or $\tilde{A}^{(0)}(a)$ (or both). This corresponds to the omission of pairing vibrations from the present considerations. Further simplification is achieved by eliminating the quasiparticle multipole operators and expressing them as polynomials in the corresponding pair operators by means of the relation (Chattopadhyay et al 1974, Li et al 1979)

$$\begin{aligned}
\tilde{B}_M^{(J)}(ab) = &(2/\hat{J}) \sum_{1,2} (-1)^{J_2 - M_2} y(J\bar{2}1) \tilde{A}_1^\dagger \tilde{A}_2 \\
&- (4/3\hat{J}) \sum_{1...6} (-1)^{J_3 + J_4 - M_3 - M_4} y(J61) y(65\bar{3}) \\
&\times y(5\bar{4}2) \tilde{A}_1^\dagger \tilde{A}_2^\dagger \tilde{A}_3 \tilde{A}_4 ,
\end{aligned} \tag{74}$$

where

$$\begin{aligned}
y(123) = &\hat{J}_1 \hat{J}_2 \hat{J}_3 \delta_{b_1 b_2} \delta_{a_1 a_3} \delta_{a_2 b_3} (-1)^{j_{a_1} + j_{b_1} + M_1} \\
&\times \begin{pmatrix} J_1 & J_2 & J_3 \\ -M_1 & M_2 & M_3 \end{pmatrix} \begin{Bmatrix} J_1 & J_2 & J_3 \\ j_{a_2} & j_{a_3} & j_{b_1} \end{Bmatrix} ,
\end{aligned} \tag{75}$$

in terms of $3 - j$ and $6 - j$ symbols, $\hat{J} = (2J + 1)^{1/2}$, $\bar{2}$ means the set $(a_2, b_2, J_2, -M_2)$, and the J in $y(J\bar{2}1)$, e.g., represents (J, M, a, b). In Eq. (74), terms with six or more operators have been neglected. We state without proof that the "correct" treatment of the Pauli principle in the current treatment is taken care of by the combination of the transformations (70) and (74).

3. Substituting Eq. (74) in H' and into the algebra yields, symbolically again,

$$H' = H'(u, v, \lambda_1, \lambda_2, \tilde{A}^{(2)}, \tilde{A}^{(2)\dagger}) , \tag{76}$$

$$[\tilde{A}, \tilde{A}^\dagger] = 1 + c\tilde{A}^\dagger \tilde{A} + ... \tag{77}$$

and c represents an appropriate matrix of constants. We now outline the calculational procedure designed to utilize the tools and ideas described above. The main point, not to be obscured by the details that follow, is that it represents a non-linear generalization of the standard BCS plus QRPA approach and includes an improved treatment of number conservation as well. The first step is to evaluate the ground state expectation value of H'. In the associated sum rule evaluation there intervene as intermediate states only 2+ states excited by the quadrupole operators. We include in the calculation one broken-pair or one phonon 2+ states (of which there will be five in this calculation = number of independent one particle quadrupole operators) that will be called $2, 2', 2''...$) and the two phonon 2+ states $2_2, 2'_2, ...$. Of the latter we shall retain only the collective one, which we identify as the unprimed state. The matrix elements that appear in this calculation of $\langle 0|H'|0\rangle$ up to the two phonon or second layer of states are, in the standard terminology (symbolically)

$$X = \langle 0|A^{(2)}|2_1\rangle, \quad \text{Tamm} - \text{Dancoff} , \tag{78}$$

$$Y = \langle 0|A^{(2)\dagger}|2_1\rangle, \quad \text{ground state correlation} , \tag{79}$$

$$\theta = \langle 2_1|A^{(2)}|2'_1\rangle, \quad \text{quadrupole matrix} , \tag{80}$$

$$\Phi = \langle 0|A^{(2)}|2_2\rangle, \quad \text{crossover matrix} . \tag{81}$$

Thus we have

$$F_0 \equiv \langle 0|H'|0\rangle = F_0(u, v, \lambda, X, Y, \theta, \Phi) . \tag{82}$$

4. To determine the quantities $u, v, \lambda_{1,2}$, as well as the amplitudes just defined, we employ various means – variational, equations of motion, and commutation relations. For example, from the conditions

$$\delta F_0/\delta u_a = \delta F_0/\delta v_a = 0 \tag{83}$$

and number conservation, we obtain a generalized form of the BCS theory which depends on the unknown amplitudes (78)-(81) in addition to the usual zero order BCS quantities. From the equations of motion for $\tilde{A}^{(2)}$, namely $[\tilde{A}^{(2)}, H']$, we obtain equations for the first order X and Y amplitudes, as well as for the second order amplitudes θ and Φ. From the commutation relations (77), we obtain normalization conditions, as well as alternative equations for computing the θ and Φ amplitudes.

5. To obtain a solution, we proceeded as follows: Setting $X = Y = \theta = \Phi = 0$, in the first cycle of iteration, we solved the BCS-like equations for the $u_a, v_a, \lambda_{1,2}$. These values were then inserted into the equations for the X, Y which constitute a non-linear version of QRPA. The latter equations were then solved, neglecting initially the θ, Φ amplitudes, and the solutions normalized. Finally the θ and Φ amplitudes were calculated by perturbation theory. This constitutes a single cycle of the calculation, and provides new input for a second cycle in which all amplitudes are non-vanishing. The steps were then repeated until convergence was achieved.

Some results of applying this theory to the isotopes of Ni will be described below. There is no reason that the calculation cannot be extended in many directions: inclusion of the two phonon states, odd nuclei, doubly open nuclei, larger shell model space.

Comparison of the results of the calculation with an exact shell model calculation for the same Hamiltonian shows that the method yields accurate solutions (Li and Klein 1979). A comparison of theory with experiment, however, shows that the Q-Q interaction is inadequate to describe the four non-collective one-phonon states, implying too compressed a spectrum. We proposed a rough-and-ready means of removing this essential disagreement: Insofar as the equations of motion have the QRPA form, the various excited states can be written in terms of phonon creation operators ($i = 1, ...5$),

$$|2q(i)> = O_{2q,i}^{\dagger}|0>, \qquad (84)$$

where 1 is the collective state. We then find that an additional term in the Hamiltonian of the form

$$\delta H = \sum_{q,i} \delta\omega_i (1 - \delta_{i,1}) O_{2q,i}^{\dagger} O_{2q,i}, \qquad (85)$$

can be used to correct the deficiency noted above. In practice we used values $\delta\omega_i = \delta\omega = $ constant. A reasonable viewpoint to take for a more elaborate study would be actually to adjust a full set of additional terms so as to obtain a perfect fit to the energies of all one phonon states, and then to see how far this would take us in fitting other data.

Some results obtained with the altered Hamiltonian are then given in Tables 1 and 2, where they are compared with experiment and with two standard shell model calculations. In these tables ω_i are the one-phonon energies, $\omega_{1,1}$ the excitation of the two-phonon 2+ state, e_n the effective charge and χ' a measure of the quadrupole interaction. In addition to energies, various B(E2) values and the quadrupole moment of the state 2_1+ have been computed. The overall agreement, with a Hamiltonian that has hardly been chosen in an optimum way, appears to be of comparable quality when compared with the results obtained with the modified surface delta interaction (MSDI) and with the adjusted surface delta interaction (ASDI) (Brussard and Glaudemans 1977). In any event the validity of the approximation scheme does not appear to be in doubt. As already remarked, work could be done in improving the Hamiltonian and in extending the calculation.

8   Other Developments

In this final section, we shall merely mention a few topics that we cannot develop within the permitted space allotment. This includes both theoretical features, applications to other topics in nuclear physics, and applications to other fields. This discussion must also be selective.

Table 1. Various results obtained for ^{60}Ni by the algebraic method are compared with shell model calculations carried out with two versions, MSDI and ASDI, of the surface delta interaction, defined in the text.

^{60}Ni	Present work $\chi' = 0.40, \delta\omega = 1.17)$	MSDI	ASDI	Experiment
ω_1 [MeV]	1.33	1.58	1.46	1.33
ω_2	3.03	2.96	2.76	3.12
ω_3	3.15	3.10	2.94	3.27
ω_4	3.39	3.33	3.34	3.39
ω_5	3.66	3.51	3.59	3.87
$\omega_{1,1}$	-	2.13	2.22	2.16
e_n	1.59	1.64	1.57	
B(E2)$[e^2 \cdot fm^4]$				
$2_1 \to 0_g$	186	186	186	186
$2_2 \to 0_g$	0.003	0.293	0.899	$0.75^{+0.52}_{-0.31}$
$2_3 \to 0_g$	0.710	0.110	0.129	$1.1^{+0.7}_{-0.4}$
$2_4 \to 0_g$	4.37	0.055	0.510	$0.49^{+0.34}_{-0.22}$
$2_5 \to 0_g$	4.83	0.786	0.167	$0.47^{+0.31}_{-0.20}$
$2_{1,1} \to 0_g$	1.68	3.19	0.001	< 1.5
$Q(2_1^+)[e \cdot f_m^2]$	-9.63	-16.8	-12.3	3 ± 7
$B(E2; 2_{1,1} \to 0_g)$	-	0.020	5×10^5	$3.2^{+0.7}_{-0.5} \times 10^{-3}$
$B(E2; 2_{1,1} \to 2_1)$				

1. *Variational principles.* Over the years, we have discussed and applied a number of variational ideas associated with our algebraic approach. The subject was discussed briefly in our earliest works. More extensive theoretical accounts can be found, e. g., in Do Dang *et al* (1968b), Klein *et al* (1980). We highlight only a few of the applications. One was already mentioned in our discussion of the Ni isotopes. Another application is to the study of the relationship between the shell model and the Interacting Boson Model (Klein and Vallières 1981). Most recently (Greenberg *et al* 1996) a variational principle was used in a study of the transition from Heisenberg matrix mechanics to a semiclassical quantum scheme.

2. *Large amplitude collective motion.* Though most of our work in this subject can be characterized (Klein *et al* 1991) as the development and application of new ways of solving time–dependent Hartree–Fock theory, the Kerman-Klein method has played a fundamental role in the derivation of a nuclear Born–Oppenheimer theory, i. e., in providing a theory for the decoupling of collective from non–collective coordinates and a classical limit of the quantum decoupling notions discussed in the body of this review (Klein and Walet 1994. Currently under development are extensions of this adiabatic theory

Table 2. Same as Table 1, except that results are for ^{62}Ni.

^{62}Ni	Present work	MSDI	ASDI	Experiment
	($\chi' = 0.5, \delta\omega = 0.999$)			
ω_1 MeV	1.22	1.44	1.42	1.17
ω_2	3.00	2.88	2.75	3.06
ω_3	3.12	2.95	3.00	3.16
ω_4	3.54	3.25	3.17	3.26
ω_5	3.67	3.44	3.49	3.27
$\omega_{1.1}$	-	2.28	2.22	2.30
e_n	1.48	1.48	1.43	-
$B(E2)[e^2 \cdot fm^4]$				
$2_1 \rightarrow 0_g$	179	179	179	179
$2_2 \rightarrow 0_g$	7.27	3.51	1.20	-
$2_3 \rightarrow 0_g$	0.200	0.621	0.851	
$2_4 \rightarrow 0_g$	8.50	0.395	0.683	
$2_5 \rightarrow 0_g$	3.18	0.018	0.073	
$2_{1,1} \rightarrow 0_g$	6.43	4.91	0.013	6.4 ± 1.9
$Q(2_1^+)[e \cdot fm^2]$	-6.78	-10.2	1.44	5 ± 12

to the diabatic case where level crossings must be taken into account and to dissipative processes.

3. *Field theories with soliton solutions.* This class of applications, both to one dimensional models (Klein and Weldon 1978) and, more recently, to the Skyrme model (Cebula *et al* 1993) can be viewed conceptually as solutions to the problem of restoring broken symmetry, since the solitons can be understood as classical solutions analogous to the mean field solutions of the many body problem. Furthermore, for each application it is clear how corrections can be included in a systematic way.

4. *Theory of effective interactions.* In a rather novel application of equation of motion methods and of the variational principle, we have suggested a new approach to the theory of the effective interactions for the nuclear many body problem (Klein and Une 1988).

Ongoing work includes the semimicroscopic theory of odd nuclei, various aspects of the problem of large amplitude collective motion, and the quantum–classical correspondence for systems with a few degrees of freedom. A major thesis of this review is that it may also be of interest to take a new look at our methods for a microscopic quantum treatment of low energy collective motion.

References

Allaart K., Boeker E., Bonsigniori G., Savoia M., Gambhir Y. K. (1988): Phys. Rep.

C169, 209

Belyaev S. T., Zelevinsky V. G. (1970): Sov. J. Nucl. Phys. **11**, 416

Belyaev S. T., Zelevinsky V. G. (1974): Sov. J. Nucl. Phys. **17**, 269

Berger J. F., Decharge J., Gogny D. (1992): *Nuclear Structure Models*, Bengtsson R., Draayer J. P., Nazarewicz W. Eds.. (World Scientific), p. 26

Bohr A., Mottelson B. R. (1975): *Nuclear Structure*, Vol.2 (Benjamin), Chap. 4

Brussard P. J., Glaudemans P. W. M. (1977): *Shell-Model Applications in Nuclear Spectroscopy* (North Holland), p. 113, 140

Cebula D., Klein A., Walet N. R. (1993): Phys. Rev. **D47**, 2113

Celenza L., Klein A., Kerman A. K. (1965): Phys. Rev. **B140**, 245

Chattopadhyay P. K., Klein A., Krejs F. (1974): Nucl. Phys. **A229**, 509

Chen Y. S., Semmes P. B., Leander G. A. (1986): Phys. Rev. **C34**, 1935

Dasso C., Klein A., Wang-Keiser C. Y., Dreiss G. J. (1973): Nucl. Phys. **A205**, 200

Do Dang G., Dreiss G. J., Dreizler R. M., Klein A., Wu C.-S. (1968a): Nucl. Phys. **A114**, 481

Do Dang G., Dreizler R. M., Klein A., Wu C. S. (1968b): Phys. Rev. **172**, 1022

Do Dang G., Dreiss G. J., Dreizler R. M., Klein A., Wu C.-S. (1968c): Nucl. Phys. **A114**, 501

Do Dang G. (1970) : *Unpublished Lectures*

Dönau F., Frauendorf S. (1977): J. Phys. Soc. Japan, (Suppl) **44**, 526

Dönau F., Frauendorf S. (1977): Phys. Lett. **71B**, 263

Dönau F., Hagemann U. (1979): Zeit. f. Physik **A293**, 31

Dönau F. (1984): Collective Phenomena in Atomic Nuclei, in *Nordic Winter School on Nuclear Physics*, Engeland T., Rekstad J., Vaagen J., S., Eds. (World Scientific)

Draayer J. P. (1992): *Nuclear Structure Models*, Bengtsson R., Draayer J. P., Nazarewicz W. eds. (World Scientific), p. 61

Dreiss G. J., Klein A. (1969): Nucl. Phys. **A139**, 81

Dreiss G. J., Dreizler R. M., Klein A., Do Dang G. (1971): Phys. Rev. **C3**, 2412

Dreizler R. M., Klein A., (1969): Phys. Lett. **30B**, 236

Elliott J. P. (1958): Proc. Roy. Soc. (London) **A245**, 128

Elliott J. P. (1958): Proc. Roy. Soc. (London) **A245**, 562

Greenberg W. R., Klein A., Li C. T. (1995): Phys. Rev. Lett. **75**, 1244

Greenberg W. R., Klein A., Zlatev I., Li C. T. (1996): Phys. Rev. **A** (to be published)

Heenen P.-H., Bonche P., Dobaczewski J., Flocard H., Krieger S. J., Meyer J., Skalski J., Tajima N., Weiss M. S. (1992): *Nuclear Structure Models*, Bengtsson R., Draayer J. P., Nazarewicz W. Eds. (World Scientific), p. 3

Jonsson S., Lyttkens J., Carlén L., Roy N., Ryde H., Walus W., Kownacki J., Hagemann G. B., Herskind B., Garrett J. D. (1984: Nucl. Phys. **A422**, 397

Kerman A. K., Klein A. (1962): Phys. Lett. **1**, 185

Kerman A. K., Klein A. (1963): Phys. Rev. **132**, 1326

Kerman A. K., Klein A. (1965): Phys. Rev. **B138**, 1323

Klein A. (1974): Revista Mexicana di Fisica **23**, 59

Klein A., Weldon A. (1978): Phys. Rev. **D17**, 1009

Klein A., Li C. T., Vassanji M. J. (1980): J. Math. Phys. **21**, 2521

Klein A., Vallières M. (1981): Phys. Lett. **98B**, 5

Klein A. (1983): Progress in Part. and Nucl. Phys., Vol. 10, Wilkinson D. Ed. (Pergamon Press) p. 39–129

Klein A., Une T. (1988): Phys. Rev. **C38**, 1897

Klein A., Marshalek E. R. (1991): Rev. Mod. Phys. **63**, 375

Klein A., Walet N. R., Do Dang G. (1991): Ann. Phys. (N.Y.) **208**, 90

Klein A., Walet N. R. (1992): *Nuclear Structure Models*, Bengtsson R., Draayer J. P., Nazarewicz W. Eds. (World Scientific), p. 229

Klein A., Walet N. R. (1994): Phys. Rev. **C49**, 1428

Kusakari H., Oshima M., Uchikura A., Sugawara M., Tomotani A., Ichikawa S., Iimura H., Morikawa T., Inamura T., Matsuzaki M. (1992): Phys. Rev. **C46**, 1257

Li C. T., Klein A., Krejs F. R. (1975): Phys. Rev. **D12**, 2311

Li C. T., Chattopadhyay P.K., Klein A., Vassanji M. J. (1979): Phys. Rev. **C19**, 2002

Li C. T., Klein A. (1979) Phys. Rev. **C19**, 2023

Nogami Y. (1964): Phys. Rev. **B134**, 313

Protopapas P., Klein A., Walet N., R. (1994): Phys. Rev. **C50**, 245

Protopapas P. (1995): Ph. D. Thesis, U. of Pennsylvania, Chap. 5

Protopapas P., Klein A., Walet N., R. (1996a): Phys. Rev. **C53**, 1655

Protopapas P., Klein A., Walet N., R. (1996b): Phys. Rev. **C54**, (to be published)

Ring P., Schuck P. (1980): *The Nuclear Many-Body Problem* (Springer-Verlag), chaps. 8–12

Rowe D. J. (1968): Rev. Mod. Phys. **40**, 153

Rowe D. J. (1970): *Nuclear Collective Motion* (Methuen), p. 229

Vallières M., Klein A., Dreizler R. M. (1973): Phys. Rev. **C7**, 2188

Vassanji M. J., Klein A. Dasso C. (1978): Phys. Rev. **C17**, 755

Wu C.-L., Feng D. H., Guidry M. (1994): *Advances in Nuclear Physics*, Vol. 21, Negele J. W., Vogt E. Eds., p. 227

Wu C. S. (1973): Phys. Rev. bf C7, 246

Zelevinsky V. G. (1983): Prog. Theor. Phys. Suppl. **74,75**, 251

Zelevinsky V., G. (1993): Nucl. Phys. **A555**, 109

Solving the Nuclear Shell Model with an Algebraic Method

Da Hsuan Feng[1], Xing-Wang Pan[1], and Mike Guidry[2,3]

[1] Drexel University, Philadelphia, PA 19104, USA
[2] University of Tennessee, Knoxville, TN 39996-1200, USA
[3] Oak Ridge National Laboratory, Oak Ridge, TN 37831, USA

Abstract. We illustrate algebraic methods in the Nuclear Shell Model through a concrete example, the Fermion Dynamical Symmetry Model (FDSM). We will use this model to introduce important concepts such as dynamical symmetry, symmetry breaking, effective symmetry, and diagonalization within a higher-symmetry basis.

1 Introduction

The shell model lies at the basis of microscopic nuclear structure physics but it can generally only be solved for light nuclei, or nuclei near closed shells. There are four modern approaches to circumventing this basic shell model problem:

1. Solution of the full problem utilizing improved algorithms and computers for traditional shell models (e. g., the shell model codes OXBASH (Brown et al. 1988), ANTOINE (Caurier 1989) and DUSM (Valliéres and Novoselsky 1993). Large-basis shell model studies using those modern shell model codes are reported in these proceedings.)
2. Stochastical path integral solutions of the shell model using Monte Carlo algorithms (e. g., Johnson et al. 1992, Ormand 1994, Koonin 1996, Honma et al. 1995).
3. Truncations of the shell model space based on guidance from mean-field models (e. g., the Hartree-Fock-Bogoliubov approach (Schmid et al. 1984, Schmid et al. 1986, Schmid et al. 1989, Bender 1996), and the Nilsson+BCS model (Hara 1996, Hara and Sun 1991))
4. Algebraic Methods in the Nuclear Shell Model (e. g., symmetry-dictated truncations of the shell model (Iachello and Arima 1987, Ginocchio 1980, Wu et al. 1994, Guidry et al. 1996)).

Approaches (1)–(3) are abundantly described in other papers in this Volume and in the literature references cited. We shall focus this discussion on symmetry-dictated truncation and its integrated solution to the shell model problem: (1) the symmetries dictate a severe truncation of the shell model space; (2) the requirement that the dominant interactions respect these symmetries provides a methodology for emphasizing a limited subset of effective interactions. Some representative examples of algebraic approaches to the shell model include

1. Elliott's SU(3) Model for the sd-shell (Elliott 1958).
2. Pseudo-Su(3) (Hecht and Adler 1969, Arima 1969) and its extensions devised for heavy deformed nuclei (Draayer 1992).
3. The Interacting Boson Model (Iachello and Arima 1987).
4. The Ginocchio Model (Ginocchio 1980).
5. The Fermion Dynamical Symmetry Model (Wu et al. 1987, Wu et al. 1994, Guidry et al. 1996)

Our discussion will concentrate on the Fermion Dynamical Symmetry Model (FDSM) as representative of a dynamical symmetry approach that can both simplify the shell model sufficiently to make calculations tractable for all heavy nuclei, and provide new conceptual insight into the nuclear many-body problem.

2   Dynamical Symmetry

A system is said to possess a dynamical symmetry if the Hamiltonian H can be expressed as a function of the Casimir invariants of a group chain $G_1 \supset G_2 \supset \ldots \supset G_n$,

$$H = f\left(C_1, C_2, \ldots, C_n\right), \tag{1}$$

where C_i is a Casimir operator of the group $G_i (i = 1, 2, \ldots n)$. If the many-body Hamiltonian exhibits an approximate dynamical symmetry, analytical solutions can be obtained that are accurate and that illustrate the physics of the problem concisely. The eigenvalues and eigenfunctions of a system having such a dynamical symmetry are given by

$$E = f\left(C_1(\nu_1), C_2(\nu_2), \ldots, C_n(\nu_n)\right) \tag{2}$$

$$\Psi = |\nu_1, \nu_2, \ldots, \nu_n\rangle \tag{3}$$

where C_i $(i = 1, 2, \ldots, n)$ are the expectation values of the Casimir operators C_i and ν_i are the quantum numbers specifying the irreducible representations (irreps) of the group G_i. Furthermore, the quantum numbers of the dynamical symmetry chain label a quantum mechanical basis for the system that is diagonal for the symmetry-limit Hamiltonian and that serves as a starting point for constructing improved solutions either perturbatively or by diagonalization of additional symmetry-breaking terms in the Hamiltonian.

Systematic comparisons of matrix elements indicate that dynamical symmetries are typically associated with collective modes of a many-body system. In many cases, the associated collective mode may be approximately identified with an excitation having geometrical significance (for example, collective rotations or vibrations). However, dynamical symmetries allow a much richer set of collective excitations to be defined since the corresponding algebraic structures need not have a geometrical counterpart.

3   Dynamical Symmetry as a Truncation Principle

A system may be truncated in such a way as to preserve a particular dynamical symmetry. We term such a truncation a *symmetry-dictated truncation*. Such symmetry-dictated truncations differ conceptually from the usual kinds of shell model truncations that we will term *energy-dictated truncations* (Wu et al. 1987, Guidry et al. 1996). An energy-dictated truncation truncates the space "spherically" in the space of symmetry generators by restricting contributing configurations according to an excitation energy prescription (energies go as the sum of the squares of the generators for compact Lie algebras; thus they select no "directions" in the symmetry space). However, a symmetry-dictated truncation reduces the space by selecting a particular "direction" (or set of directions) in the space of symmetry generators. Because such an approach involves a choice of preferred directions in the symmetry space, it is typically associated with spontaneous symmetry breaking, phase transitions, and the emergence of corresponding collective modes.

In realistic symmetry-dictated calculations both methods of truncation are used: energy-dictated truncation selects a valence space with an associated effective interaction, and symmetry-dictated truncation is then applied to that valence space to further reduce (often dramatically) the contributing configurations. Stated in other language, energy-dictated truncation selects shells and symmetry-dictated truncation selects a set of collective modes (defined algebraically in terms of the particle operators) for special emphasis in that space. The goal of such an approach is to describe a certain subset of states that can arise from interacting particles distributed in a valence space, with the effect of the remainder of the space absorbed into a renormalization of these modes. The hope is that this limited set of states dominates the low-energy configurations of realistic nuclei; this hope is motivated by the observation that such low-lying states are extremely simple in realistic nuclei when compared with the available level of complexity for a large space of interacting fermions.

4   Fermion Dynamical Symmetries in Quasi-LS Coupling

The recognition and implementation of dynamical symmetries is often facilitated by a unitary transformation to a new basis in which the symmetry and associated truncation are more easily identified than in the standard basis. This is the case for the FDSM, which is most conveniently formulated in terms of an angular momentum coupling basis that generalizes the shell model LS coupling scheme: the total single-particle angular momentum $\mathbf{j}$ is decomposed vectorially into an integer pseudoorbital part $\mathbf{k}$ and a half-integer pseudospin part $\mathbf{i}$ (Ginocchio 1980).

4.1   Hamiltonian

Let us illustrate the FDSM coupling scheme using the 6th nuclear major shell, which consists of the single-particle levels $s_{1/2}$, $d_{3/2}$, $d_{5/2}$, $g_{7/2}$, and $h_{11/2}$. Then we can consider the normal parity single-particle levels $\frac{1}{2}$, $\frac{3}{2}$, $\frac{5}{2}$, and $\frac{7}{2}$ as resulting from the coupling of a pseudospin $i = 3/2$ part and pseudoorbital $k = 2$, with an associated Lie algebra $SO(2(2i+1)) \times SO(2k+1)$. The remaining (unique-parity) orbital can then be associated with the quasispin group $\mathcal{SU}_2$. If both proton and neutron degrees of freedom are taken into account explicitly, the total symmetry coupling can be written as

$$[SO_i(8) \times SO_k(5) \times \mathcal{SU}_j(2)]_\pi \times [SO_i(8) \times SO_k(5) \times \mathcal{SU}_j(2)]_\nu$$

for protons (π) and neutrons (ν) filling the $N = 6$ valence shell (Wu et al. 1994).

In the symmetry limits of the FDSM the numbers of nucleons in the normal and abnormal parity orbits are fixed for a given nucleus; therefore the quasispin group $\mathcal{SU}_2$ associated with the abnormal levels plays no explicit dynamical role for low-lying states in this limit. (It enters implicitly through the conservation of particle number and through effective interaction parameters in the symmetry limit, and explicitly through symmetry-breaking terms in a realistic Hamiltonian.) The wavefunctions for both even and odd nuclei are given by the dynamical group chain

$$(SO_8^i \supset SO_6^i \supset SO_5^i) \times SO_5^k \supset SO_5^{i+k} \supset SO_3^{k+i} \qquad (4)$$
$$[l_1 l_2 l_3 l_4] \quad [\sigma_1 \sigma_2 \sigma_3] \quad [\tau_1 \tau_2] \quad [\tau] \qquad [\omega_1 \omega_2] \qquad J$$

where $[l_1 l_2 l_3 l_4]$, $[\sigma_1 \sigma_2 \sigma_3]$, and $[\tau_1 \tau_2]$ are the Cartan–Weyl labels for the groups SO_8, SO_6, and SO_5, respectively, $\tau = 0(1)$ for even (odd) nuclei, and k and i indicate pseudoorbital and pseudospin parts of the groups, respectively. We note the resemblance between Eq. (4) and a nuclear supersymmetry (NSUSY) group chain (see Pan et al. 1996c). The FDSM Hamiltonian is

$$H_{\text{FDSM}} = \varepsilon_1 n_1 + G_0 S^\dagger S + G_2 D^\dagger \cdot D + \sum_{r=1}^{3} B_r P^r(i) \cdot P^r(i)$$
$$+ \sum_{r=1,3} [B_r P^r(k) \cdot P^r(k) + 2b_r P^r(i) \cdot P^r(k)], \qquad (5)$$

where ε_1 is the energy for the normal parity orbits (assumed degenerate) and n_1 is the number of nucleons in the normal parity orbits,

$$S^\dagger = A^{0\dagger} \qquad D_\mu^\dagger = A_\mu^{2\dagger}$$
$$A_\mu^{r\dagger} = \sqrt{\Omega_1/2} \left[b_{ki}^\dagger b_{ki}^\dagger \right]_{0\mu}^{0r} \qquad (r = 0, 2) \qquad (6)$$

where $k = 2$, $i = \frac{3}{2}$ and $\Omega_1 \equiv \Omega_{ki} = (2k + 1)(2i + 1)/2$. Similarly,

$$P_\mu^r(i) = \sqrt{\Omega_1/2} \left[b_{ki}^\dagger \tilde{b}_{ki} \right]_{0\mu}^{0r} \qquad (r = 0, 1, 2, 3) \qquad (7)$$

$$\bar{P}_\mu^r(k) = (-)^{[\frac{r}{2}]} \sqrt{8/5} P_\mu^r(k) \qquad P_\mu^r(k) = \sqrt{\Omega_1/2} \left[b_{ki}^\dagger \tilde{b}_{ki} \right]_{\mu 0}^{r0} \qquad (r = 0, 1, 2, 3), \qquad (8)$$

where $\left[\frac{r}{2}\right]$ is the integer part of $\frac{r}{2}$. The operators $P_\mu^r(i)$ and $\bar{P}_\mu^r(k)$ for $r = 1$ and 3 form the Lie algebras SO_5^i and SO_5^k, respectively. The commutators among the $P_\mu^r(i)$ are given by Eq. (3.12) of ref. (Wu et al. 1987) for the i-active case and those for $\sqrt{\Omega_1/2}[b_{ki}^\dagger \tilde{b}_{ki}]_{\mu 0}^{r0}$ can be obtained from Eq. (3.12) of ref.(Wu et al. 1987) for the k-active case with

$$\sqrt{3}\begin{Bmatrix} r & s & t \\ 1 & 1 & 1 \end{Bmatrix} \longrightarrow \sqrt{5}\begin{Bmatrix} r & s & t \\ 2 & 2 & 2 \end{Bmatrix}.$$

In Eq. (8) we have renormalized the multipole operators $\bar{P}_\mu^r(k)$ so that they are isomorphic with $P_\mu^r(i)$. Furthermore, $P_\mu^1(i)$ and $\bar{P}_\mu^1(k)$ are related to the total pseudoorbital angular momentum and pseudospin by

$$P_\mu^1(i) = \frac{1}{\sqrt{5}}I_\mu, \qquad \bar{P}_\mu^1(k) = \frac{1}{\sqrt{5}}L_\mu. \tag{9}$$

By using the Casimir operators of SO_8, SO_6 and SO_5, the Hamiltonian of Eq. (5) can be rewritten as

$$\begin{aligned} H_{FDSM} = H_0 + \epsilon_1 n_1 + g_S S^\dagger \cdot S + g_6 C_{SO_6^i} + g_5 C_{SO_5^{k+i}} \\ + g_5^i C_{SO_5^i} + g_5^k C_{SO_5^k} + g_I \mathbf{I}^2 + g_L \mathbf{L}^2 + g_J \mathbf{J}^2, \end{aligned} \tag{10}$$

where the total angular momentum is $\mathbf{J} = \mathbf{I} + \mathbf{L}$.

In Fig. 1, the transitions between symmetry limits are presented by a pairing plus quadrupole hamiltonian. For the $SO(6)$ limit, which is realized with certain conditions in the Hamiltonian parameters in eq. (5) (Pan et al. 1996c), the analytical eigenvalues of energy for even and odd systems are

$$E^{\text{even}} = E_0^{(e)} + g_6\sigma(\sigma + 4) + g_5'\tau(\tau + 3) + g_I' J(J + 1), \tag{11}$$

where $J(J + 1)$ is used instead of $I(I + 1)$ [since $L = 0$, and thus $J = I$], and

$$\begin{aligned} E^{\text{odd}} = E_0^{(o)} + g_6[\sigma_1(\sigma_1 + 4) + \sigma_2(\sigma_2 + 2) + (\sigma_3)^2] \\ + g_J J(J + 1) + (g_5' - g_5)[\tau_1(\tau_1 + 3) + \tau_2(\tau_2 + 1)] \\ + g_5[\omega_1(\omega_1 + 3) + \omega_2(\omega_2 + 1)]. \end{aligned} \tag{12}$$

Reduction rules for the quantum numbers are given in (Pan et al. 1996c) and vector and spinor group representations are for even and odd system, respectively. In (Pan et al. 1996c), this unified description of even and odd systems is applied to the Xe–Ba region.

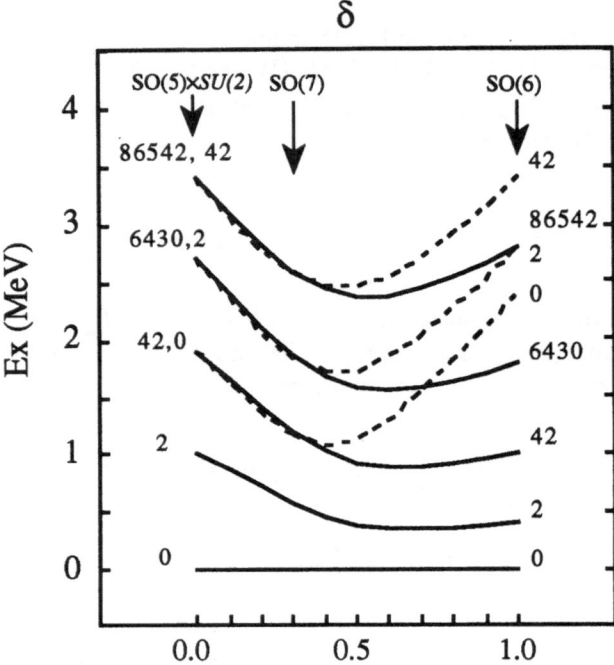

Fig. 1. Transition of the spectrum from vibrational (SO(5)×$SU(2)$) via SO(7) to γ-soft (SO(6)). The numbers inside the box are the permissable angular momenta of a given SO(5) multiplet. The Hamiltonian is of pairing plus quadrupole form: $H = -0.05(1 - \delta)S^\dagger S - 0.1\delta P^2 \cdot P^2$, and the major shell degeneracy $\Omega = \sum j(j + \frac{1}{2})$ is taken to be 20. The figure is from (Pan and Feng 1994a).

4.2 Wave Functions

For even systems the wave functions are

$$|N_1\sigma\tau n_\Delta IM\rangle = \mathcal{P}_{N_1\sigma\tau}|N_1\sigma\tau n_\Delta IM\rangle^{\text{IBM}}_{b\to f}, \qquad (13)$$

where $\mathcal{P}_{N_1\sigma\tau}$ is a Pauli factor,

$$\mathcal{P}_{N_1\sigma\tau} = \left[\frac{(\Omega_1 - N_1 - \sigma)!!(\Omega_1 - N_1 + \sigma + 4)!!}{\Omega_1!!(\Omega_1 + 4)!!}\right]^{\frac{1}{2}}, \qquad (14)$$

and $|N_1\sigma\tau n_\Delta IM\rangle^{\text{IBM}}_{b\to f}$ denotes a wave function resulting from replacing the boson operators $s^\dagger$ and $d^\dagger_\mu$ by the fermion operators $S^\dagger$ and $D^\dagger_\mu$, in the $U_6 \supset O_6 \supset O_5 \supset O_3$ IBM wave function $|N_1\sigma\tau n_\Delta IM\rangle^{\text{IBM}}$.

For odd systems the wave functions are

$$|2N_1 + 1, \{\sigma_1 + \tfrac{1}{2}, \tau_1\} n_\Delta IM, [10]k = 2, m\rangle$$

$$= \frac{1}{\sqrt{2^{N_1}}} K_l^{-1} \left(N_1, \langle \sigma + \tfrac{1}{2} \rangle \right)$$

$$\times \sum_{\tau I' n'_\Delta} \xi_{\sigma + \frac{1}{2} \tau_1 n_\Delta I}^{\sigma \tau n'_\Delta I'} \left[b_{2m,3/2}^\dagger | N_1 \sigma \tau n'_\Delta I' \rangle_{b \to f}^{\mathrm{IBM}} \right]_M^I, \tag{15}$$

where the shorthand notation stands for

$$\left\{ \sigma + \tfrac{1}{2}, \tau_1 \right\} \equiv \left\{ \langle \sigma + \tfrac{1}{2} \rangle, [\tau_1 \tfrac{1}{2}] \right\}, \qquad \langle \sigma + \tfrac{1}{2} \rangle \equiv \left[\sigma + \tfrac{1}{2}, \tfrac{1}{2} \tfrac{1}{2} \right], \tag{16}$$

and the matrix K_l^{-1} contains a Pauli factor,

$$K_l^{-1}(N_1, \langle \sigma + \tfrac{1}{2} \rangle) = \left[\frac{2^{-N_1} \Omega_1}{\Omega_1 - N_1 - \sigma} \right]^{\frac{1}{2}} \mathcal{P}_{N_1 \sigma \tau}. \tag{17}$$

4.3 Electromagnetic Transitions

In Ref. (Wu et al. 1994), the $E2$ transition operator in the FDSM is defined as

$$T(E2)_\mu^2 = q P_\mu^2(i), \tag{18}$$

while the $E2$ transition operator for the IBM SO_6 limit is (Iachello and Kuyucak 1981)

$$T(E2)_\mu^2 = q B_\mu^2, \qquad B_\mu^2 = (d^\dagger \tilde{s} + s^\dagger \tilde{d})_\mu^2. \tag{19}$$

Owing to the isomorphism between the commutators for the FDSM and IBM:

$$[P_\mu^2(i), S^\dagger] \longleftrightarrow [B_\mu^2, s^\dagger], \tag{20}$$

$$[P_\mu^2(i), D_\nu^\dagger] \longleftrightarrow [B_\mu^2, d_\nu^\dagger], \tag{21}$$

the formula for the reduced matrix elements of the $E2$ transition operator in the FDSM is identical to that in the IBM,

$$\langle N\sigma\tau'n'_\Delta J' \parallel P^2(i) \parallel N\sigma\tau n_\Delta J \rangle^{\mathrm{FDSM}} = (N\sigma\tau'n'_\Delta J' \parallel B^2 \parallel N\sigma\tau n_\Delta J)^{\mathrm{IBM}} \tag{22}$$

Note that there are no Pauli factors in this symmetry limit for the $E2$ transition rate when the $E2$ operator is defined through the quadrupole operator. A Pauli effect will appear in the $E2$ transitions if one supplements this minimal $E2$ operator with an additional term $(D^\dagger \tilde{D})_\mu^2$ (Pan et al. 1996c).

For the $u = 1$ case, the $E2$ transition operator can be defined as

$$T(E2)_\mu^2 = q P_\mu^2(i) + q'' P_\mu^2(k). \tag{23}$$

The reduced matrix element is

$$\langle\{N_1 + \frac{1}{2}, \tau_1'\}[\omega_1'\omega_2']J' \| qP^2(i) + q''P^2(k) \| \{N_1 + \frac{1}{2}, \tau_1\}[\omega_1\omega_2]J\rangle$$

$$= \sum_{II'} \left(\begin{array}{c|c}[\tau'\frac{1}{2}] \ [10] & [\omega_1'\omega_2'] \\ I' & 2 & J'\end{array}\right)\left(\begin{array}{c|c}[\tau_1\frac{1}{2}] \ [10] & [\omega_1\omega_2] \\ I & 2 & J\end{array}\right) M, \qquad (24)$$

where M is given by

$$M = \langle\{N_1 + \frac{1}{2}, \tau_1'\}(I', [10]k = 2)J' \| qP^2(i) \| \{N_1 + \frac{1}{2}, \tau_1\}(I, [10]k = 2)J\rangle$$

$$+ \langle\{N_1 + \frac{1}{2}, \tau_1'\}(I', [10]k = 2)J' \| q''P^2(k) \| \{N_1 + \frac{1}{2}, \tau_1\}(I, [10]k = 2)J\rangle.$$

According to Eq. (6.8) in (Judd 1963),

$$[b_{ki}^\dagger b_{ki}]_{\mu 0}^{K0} = \frac{1}{\sqrt{2i+1}}\sum_{p=1}^{n}[b_k^\dagger(p)b_k(p)]_\mu^K. \qquad (25)$$

Therefore, in computing the matrix elements of $P_\mu^2(k)$ the operator can be replaced by

$$P_\mu^2(k) = \sqrt{\frac{\Omega_1}{8}}\sum_{p=1}^{n}[b_k^\dagger(p)b_k(p)]_\mu^2. \qquad (26)$$

Using (26) we have

$$M = \hat{J}\hat{J}'(-)^{I'+J}\left\{\begin{array}{ccc}J' & J & 2 \\ I & I' & 2\end{array}\right\}\langle\{N_1 + \frac{1}{2}, \tau_1'\}I' \| qP^2(i) \| \{N + \frac{1}{2}, \tau_1\}I\rangle$$

$$+ q''\delta_{\tau_1\tau_1'}\delta_{II'}(-)^{I+J'}\hat{J}\hat{J}'\sqrt{\frac{5\Omega_1}{2}}\frac{\hat{I}}{\hat{i}}\frac{(\Omega_1 - 2N_1 - 1)}{(\Omega_1 - 1)}\left\{\begin{array}{ccc}J' & J & 2 \\ 2 & 2 & I\end{array}\right\}. \qquad (27)$$

Now only the matrix elements of $P_\mu^2(i)$ remain to be calculated. The generators of $Spin(6)$ for the IBFM are

$$G_\mu^2 = B_\mu^2 + F_\mu^2, \qquad F_\mu^2 = (a_{\frac{3}{2}}^\dagger \tilde{a}_{\frac{3}{2}})_\mu^2 \qquad (28)$$

corresponding to the commutator for the IBFM

$$[F_\mu^2, a_{\frac{3}{2}m_i}^\dagger] = (-1)^{\frac{3}{2}-m_i}\langle im_i + \mu, i - m_i|2\mu\rangle a_{\frac{3}{2}m_i+\mu}^\dagger. \qquad (29)$$

There is a similar commutator in the FDSM

$$[P_\mu^2(i), b_{22\frac{3}{2}m_i}^\dagger] = \sqrt{\frac{\Omega_1}{2(2k+1)}}(-1)^{\frac{3}{2}-m_i}\langle im_i + \mu, i - m_i|2\mu\rangle b_{22\frac{3}{2}m_i+\mu}^\dagger, \qquad (30)$$

where $\sigma = \pi$ or ν, and the factor $[\Omega_1/(2(2k+1))]^{1/2}$ is always equal to 1 for the 6th shell.

Because of eqs. (20–21) and eqs. (29–30), we have the following isomorphism between the commutators in the FDSM and IBFM,

$$[P_\mu^2, S^\dagger] \longleftrightarrow [B_\mu^2, s^\dagger] = [G_\mu^2, s^\dagger], \tag{31}$$

$$[P_\mu^2(i), D_\nu^\dagger] \longleftrightarrow [B_\mu^2, d_\nu^\dagger] = [G_\mu^2, d_\nu^\dagger], \tag{32}$$

$$[P_\mu^2(i), b_{22\frac{3}{2}m_i}^\dagger] \longleftrightarrow [F_\mu^2, a_{\frac{3}{2}m_i}^\dagger] = [G_\mu^2, a_{\frac{3}{2}m_i}^\dagger]. \tag{33}$$

Therefore we establish the following identity:

$$\langle \{N + \frac{1}{2}, \tau_1'\} I' \| P^2(i) \| \{N + \frac{1}{2}, \tau_1\} I \rangle^{\text{FDSM}}$$

$$= \left(\{N + \frac{1}{2}, \tau_1'\} I' \| G^2 \| \{N + \frac{1}{2}, \tau_1\} I \right)^{\text{IBFM}}. \tag{34}$$

The reduced matrix element of G_μ^2 is derived in Ref. (Iachello and Kuyucak 1981). With these results we can calculate the $B(E2)$ values and the quadrupole moments for odd-mass nuclei. In Fig. 2, typical E2 transitions in three symmetry limits are illustrated.

5 Symmetry Breaking

Although dynamical symmetries are often realized rather well in the properties of low-lying states, the realistic case generally involves some level of symmetry breaking.

5.1 Deviation from the Symmetry Limit

The condition for realization of the $SO(6)$ limit is that the pairing strength g_S in eq. (10) should vanish. First, let us assume that g_S is non-zero and small, so that the term $g_S S^\dagger \cdot S$ is perturbative. Thus the spectrum

$$E^{\text{even}} = E_0^{(e)} + g_6 \sigma(\sigma + 4) + g_5' \tau(\tau + 3) + g_I' J(J + 1) + g_S \langle S^\dagger S \rangle, \tag{35}$$

deviates slightly from the $SO(6)$ spectrum of eq. (11). From the reduced matrix elements

$$\langle N + 1, \sigma + 1, \tau \| S^\dagger \| N\sigma\tau \rangle =$$

$$\left[\frac{(\Omega_1 - \sigma - N)(\sigma - \tau + 1)(\sigma + \tau + 4)(N + \sigma + 6)}{4(\sigma + 2)(\sigma + 3)} \right]^{\frac{1}{2}} \tag{36}$$

and

$$\langle N + 1, \sigma - 1, \tau \| S^\dagger \| N\sigma\tau \rangle =$$

$$- \left[\frac{(\Omega_1 + \sigma - N + 4)(\sigma - \tau)(\sigma + \tau + 3)(N - \sigma + 2)}{4(\sigma + 1)(\sigma + 2)} \right]^{\frac{1}{2}} \tag{37}$$

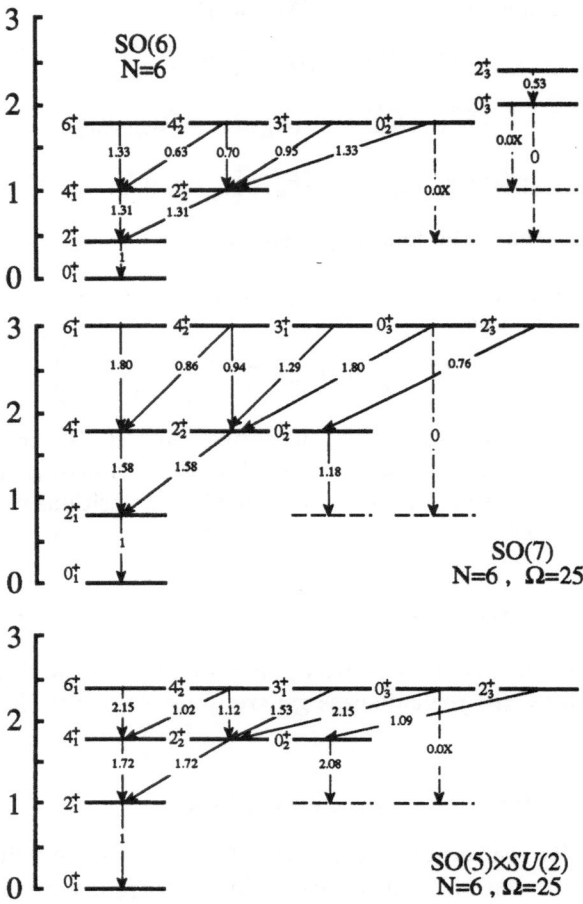

Fig. 2. Typical level schemes and E2 transitions for SO(6), SO(7) and SO(5)×SU(2) symmetries. The B(E2) values are in units of $B(E2; 2_1^+ \rightarrow 0_1^+)$.

we can expect that the resulting spectrum of the $SO(6)$ + Pairing Hamiltonian will not exactly obey the $\tau(\tau+3)$ rule, since the reduced matrix elements depends on the $SO(5)$ quantum number τ but not the angular momentum J. On the other hand, the pairing interaction $S^\dagger \cdot S$ is a scalar in $SO(5)$ and therefore the $SO(6)$ + Pairing Hamiltonian will still preserve the basic $SO(5)$ scheme with τ still a good quantum number. The term $S^\dagger S$ transforms as a tensor $T^{\langle 2 \rangle}$ under $SO(6)$. Thus for the symmetric irrep of $SO(6)$, the Knoenecker product contains the $SO(6)$ irreps $\langle \sigma + 2 \rangle$, $\langle \sigma \rangle$ and $\langle \sigma - 2 \rangle$, which implies that the matrix elements of $S^\dagger \cdot S$ are non-vanishing only when $\Delta\sigma = 0, \pm 2$. When the pairing interaction is included perturbatively up to second order, the approximate $SO(6)$ + Pairing spectrum assumes a simple form (Pan et al. 1992, Pan et al. 1994):

$$E'^{\text{even}} \cong E_0^{(e)} + g_6\sigma(\sigma + 4) + A'\tau(\tau + 3) - B'[\tau(\tau + 3)]^2 + g_I' J(J + 1). \quad (38)$$

Fig. 3 illustrates this spectrum for ^{128}Xe . Clearly the theoretical results in Fig. 3c are in good agreement with the data; in particular, the observed τ compression is well reproduced.

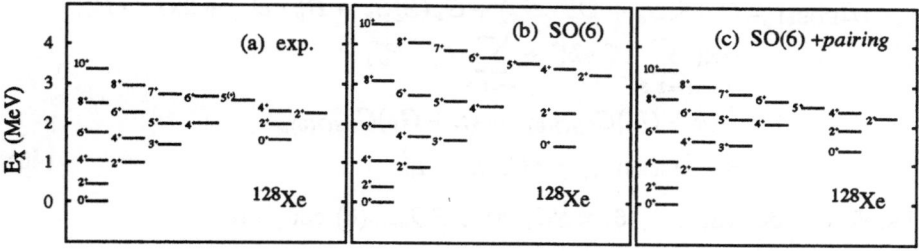

Fig. 3. Comparison between the experimental spectrum of ^{128}Xe and the spectra of $SO(6)$ and $SO(6)$ + Pairing Hamiltonians. The figure is taken from (Pan et al. 1994).

5.2 Mixed Symmetries

In the interacting boson model (IBM) (Iachello and Arima 1987), nuclear low-lying quadupole collective states are treated as totally symmetric states in terms of the $SU(6)$ algebra of IBM-1; thus, proton and neutron bosons are not distinguished. Proton and neutron boson degrees of freedom are distinguished in the IBM-2 (Arima et al. 1977, Otsuka 1981) and both symmetric and mixed-symmetry states will appear. States are of symmetric or mixed-symmetry character depending on whether they are symmetric or non-symmetric representations of a given $G_{\pi+\nu}$ symmetry for the proton–neutron system. In IBM-2, F-spin (Iachello 1984) provides a simple classification of the low-lying symmetric and mixed-symmetry states. The IBM-2 Hamiltonian under this coupling is F-spin invariant and the phemomenological Majorana interaction,

$$\langle \xi M \rangle = \left\langle \tfrac{\xi}{2} \left(\hat{N}(\hat{N} + 5) - \hat{C}_{2U(6)} \right) \right\rangle = \xi f(N - f + 1) \quad (39)$$

(where $C_{2U(6)}$ is the quadratic Casimir operator of $U_{\pi+\nu}(6)$), which is diagonal for all three symmetry limits discussed above, is introduced to adjust mixed-symmetry states to their correct positions.

However, if one starts from a truncated fermion subspace the F-spin formalism is not available to classify the symmetric and mixed symmetry states. Physically, the Hamiltonian for the proton part is not necessarily symmetric to the neutron part in realistic cases. Furthermore, there is no obvious correspondence for the Majorana interaction in a truncated fermion shell model because the microscopic origin of the Majorana force is not clear. Therefore, one may ask a question: How is one to treat low-lying symmetric and mixed-symmetry states in a truncated fermion shell model?

In more realistic cases the contributions of $P_\pi^2 \cdot P_\pi^2$ and $P_\nu^2 \cdot P_\nu^2$ are much smaller than $P_\pi^2 \cdot P_\nu^2$ and an $SO(6)$ dynamical symmetry is still obtained,

$$
\begin{aligned}
H_{\mathrm{FDSM}} =\ & G_\pi(S_\pi^\dagger S_\pi + D_\pi^\dagger \cdot D_\pi) + G_\nu(S_\nu^\dagger S_\nu + D_\nu^\dagger \cdot D_\nu) + 2\kappa P_\pi^2 \cdot P_\nu^2 \\
& -\kappa\Big(\sum_{r=1,3} P_\pi^r \cdot P_\pi^r + \sum_{r=1,3} P_\nu^r \cdot P_\nu^r\Big) \\
=\ & -(\kappa + G_\pi)C_{2SO(6)_\pi} - (\kappa + G_\nu)C_{2SO(6)_\nu} \\
& +\kappa C_{2SO(6)_{\pi+\nu}} - \kappa C_{2SO(5)_{\pi+\nu}}
\end{aligned}
\tag{40}
$$

Therefore under the $SO_\pi(6) \times SO_\nu(6) \supset SO_{\pi+\nu}(6)$ coupling

$$
\langle \sigma_\pi 00\rangle \otimes \langle \sigma_\nu 00\rangle = \langle \sigma_1, \sigma_2\rangle = \sum_{l=0}^{\min(\sigma_\pi,\sigma_\nu)} \sum_k \langle \sigma_\pi + \sigma_\nu - 2l - k, k\rangle ,
\tag{41}
$$

where k satisfies $\sigma_\pi + \sigma_\nu - 2l \geq 2k \geq 0$ and $\min(\sigma_\pi, \sigma_\nu) - l \geq k$, the eigenvalues of eq. (40) are

$$
\begin{aligned}
E =\ & -(\kappa + G_\pi)\sigma_\pi(\sigma_\pi + 4) - (\kappa + G_\nu)\sigma_\nu(\sigma_\nu + 4) \\
& +\kappa(\sigma_1(\sigma_1 + 4) + \sigma_2(\sigma_2 + 2)) - \kappa(\tau_1(\tau + 3) + \tau_2(\tau + 1)) .
\end{aligned}
\tag{42}
$$

The reduction rules for some lower representations from $\langle \sigma_1, \sigma_2\rangle$ to $\langle \tau_1, \tau_2\rangle$ of $SO_{\pi+\nu}(5)$ are given in (Iachello and Arima 1987). Note that in the above spectrum the $J(J+1)$ term is ignored for simplicity. It can be added to the Hamiltonian through the terms $(P_\pi^1 \cdot P_\pi^1 + P_\nu^1 \cdot P_\nu^1 + P_\pi^1 \cdot P_\nu^1)$. The spectrum typical of eq. (42) is shown in Fig. 4.

From eq. (42) and Fig. 4, one can see that in contrast to the $\langle \sigma_\pi + \sigma_\nu, 0\rangle$ irrep, which lies lowest because of the attractive coupling (i.e., $\kappa < 0$), $\langle \sigma_\pi + \sigma_\nu - 1, 1\rangle$ and $\langle \sigma_\pi + \sigma_\nu - 2, 0\rangle$ are lower than other $SO_{\pi+\nu}(6)$ irreps. Note that here $\langle \sigma_\pi + \sigma_\nu - 2, 0\rangle$ means $|\langle \sigma_\pi - 2\rangle \otimes \langle \sigma_\nu\rangle : \langle \sigma_\pi + \sigma_\nu - 2, 0\rangle > $ or $|\langle \sigma_\pi\rangle \otimes \langle \sigma_\nu - 2\rangle : \langle \sigma_\pi + \sigma_\nu - 2, 0\rangle > $ because $|\langle \sigma_\pi\rangle \otimes \langle \sigma_\nu\rangle > : \langle \sigma_\pi + \sigma_\nu - 2, 0\rangle$ always lies above the $\langle \sigma_\pi + \sigma_\nu - 1, 1\rangle$ irrep as shown in Fig. 4. Their relative energies are

$$
\Delta E(\langle N - 1, 1\rangle) = -2\kappa(N + \frac{1}{2}) ,
\tag{43}
$$

$$
\Delta E(\langle N - 2, 0\rangle_\pi) = -4\kappa\sigma_\nu + 4G_\pi(\sigma + 1) ,
\tag{44}
$$

$$
\Delta E(\langle N - 2, 0\rangle_\nu) = -4\kappa\sigma_\pi + 4G_\nu(\sigma + 1) .
\tag{45}
$$

From these equations it is easy to understand that the the mixed-symmetry irreps like $\langle N-1,1\rangle$ can be pushed up by increasing the strength of κ, while the symmetric irrep can be shifted up by increasing the pairing strengths. In (Pan and Feng 1994b), the above symmetry argument is applied to describe mixed symmetry states in ^{132}Ba.

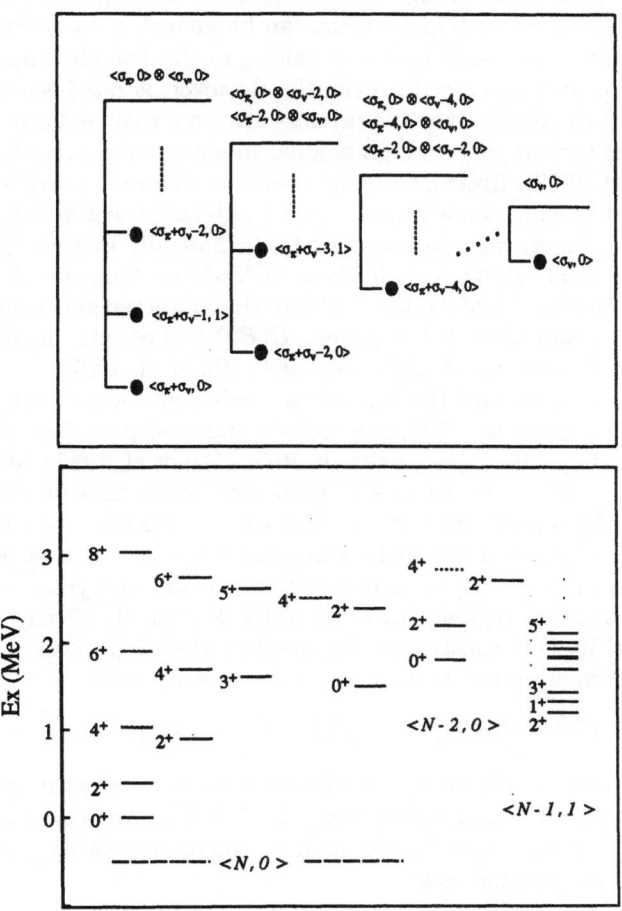

Fig. 4. Upper Panel is the $SO(6)_\pi \times SO(6)_\nu$ coupling scheme. Lower Panel is a typical dynamical symmetry limit spectrum of $SO(6)_{\pi+\nu}$ with $E = -67C_{SO(6)} + 83C_{SO(5)} + 10J(J+1)$. Since the Hamiltonian has no Majorana term, the states in the $\langle N-1,1\rangle$ representation lie quite low. Thus, identification of the symmetric and mixed-symmetry states is difficult.

5.3   Broken Pairs

The space of no broken pairs (technically the heritage $u = 0$ space) that we
have discussed to this point gives a good description for even–even nuclei in
states of low angular momentum. However, numerical calculations within the
$u = 0$ space consistently overestimate the energies and underestimate the $B(E2)$
values for higher angular momentum states, and these discrepancies increase
with angular momentum. Some improvement can be gained in the energies by
correcting perturbatively for the influence of pairing in the same manner as the
$SO(5)$ + Pairing approach discussed previously. However, it has been realized
since the inception of the FDSM that the primary reason for these discrepancies
is the contribution of broken pairs to high angular momentum states (Guidry et
al. 1986, Guidry et al. 1987). Broken pairs in normal or abnormal parity orbitals
may carry significant angular momentum, thereby reducing the amount carried
by the S–D condensate. As was demonstrated schematically in Refs. (Guidry
et al. 1986, Guidry et al. 1987), this leads to a Variable Moment of Inertia
(VMI) behavior for the moment of inertia similar to that observed experimentally,
and brings calculated and observed high-spin $B(E2)$ values into quantitative
agreement with data (Guidry et al. 1987, Wu 1990, Wu et al. 1987).

   Thus, it is natural to extend the numerical implementation of symmetry-
dictated truncation by using the FDSM to include unpaired particles. This has
been discussed for a single unpaired particle in Refs. (Wu et al. 1987a, Wu et al.
1987b, Wu et al. 1988, Wu et al. 1987). We shall not discuss that further here,
but instead conclude by summarizing recent work that incorporates broken pairs
in such calculations. We have developed a computer code SU3su2 that includes
explicit broken pairs in an $SU(3)$ limit in order to examine the yrast states in
the rare earth and actinide regions (Di et al. 1993, Pan et al. 1996a). In this
extension, the model basis is constructed by coupling the $SU(3)$ basis and one
broken pair in the normal parity levels or the unique parity level

$$(S, D)^{N-1} \otimes A'(i) \quad \text{or} \quad (S, D)^{N-1} \otimes A'(j_0). \tag{46}$$

where S and D are the usual symmetry-dictated coherent fermion pairs with coupled angular momentum 0 and 2, respectively, $A'(i)$ designates broken normal-
parity pairs, and $A'(j_0)$ designates broken pairs in the unique parity level. The
corresponding creation operators are

$$A'^{\dagger}(i) = \sum_{ki} \sqrt{\Omega_1/2} \left[b_{ki}^{\dagger} b_{ki}^{\dagger} \right]_{(M_K M_I) M_{L'}}^{(KI) L'}, \tag{47}$$

$$A'^{\dagger}(j_0) = \sum_{ki} \sqrt{\Omega_1/2} \left[b_{j_0}^{\dagger} b_{j_0}^{\dagger} \right]_{M_{I_0}}^{I_0}. \tag{48}$$

The basis for the $SU(3)$ core is

$$\begin{array}{cccc} Sp(6) & \supset & SU(3) & \supset SO(3) \\ N_1, u = 0 & & (\lambda, \mu) & \kappa L \end{array} \tag{49}$$

where κ is an additional quantum number to distinguish orthogonal states having the same (λ, μ) and J. For an $SU(3)$ core and one broken pair, the following coupling schemes are possible:

$$|N_1 u = 0(\lambda_1, \mu_1) \otimes (2,0); (\lambda_2, \mu_2)KIJM_J\rangle,$$

$$|N_1 u = 0(\lambda_1, \mu_1)\kappa L \otimes K_i; KIJM_J\rangle, \qquad (50)$$

$$|N_1 u = 0(\lambda_1, \mu_1)\kappa L \otimes I_0; JM_J\rangle.$$

and the total Hamiltonian is

$$H = H_{SD} + H_{A'} + H_{\text{mix}}, \qquad (51)$$

where H_{SD} is the $SU(3)$-plus-Pairing Hamiltonian, $H_{A'}$ corresponds to the broken pair, and H_{mix} is the interaction between the (S, D) core and the broken pair that leads to the mixing of heritage. The Hamiltonian can be expressed explicitly as

$$
\begin{aligned}
H = {} & \alpha K \cdot K + \gamma I \cdot I - \delta I \cdot K + \gamma_0 I_0 \cdot I_0 - \delta_0 I_0 \cdot K \\
& - \kappa P^2(k) \cdot P^2(i) + \kappa'(D^\dagger D)^2 \cdot P^2(i) - \kappa_0 P^2(k) \cdot P^2(i_0) \\
& + \kappa_0'(D^\dagger D)^2 \cdot P^2(i_0) + \Delta B ,
\end{aligned}
\qquad (52)
$$

where ΔB is the energy required to break one pair in either normal parity or abnormal parity levels.

In Fig. 5, we present the results of a $u = 2$ calculation of the energy levels up to $J = 20$ for the even–even isotopes $^{160-166}$Er (Pan et al. 1996a). The inclusion of a single broken pair in the $SU(3)$ symmetry limit leads to a quantitative description of the spectrum. Preliminary indications are that the inclusion of a broken pair leads to a substantial increase in the $B(E2)$ strengths at higher spins, but the calculated values are still low in the angular momentum 10–20 region, suggesting that configurations having two broken pairs may be important for a quantitative description of $B(E2)$ values near angular momentum 20 for rare earth nuclei.

6   Effective Dynamical Symmetries

Symmetry limit calculations are attractive because of their algebraic brevity and simple physical insight, but we often encounter non-symmetric cases in realistic situations. As we have already discussed, it is then necessary to resort to inclusion of symmetry-breaking terms. However, in some cases, the opposite situation occurs: a system may exhibit a very good effective dynamical symmetry, even for cases where formally the symmetry would not be expected to exist. We now give an example to show that a very accurate fermion $SO(6)$ dynamical symmetry exists for ^{196}Pt (and other nuclei in that region), even though the coupled system has no such formal dynamical symmetry in the Fermion Dynamical Symmetry Model.

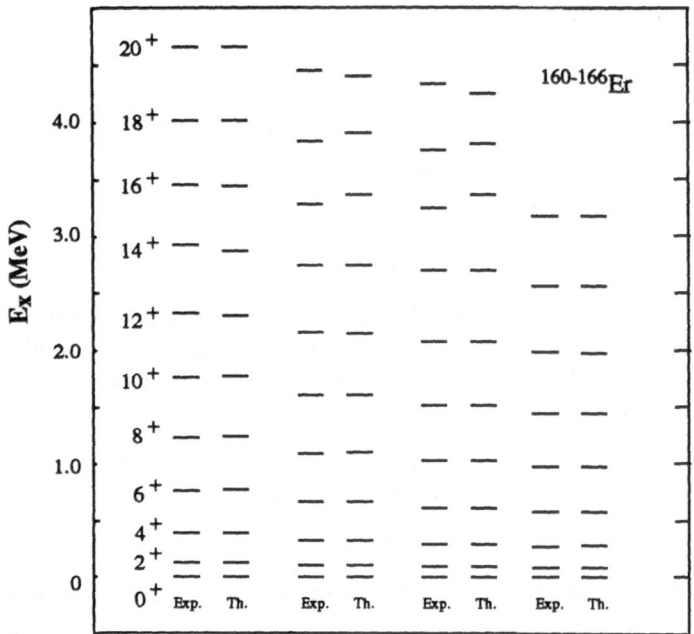

Fig. 5. Experimental and theoretical energy levels in some even–even Er isotopes calculated with a single broken pair using the code SU3su2 (Di et al. 1993, Pan et al. 1996a). The spectrum of ^{160}Er is to the left and that of ^{166}Er to the right. The figure is taken from (Guidry et al. 1996).

In Sect. 2, we discussed the $SO(6)$ dynamical symmetry limit. This limit is applicable to the xenon and barium regions since the 50–82 shell is the valence shell for both the protons and the neutrons, and an $SO(8) \supset SO(6)$ dynamical symmetry chain can be constructed algebraically in this shell. This is consistent with observation of clear $SO(6)$ behavior in this region. However, it is not obvious that ^{196}Pt can be $SO(6)$-like in the FDSM because the shell symmetry for rare earth nuclei is $Sp^\nu(6) \times SO^\pi(8)$, which has no n–p coupled $SO(6)$ dynamical symmetry chain. Thus, it is a challenge for the FDSM to explain why there is an $SO(6)$ dynamical symmetry for the platinum isotopes, as indicated by the data, and as obtained in the IBM. (The IBM has no inherent difficulty in obtaining the $SO(6)$ limit from $SU^\nu(6) \times SU^\pi(6)$ coupling since the IBM symmetry is empirical and not constrained by the underlying shell structure as is the case for the FDSM).

In (Feng et al. 1993, Ping et al. 1996), we reported the discovery of a very well-defined effective $SO(6)$ fermion dynamical symmetry for this region with a Pairing + Quadrupole Hamiltonian:

$$H = G'_{0\pi} S^\dagger_\pi S_\pi + G'_{0\nu} S^\dagger_\nu S_\nu + B'_{2\pi} P^2_\pi \cdot P^2_\pi + B'_{2\nu} P^2_\nu \cdot P^2_\nu + B_{2\pi\nu} P^2_\pi \cdot P^2_\nu, \qquad (53)$$

where the quadrupole interactions between neutrons and protons are significantly larger than those among protons or among neutrons. (In the FDSM the explicit quadrupole pairing interactions can always be ignored if the spectrum is our only concern (Wu et al. 1987); the quadrupole pairing is taken into account by redefining the parameters in eq. (1): $G'_{0\sigma} = G_{0\sigma} - G_{2\sigma}$ and $B'_{r\sigma} = B_{r\sigma} - G_{2\sigma}$ ($\sigma = \pi, \nu$).)

In Fig. 6, energy levels and branching ratios are compared with observations and with IBM calculations for ^{196}Pt. Overall the FDSM results are in quantitative agreement with the IBM-1 $O(6)$ level pattern and branching ratios. This agreement confirms that while the FDSM does not have an explicit mathematical $SO(6)$ dynamical symmetry, it has a remarkably accurate practical one for this case.

Because of the different shell symmetries in the FDSM, the effective $SO(6)$ dynamical symmetry in the platinum region and the $SO(6)$ symmetry of the mass-130 region differ fundamentally at the microscopic level. The Xe–Ba region corresponds to an $SO^\nu(8) \times SO^\pi(8)$ coupling scheme in the FDSM; this admits a coupled $SO(6)^{\pi+\nu}$ dynamical symmetry and a formal $SO(6)$ symmetry is possible for these nuclei. For Pt isotopes, the realization of the $SO(6)$-like structure is critically dependent on a specific Hamiltonian and the number of valence pairs. Thus, in this view the $SO(6)$ dynamical symmetry of the mass-130 region is required by the shell structure, while (ironically) the original example of $O(6)$ symmetry in the Pt nuclei is more accidental. The shell structure permits it, but does not demand it. This would explain the more ubiquitous appearance of $SO(6)$ symmetry in the Xe–Ba region. We note in this connection that the Xe–Ba region not only exhibits an $SO(6)$ symmetry in the even–even isotopes, but in the even–odd ones as well, and that the level pattern for the latter is quite different from the Pt even–odd ones.

7   Diagonalization within a Higher-Symmetry Basis

In this section we use the Pairing + Quadrupole Hamiltonian given in the previous section to carry out systematic numerical calculations in rare-earth nuclei that are not restricted to symmetry limits. The five effective interaction parameters of Eq. (52) were determined numerically using a gradient search within the FDU0 code (Wu and Valliéres 1989, Valliéres 1991) to best reproduce the experimental spectrum of the nuclides in question. The experimental energies used in the fit for the systematics are those for the 2^+_1, 4^+_1, 6^+_1, 1^+, 2^+_2, 0^+_2 states in Nd, Sm, Gd, Dy, and Er (some 1^+ states are not known experimentally for Sm and Nd isotopes). This procedure was interated until a good match to the experimental spectra and a smooth trend in particle number were obtained.

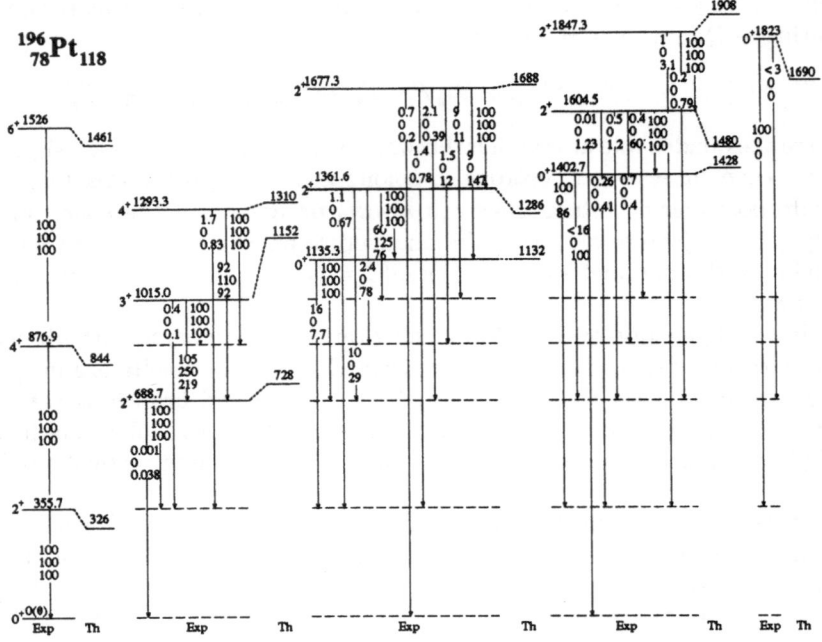

Fig. 6. Level scheme for positive-parity states in ^{196}Pt. Experimental levels are taken from (Ping et al. 1996). The theoretical levels are from an FDSM calculation using the FDU0 code (Wu and Valliéres 1989, Valliéres 1991). The parameters are $G_{0\nu} = -49$, $G_{0\pi} = -18$, $B_{2n p} = -386$, $B_{2\nu} = 97$, $B_{2\pi} = 48$, with all units in keV. The upper number on the transition arrows is the measured relative $B(E2)$ value; the middle number is the IBM-1 predicted value; the lowest number on each transition arrow is the FDSM prediction. There are only two transitions where the FDSM calculation does not agree with the IBM-1 prediction: the $2_4^+ \rightarrow 3_1^+$ and $0_3^+ \rightarrow 2_2^+$ transitions are forbidden in the IBM-1, but not in the FDSM. For weak $\Delta \tau = 0, \pm 2$ transitions the FDSM results agree with data quite well. An additional $d^\dagger \tilde{d}$ term is required in the $E2$ operator to reproduce these transitions with IBM-1.

Similar calculations have been described in (Wu et al. 1987, Guidry et al. 1996) and details of the present fitting process will be discussed in a forthcoming paper (Pan et al. 1996b). The mass dependences of the five parameters as determined from the global fitting are shown in Fig. 7.

When a suitable fit is found for a particular nuclide, the correlation of the symmetric and mixed-symmetry states is given by the unified Hamiltonian of Eq. (52). We find that the QQ-interaction plays the crucial role in correlating the 2_1^+ and 1^+ states because the excitation energies of these states depend sensitively

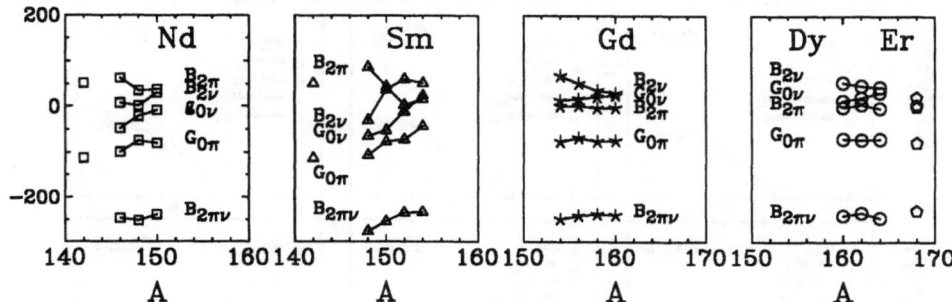

Fig. 7. The mass dependence of Hamiltonian parameters (in keV).

on this term. Once a suitable spectrum has been determined, the next step is to see whether there is correlation between symmetric and mixed-symmetry states for the electromagnetic transitions that is associated with the same $Q_\pi \cdot Q_\nu$ strength.

To compute the electromagnetic transition rates, one needs the wavefunctions for the states in question and the effective transitional operators. The calculation described above yields these wavefunctions; the $M1$ and $E2$ effective transition operators in the FDSM are

$$T(M1)^1_\mu = \sqrt{\frac{3}{4\pi}}\,(g_\pi L_\pi + g_\nu L_\nu) \qquad T(E2)^2_\mu = e_\pi P^2_\mu(i)_\pi + e_\nu P^2_\mu(k)_\nu, \quad (54)$$

respectively, where

$$P^r_\mu(i) = \sqrt{5}\left[b^\dagger_{ki}\tilde{b}_{ki}\right]^{0r}_{0\mu} \qquad P^r_\mu(k) = \sqrt{15/2}\left[b^\dagger_{ki}\tilde{b}_{ki}\right]^{r0}_{\mu0} \qquad (r = 1,2)$$

$$L_\pi = \sqrt{5}P^1_\sigma(i) \qquad L_\nu = \sqrt{8/3}P^1_\sigma(k) \qquad (55)$$

The proton and neutron effective charges e_π and e_ν were fixed at global values of 0.24 eb and 0.20 eb, respectively, for all cases. The g's are the g-factors in the S-D subspace; we take $g_\pi=1.0\ \mu_N$ and $g_\nu=0$ for all nuclei examined here.

In Fig. 8, we plot the experimental and calculated energies of the 2^+_1 and 1^+_1 states as a function of the factor $P \equiv N_p N_n/(N_p + N_n)$ where N_p (N_n) are the valence proton (neutron) numbers, respectively. The P scheme was introduced by in Casten et al. 1987, Rangacharyulu et al. 1991) to facilitate the display of global systematics. In a separate paper (Pan et al. 1996b), we shall discuss the detailed level structures of the rare-earth nuclei. There we will show that nuclei with the same P factor exhibit a correlation between the ground-band structure and that of mixed-symmetry states because they have the same $Q_\pi \cdot Q_\nu$

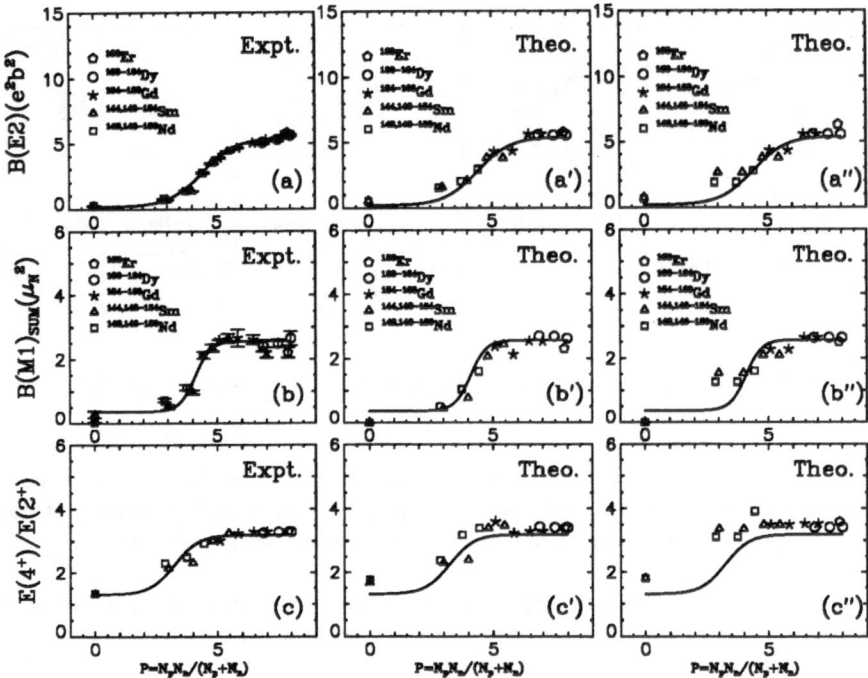

Fig. 8. Comparison of experimental and theoretical $B(E2)$ and summed $B(M1)$ values for rare-earth nuclei as a function of P. The experimental results in the left column are taken from the compilation given in (Rangacharyulu et al. 1991) and (Raman et al. 1989). For comparison, curves from the same empirical relation employed there, $B(E2, M1) = a_1 + a_2/[1 + \exp((c - P)/d)]$ are also plotted. The symbols in (c)–(c″) have the same meaning as in the $E2$ and $M1$ cases. The theoretical results in the second column of figures correspond to best adjustment of effective interaction parameters to reproduce spectra. The theoretical results in the third column of figures correspond to constant values of the effective interaction parameters for all nuclei examined. The figure is from (Smith 1995).

contribution, while prominent discrepancies in the β and γ excitation states may be largely attributed to the different pairing for protons and neutrons.

The corresponding wavefunctions are used to compute the $E2$ and $M1$ transition strengths. The $B(E2)$ values and the summed $B(M1)$ strengths from this calculation are shown in Figs. 8a′ and 8b′, while the corresponding data are shown in Figs. 8a and 8b. The curves are the empirical relations presented in Ref. (Rangacharyulu et al. 1991) that summarize the approximate behavior of the data. The $B(E2)$ and $B(M1)$ strengths are reproduced quantitatively by the calculations. Thus, we find theoretical evidence for the approximate universal behavior of $E2$ and $M1$ strengths exhibited by the data. Furthermore, we observe that even the deviations from universality exhibited by the data [for example, the $M1$ saturation is sharper and occurs at least 1 unit of P lower than that for the $E2$ strength (which never completely saturates)], is reproduced *quantitatively* by the calculations, without parameter adjustment.

In Figs. 8c and Figs. 8c$'$ we have plotted the ratio $E(4_1^+)/E(2_1^+)$ as a function of P. This quantity is also seen to exhibit an empirical variation with P that is similar to that of the $E2$ and $M1$ strengths, and it is also quantitatively reproduced by these calculations. We expect this ratio to be sensitive to the $Q_\pi \cdot Q_\nu$ interaction so this is an expected result, given the success of the preceding calculations and our previous assertion that the $Q_\pi \cdot Q_\nu$ term is the most important factor governing the relationship between the properties of the symmetric and mixed-symmetry states.

In Figs. 8a$''$–8c$''$ we repeat the calculations of Figs. 8a$'$–8c$'$, but with a *fixed set of parameters for all nuclei*: $G'_{0\pi} = -0.074$ MeV, $G'_{0\nu} = 0.020$ MeV, $B'_{2\pi} = -0.001$ MeV, $B'_{2\nu} = 0.047$ MeV, and $B_{2\pi\nu} = -0.243$ MeV. These calculations are also in good agreement with observations. Thus, the quantitative reproduction of $E2$ and $M1$ strengths in the rare earth nuclei is an inherent feature of the FDSM, not a consequence of parameter adjustment. It should be noted that the small positive value of the renormalised $G'_{0\nu}$ means that the neutron quadrupole pairing is strong, thus implying that higher angular momentum pairs may also play some role.

8   Conclusions

The Fermion Dynamical Symmetry Model (FDSM) has been discussed as an example of a systematic method for truncating the spherical shell model according to principles of dynamical symmetry. In the resulting symmetry-dictated truncation, a valence space is selected using energy considerations and principles of dynamical symmetry are then used to radically truncate the shell model space. The resulting truncated space permits systematic shell model calculations to be implemented for all heavy nuclei, but since the space has been severely truncated the corresponding interactions are highly effective with respect to the original shell model; thus it is necessary to determine the appropriate FDSM effective interaction for each valence space.

We have presented examples of systematic FDSM calculations that have been used to determine an effective interaction appropriate for heavy nuclei having no broken pairs. This interaction is simple and has a weak dependence on particle number. Calculations using this interaction reproduce low-lying spectra, moments, and transition rates for broad ranges of collective behavior. Finally, we have presented an initial extension of this approach to include broken pairs in the configuration space.

There remains the task of systematizing and refining the interaction over all mass numbers through a series of numerical calculations, and investigating in detail the contributions of additional symmetry breaking associated with effects like single-particle splitting, model-space dependent pair structure, and contributions from higher angular momentum pairs. Neverthless, the results summarized here indicate that truncation of shell model spaces based on the maxim of preserving certain algebraically defined dynamical symmetries can be a powerful method of solving the nuclear many-body problem.

9 Acknowledgement

This work was supported by the NSF (Drexel). Theoretical nuclear physics research at the University of Tennessee is supported by the U. S. Department of Energy through Contract No. DE–FG05–87ER40361 and DE–FG05–93ER40770. Oak Ridge National Laboratory is managed by Lockheed Martin Energy Research Corp. for the U. S. Department of Energy under Contract No. DE–AC05–96OR22464. We are grateful to Drs. C. L. Wu and J. Q. Chen for useful discussions.

References

Arima A., Harvey M., Shimizu K. (1969): Phys. Lett. **B30**, 517

Arima A., Otsuka T., Iachello F., Talmi I. (1977): Phys. Lett. **B66**, 205

Bender E., Schmid K.W., Faessler A. (1996): Nucl. Phys. Rev. **A596**, 1

Brown B.A., Etchegoyen A., Rae W.D.M., Ormand W.E., Winfield J.S., Zhao L.(1988): *OXBASH code*, MSUNSCL Report **524**

Casten R.F., Brenner D.S., Haustein P.E. (1987): Phys. Rev. Lett. **58**, 658

Caurier E. (1989): code ANTOINE, Strasbourg

Di Y.M., Yoshida N., Pan X.W. (1993): *SU3u2 code* (unpublished)

Draayer J. P. (1992): *Nuclear Structure Models*, Bengtsson R., Draayer J. P., Nazrrewicz W. eds. (World Scientific), P.61

Elliott J.P. (1958): Proc. Roy. Soc. **A245**, 128, 562

Feng D.H., Guidry M.W., Pan X.W., Wu C.L., Zlatev I. (1993): Phys. Rev. **48**, R1488

Ginocchio J.N. (1980): Ann. Phys. **126**, 234

Guidry M.W., *et al.*, Phys. Lett. **B176**, 1

Guidry M.W., *et al.* (1987): Phys. Lett. **B187**, 210

Guidry M.W., Feng D.H., Pan X.W., Wu C. L. (1996): J. Phys. **G22**, 425 (1996).

Hara K. (1996): these proceedings

Hara K. Sun Y. (1991): Nucl. Phys. **A529**, 445

Hecht K.T., Adler A. (1969): Nucl. Phys. **A137**, 129

Honma M, Mizusaki T., Otsuka T. (1995): Phys. Rev. Lett. **75**, 1284

Iachello F., Kuyucak S. (1981): Ann. Phys. **136**, 19

Iachello F., (1984): Phys. Rev. Lett. **53**, 14271

Iachello F., Arima A. (1987): *The interacting boson model* (Cambridge U.P., Cambridge)

Johnson C.W., *et al.* (1992): Phys. Rev. Lett. **69**, 3147

Judd B.R. (1963): *Operator Techniques In Atomic Spectroscopy*, (Mcgraw-Hill)

Koonin S.E., Dean D.J., Langanke K. (1996): Physics Reports (in press)

Ormand E. , *et al.* (1994): Phys. Rev. **C49**, 1422

Otsuka T. (1981): Nucl. Phys. **A368**, 244

Ping J.L., Zlatev I., Pan X.W., Feng D.H., Guidry M.W. (1995): Phys. Rev. **C** (Submitted)

Pan X.W., Otsuka T., Chen J.Q., Arima A. (1992): Phys. Lett. **B287**, 1

Pan X.W., Feng D.H., Chen J.Q., Guidry M.W. (1994): Phys. Rev. **C49**, 2493

Pan X.W., Feng D.H. (1994): Phys. Lett. **B336**, 285

Pan X.W., Feng D.H. (1994): Phys. Rev. **C50**, 818

Pan X.W., Di Y.M., Yoshida N., Guidry M.W. (1996): *The Broken Pair Effect on the Moment of Inertia in the Fermion Dynamical Symmetry Model* (preprint)

Pan X.W., Smith B.H., Feng D. H., Guidry M.W. (1996): (in preparation)

Pan X.W., Ping J.L., Feng D.H., Chen J.Q., Wu C.L., Guidry M.W. (1996): Phys. Rev. **C53**, 715

Raman S., Nestor C.W. Jr., Kahane S., Bhatt K.H. (1989): At. Data Nucl. Data Tables **42**, 1

Rangacharyulu C., et al. (1991): Phys. Rev. **C43**, R949.

Schmid K.W., Grummer F., Faessler A. (1984): Nucl. Phys. **A431**, 205; Phys. Rev. **C29**, 308

Schmid K.W., Grummer F., Kyotoku M., Faessler A. (1986): Nucl. Phys. **A452**, 493

Schmid K.W., Zheng R.R., Grummer F., Faessler A. (1989): Nucl. Phys. **A499**, 63

Smith B.H., Pan X.W., Feng D.H., Guidry M.W. (1995): Phys. Rev. Lett. **75** , 3086

Valliéres M., Wu H. (1991): in *Computational Nuclear Physics 1*, ed. by Langanke K., Maruhn J.A., Koonin S.E. (Springer–Verlag)

Valliéres M., Novoselsky A. (1993): Nucl. Phys.**A570**, 345c

Wu H., *et al.* (1987): Phys. Lett. **B193**, 163

Wu H., Feng D.H., Wu C.L., Guidry M.W. (1987): Phys. Lett. **B198**, 119

Wu H., Wu C.L., Feng D.H., Guidry M.W. (1988): Phys. Rev. **C37**, 1739

Wu H., Valliéres M. (1989): Phys. Rev. **C39**, 1066

Wu C.L., Feng D.H., Chen X.G., Chen J.Q., Guidry M. (1987): Phys. Rev. **C36**, 1157

Wu C.L. (1990): Nucl. Phys. **A520**, 459c

Wu C.L., Feng D.H., Guidry M.W. (1994): Advances in Nuclear Physics **21**, 227

Challenges to Microscopic Theories
of Nuclear Structure

R. F. Casten

Wright Nuclear Structure Laboratory, Yale University, New Haven CT 06520, USA

Abstract. Several recent developments in nuclear structure] research and technology that provide current challenges to microscopic theories are briefly discussed.

1 Introduction

In these brief comments, we would like to raise several issues, relating to major new directions in nuclear structure research, that pose significant challenges to microscopic nuclear theories. At the outset, we emphasize two points. This is only a small sampling of current issues that could be raised. The emphasis is on areas of interest to the author: They focus mostly on low spin phenomena. Even within that area, a number of other equally interesting topics could have been raised. Secondly, the use of the phrase "microscopic theories" rather than "Shell Model" is intentional since it is realized that many of these issues are beyond current applications of the Shell Model. Rather, to be computationally tractable, they entail truncations, constraints, paradigm-shifts, or approximate methods of calculation. We will focus on three specific but interrelated topics. To emphasize both their individuality and their interrelationships it is useful (and mnemonic) to link them through the acronym

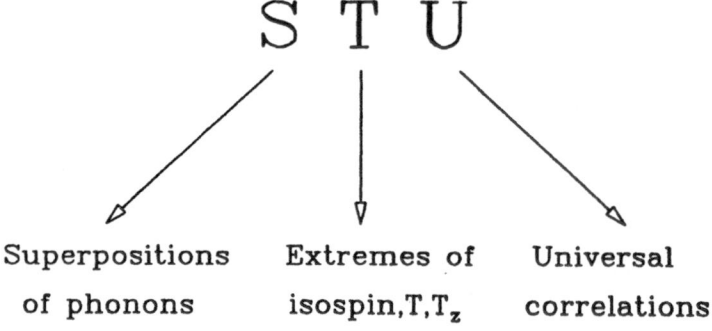

$$S \quad T \quad U$$

| Superpositions of phonons | Extremes of isospin, T, T_z | Universal correlations |

2   Multi-phonon States

Ever since the recognition of vibrational modes in spherical and deformed nuclei in the 1950's [1], there has been debate as to the viability of multi-phonon excitations. The question is intimately related to the effects of the Pauli Principle in a quantal finite-body environment and to our microscopic understanding of these collective modes. A number of calculations [2] concluded that multi-phonon states would be almost completely fragmented. In the last few years this general skepticism has come face to face with new experimental results, in several structural regions, that revealed multi-phonon states with largely intact collectivity.

A first hint of this occurred a decade ago in a study [3] of ^{118}Cd which revealed, for the first time, candidates for the full 3-phonon close lying quintuplet of spherical quadrupole vibrational states. Though not all spins were fixed in that experiment, and though subsequent work has revealed additional complexity in the level scheme, this work awakened a new attitude to the possibility of multi-phonon states and spurred further studies. One such study [4] of the more accessible nucleus ^{114}Cd revealed a wealth of multi-phonon levels. As illustrated in Fig. 1, the B(E2) values agree remarkably well with the phonon picture even up through candidates for 4 or 5 phonon states. Although the energies are highly anharmonic, and although competitive interpretations exist involving mixing with intruder states, all intepretations of ^{114}Cd include at least some states (including *non*-yrast levels) up to 4 phonons.

In deformed nuclei, where the lowest $K = 0$ and γ vibrations are typically at ~ 1 MeV, and where the Pauli Principle acts on 2-fold, rather than $(2j + 1)$-fold degenerate sub-states, it was long thought much less likely that even 2-phonon states would survive. It came as a surprise then when a $K = 4$ band in ^{168}Er was shown [5] to have $B(E2)$ values for decay to the γ band with about half the full 2-phonon collectivity. Subsequently, a Coulomb excitation study [6] of ^{232}Th concluded that a $K = 4$ band has the full 2-phonon collectivity. The situation with respect to these K = 4 excitations has recently been reviewed [7].

The situation with respect to the lowest $K = 0$ band is murky but intriguing. Long thought to be β vibrations, such an interpretation is now more or less discarded but their true nature is still actively debated. [8,9,10]. The discussion is based on the puzzling experimental results shown in Table 1, namely that the $B(E2 : 0_2^+ \rightarrow 2_\gamma^+)$ values are often 2-3 *orders of magnitude* larger than $B(E2 : 0_2^+ \rightarrow 2_g^+)$ values. Although there are no *absolute* $B(E2 : 0_2^+ \rightarrow 2_\gamma^+)$ values known, estimates from the branching ratios in Table 1, using typical $B(E2 : 0_2^+ \rightarrow 2_g^+)$ values suggest that these $K = 0^+ \rightarrow \gamma$ vibrational transitions may even be within a factor of 3-10 of *rotational* transition strengths. Clearly, this suggests the $K = 0_2^+$ excitation is a very collective phonon mode built on the γ vibration. Interestingly, this phenomenon is predicted (automatically) [8] by the IBA [11] but this does not yet provide a microscopic understanding.

Two caveats are important here. First, the errors in Table 1 are quite large. Some of the $B(E2)$ ratios are obtained indirectly via other $K = 0_2^+ \rightarrow \gamma$ tran-

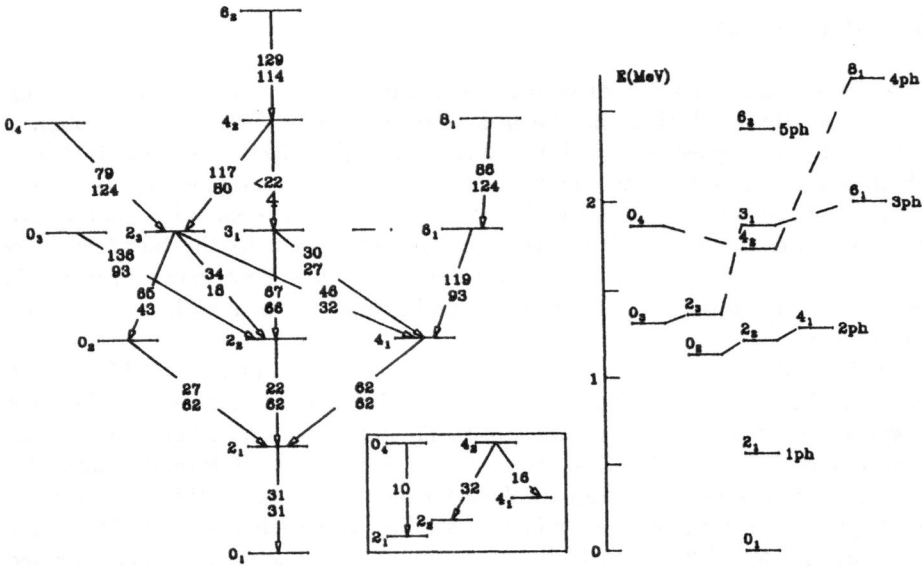

Fig. 1. B(E2) values (in W.u.) in ^{114}Cd. On the left the energies are aligned according to the assigned phonon number. The upper numbers on each arrow are the observed values, the lower are the predictions of the harmonic vibrator model. The agreement is impressive. Allowed transitions are strong, with strengths up to and exceeding 100 W.u. compared to 31 for the ground state transition, and forbidden transitions are weak (only one of 14 spin- allowed but structure-forbidden transitions is above 16 W.u). The few most significant disagreements are highlighted in the box. The right side shows the actual level energies, with dashed lines connecting levels of given phonon number. The strong anharmonicities are evident. (Based on [4].)

sitions which are converted to $0_2^+ \rightarrow 2_\gamma^+$ values by appropriate Clebsch Gordon coefficients. Band mixing will affect these values, possibly by large amounts if the K = 0 and γ bands are close lying. Nevertheless, it is unlikely that the extraordinary dominance of transitions to the γ band can be explained by plausible K = 0 - γ mixing. Secondly, not all deformed nuclei display this dominance. There are two known cases (see Table) where the ratio is on the order of unity and new measurements [12] in several rare earth nuclei have also not found large ratios. It could be that the phenomenon shown in Table 1 only occurs in some nuclei. This is not unlikely. The 0_2^+ states in the Table are the first excited 0^+ levels. Other 0^+ excitations often occur nearby in energy. The relative energies of these intrinsic excitations can fluctuate from nucleus to nucleus and, in some cases, a non-collective (e.g., quasi-particle) 0^+ state may be the lowest excited 0^+ state. Another interpretation could be that there are problems with the data used to construct Table 1. These K = $0_2^+ \rightarrow \gamma$ transitions are extremely weak,

Table 1. Ratios of B(E2) values for decay of the 0_2^+ level to the 2_γ^+ and 2_g^+ states. Note that only a few of these transitions are *directly* measured. In most cases, the ratios are measured for higher spin $K = 0^+$ band members (J = 2^+, 4^+ or 6^+) and these ratios are connected to the indicated ratio via appropriate Clebsch Gordon coefficients (Alaga rules). This procedure, sometimes possible for more than one branching ratio in a given nucleus, ignores the (possibly large) effects of band mixing but it is thought unlikely that the qualitative dominance of strength to the γ-band would be altered by this consideration. (Based on [10].)

Nucleus	$\dfrac{B(E2:K=0_2^+\to2_\gamma^+)}{B(E2:K=0_2^+\to2_g^+)}$
^{158}Gd	360(80)
^{160}Gd	610(410)
^{160}Gd	1600(800)
^{160}Dy	300(150)
^{162}Dy	120(20)
^{162}Dy	1600(1100)
^{164}Er	1.9 (1.0)
^{166}Er	0.9 (0.1)
^{168}Er	56(9)
^{168}Er	110(40)

and typically in high background regions of the spectrum. In any case, the data are certainly striking and deserve further study from both the experimental and theoretical points of view.

Multi-phonon states of another type have also been found in or near closed shell regions. These are typically 2-phonon modes of the type $2^+ \otimes 3^-$. Examples are known in ^{96}Zr [13] and N = 82 [14].

Finally, in the last few years, heavy ion scattering reactions have revealed 2-phonon E1 giant resonances with substantial collectivity. Examples are found [15,16,17] in ^{40}Ca and ^{208}Pb.

All in all, this proliferation of 2- and multi-phonon states, in a number of different collective regimes, represents one of the most interesting developments in nuclear structure in recent years and, especially given the *a priori* scepticism concerning such modes, calls for renewed efforts at a microscopic understanding.

3 Nuclei at Extremes of Isospin, T

Recent technological developments offer a new opportunity for nuclear structure research in nuclei with extreme ratios of proton and neutron number. Radioactive beam facilities, either of Fragmentation or ISOL type, are beginning to provide access to exotic nuclei on both sides of stability. Of course, these beams are orders of magnitude weaker than the stable beams we are accustomed to and new and more efficient detector systems will have to be developed. Fortunately, there have been major recent advances in mega-instrumentation such as arrays of Compton suppressed γ-ray detectors, Ge detector segmentation techniques, sophisticated recoil mass separators (such as the FMA at ANL), and high resolution magnetic spectrometers with advanced focal plane detectors. Many of these were originally developed for studies of nuclei at extremes of angular momentum but they present a new opportunity for low spin studies, especially in nuclei far from stability. Coupled with new empirical signatures of structure (See Section 4), it will soon be possible to study $N = Z$ nuclei all the way to ^{100}Sn on the proton rich side, and on the neutron rich side, to extend far enough off stability to see nuclei with low density outer skins of weakly bound, nearly pure, neutron matter.

Recent calculations by Dobaczewski, Nazarewicz, and colleagues [18] have suggested that near-neutron drip line nuclei may have structures radically different from anything observed near stability. Figure 2 illustrates one such possible scenario in which the underlying shell structure itself changes so that the single particle levels resemble those obtained with a harmonic oscillator potential with spin-orbit interaction. Such a potential has a more rounded bottom than the usual Woods-Saxon potential and is roughly equivalent to the Nilsson potential without the ℓ^2 term. As a result, the magic numbers themselves change and shell gaps may weaken as well. Such changes are fundamental. In the usual shell model typical of medium mass and heavy nuclei, the normal parity orbits exhibit a nearly monotonic decrease of spin through a shell, and a $\Delta j = -1$ relationship of successive orbits. In the Dobaczewski-Nazarewicz scenario, the j-values are nested, with the highest values surrounding lower ones and with a uniform sequence $\Delta \ell = \Delta j = -2$. (Note that j = 3/2 and 1/2 differ by $\Delta j = 2$ since the wave functions are symmetrized with respect to $\pm$ m substates). Since a quadrupole residual interaction strongly couples such $\Delta \ell = \Delta j = 2$ pairs of orbits, the onset, evolution and phenomenology of collectivity and collective excitations may be strongly altered.

The fragility of magic numbers should not, of course, come as a surprise. Many examples are known, such as the disappearance of the $Z = 64$ gap for $N \geq 90$, $Z = 38$ or 40 for $N \geq 60$, and $N = 20$ in proton rich ^{32}Mg.

Clearly, if shell structure as we know it applies only near stability and if completely new facets of structure and its microscopic underpinnings appear in very neutron rich nuclei, a proper understanding of this from a microscopic perspective will be essential. Initial studies such as those of Dobaczewski and Nazarewicz have not gone unchallenged, nor have they been tested experimentally. Much more microscopic work in this area is urgently needed.

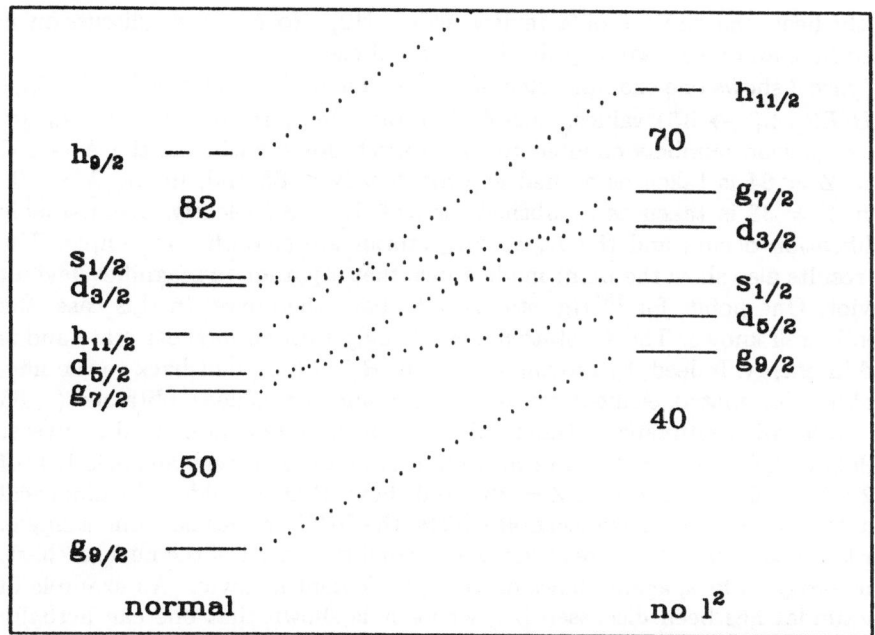

Fig. 2. Left: Normal shell model levels for the 50-82 nucleon region. Right: the level sequence for a shell model Hamiltonian consisting of harmonic oscillator and spin-orbit terms only. This is similar to the sequence that results from calculations of Dobaczewski et al.[18] for exotic nuclei near the neutron drip line.

4 Universal Correlations and New Signatures of Structure

A third recent development in nuclear structure has been the recognition of re-markable simplicities in the phenomenology of observables reflecting equilibrium structure and collectivity. In most cases these phenomenological correlations are extraordinarily (but unexpectedly) smooth and compact. As such, they can pro-vide signatures of structure, often heretofore unrecognized, and can give new paradigms or benchmarks of structure. By acting as a magnifying glass, such correlations provide a highly sensitive way to spot deviant nuclei and, therefore, to study the interactions and degrees of freedom that lead to such deviations. Moreover, these phenomenologies relate to observables, such as $E(2_1^+)$, $E(4_1^+)$, $R_{4/2} = E(4_1^+)/E(2_1^+)$, $B(E2 : 2_1^+ \rightarrow 0_1^+)$ values, separation energies and the like that are particularly easy to measure. Therefore, they will be of value with radioactive beam experiments where we will never have the quantity of data we are accustomed to.

These correlations have been extensively discussed in the literature and have recently been the subject of a review article [19]. No extensive discussion is needed here and hence we only illustrate two of them briefly.

Figure 3 shows one example each of a normal and $N_p N_n$ plot of $E(2_1^+)$, $R_{4/2}$ and $B(E2 : 2_1^+ \rightarrow 0_1^+)$ values, each for a typical mass region. N_p and N_n are valence nucleon numbers counted to the nearest closed shell. (In the A $\sim$ 150 region, Z = 64 is taken as a shell gap for 84$\leq$ N $\leq$ 88 and, in the A $\sim$ 100 region, Z = 38 is taken as a subshell for 52$\leq$ N$\leq$ 58.) Clearly, a remarkable simplification occurs and the $N_p N_n$ trajectories are smooth and simple. The $R_{4/2}$ results also show the point made above that a paradigm magnifies deviant behavior. One point, for ^{184}Hg, stands out above the curve. In this case, the reason is well known. The 4^+ state mixes strongly with an intruder state and is raised in energy. Indeed, by moving the errant $R_{4/2}$ data point back to the line, the interaction matrix element can be deduced and the value so obtained ($\sim$ 85 keV) agrees with estimates obtained through much more complicated analyses.

Global $N_p N_n$ systematics are also often smooth. A recent analysis [20] of $B(E2 : 2_1^+ \rightarrow 0_1^+)$ values from Z = 38 - 100 shows that, provided the empirical values are corrected for hexadecapole effects, the B(E2) values lie along a single, linear trajectory for this entire region which comprises most of the nuclear chart. The linearity, in fact, again allows one to spot deviant behavior. An example in the actinides has been discussed [20] where it is shown that one can actually extract a β_4 deformation from the $B(E2 : 2_1^+ \rightarrow 0_1^+)$ value alone and, moreover, this leads to suggested refinements to macroscopic-microscopic models of the stability of the heaviest nuclei.

The second correlation we illustrate is that of one collective observable against another. Figure 4 shows $E(4_1^+)$ against $E(2_1^+)$ for all nuclei from Z = 50-82 and N = 82-126. The normally very complex behavior of $E(4_1^+)$ is transformed by such a plot into a remarkably simple tri-linear trajectory that leads to a tri-partite classification of the evolution of structure into rotor, anharmonic vibrator (AHV), and pre-collective regimes with slopes of 3.33, 2.00, and 1.0, respectively.[21] The AHV region is particularly interesting and unexpected. The nuclei involved here have $R_{4/2}$ values from $\sim$ 2.0 to $\sim$ 3.15 and therefore span an enormous variety of structures. Yet they all lie along a single, compact (half width $\sim$ 40 keV) path against $E(2_1^+)$. This path, satisfying the equation

$$E(4_1^+) = 2.0E(2_1^+) + \epsilon_4 \tag{1}$$

is that of an anharmonic vibrator, with constant anharmonicity, ϵ_4. That is, in terms of a phonon picture, one has here another example of 2-phonon behavior, but one in which, despite widely varying internal phonon structure, the phonon-phonon interaction (ϵ_4) is constant. Why this is, is surely a major challenge to microscopic theory. Higher lying yrast states also satisfy the generalization of Eq. 1. Finally, the sharp kinks at the junctions of the three regions in Fig. 4 suggest true phase transitional behavior and point to the intriguing possibility of viewing $E(2_1^+)$ itself as a kind of critical parameter. Again, an obvious question for microscopy is raised by this result. We note that, very recently, a similar

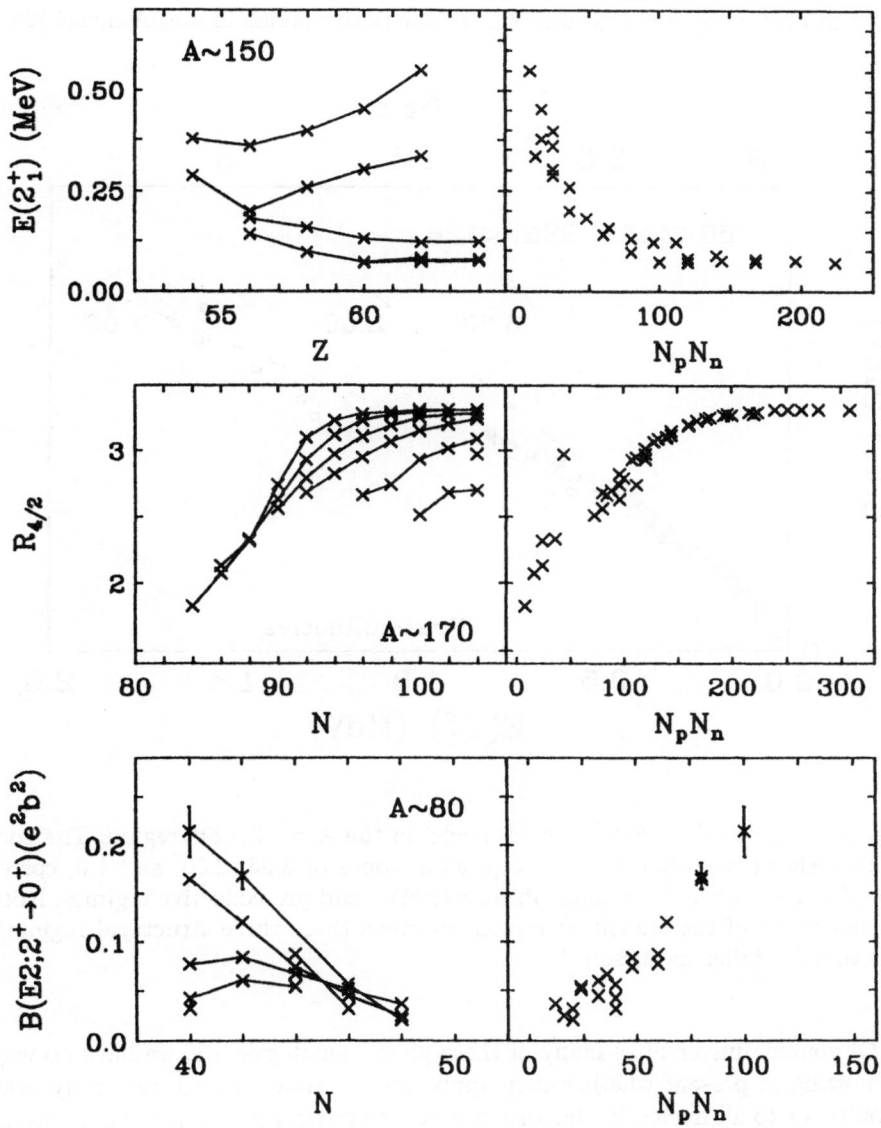

Fig. 3. Normal and $N_p N_n$ plots for three observables, in three different mass regions. The simplification that results in the $N_p N_n$ scheme are evident. In the middle right panel the one anomalous (high) point is that for ^{184}Hg and is discussed in the text. (Based on [19].)

AHV behavior, and a tri-partite classification of structure, has been shown to apply to both unique parity and non-unique parity bands in odd-A nuclei [22].

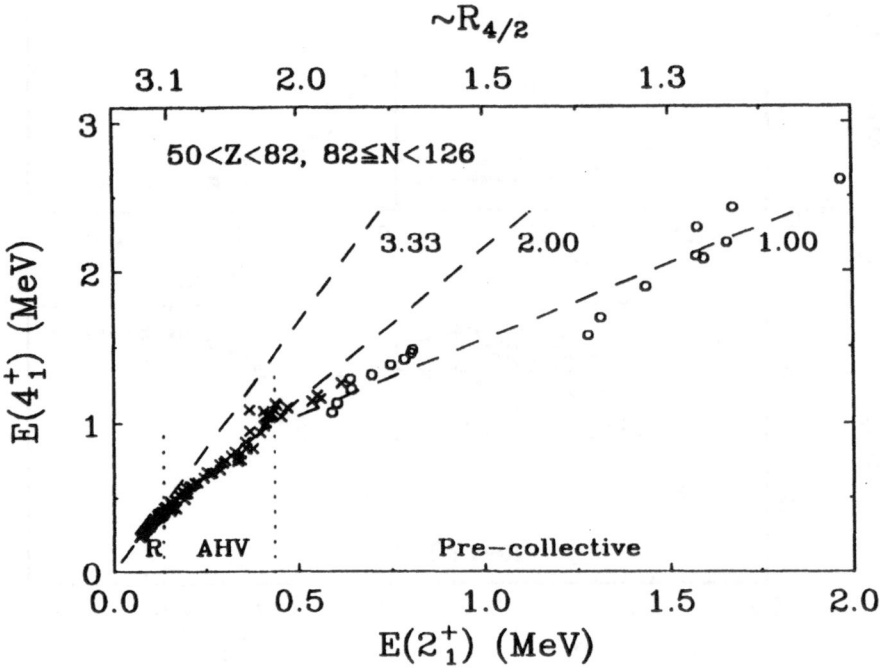

Fig. 4. $E(4_1^+)$ against $E(2_1^+)$ for *all* nuclei in the A = 132-208 region. The correlation shows a tri-linear trajectory, with slopes of 3.33, 2.00, and 1.0, corresponding to rotor, anharmonic vibrator (AHV) and pre-collective regimes. Note the sharpness of the transition regions between these three structural regimes. (See text for fuller discussion.)

Of course, universal as many of these phenomenologies are, we have no way of knowing at present whether they apply only to known nuclei, relatively near stability, or to all nuclei. In the former case, the critical question is what special aspects of shell structure and interactions are responsible and what changes in underlying structure can lead to disruptions in the smooth phenomenology. In the latter case, the critical question is what generic features of shell structure and interactions are at work. While shell structure in known nuclei has a certain consistency (e.g., above N,Z = 50, a more or less monotonic, $\Delta j = -1$ decrease of j across the normal parity orbits of a major shell), it also has an obvious and important variability, namely, the position of the unique parity orbit, which ranges from a location at the top of a shell ($1g_{9/2}$ orbit in the 20-50 shell) to the bottom ($1j_{15/2}$ orbit in the actinides). The simple correlations recently discovered persist despite such variability and, in some cases, encompass several major shell regions in a single universal pattern of behavior. Therefore, they are quite robust and would seem to require truly major alterations of underlying shell structure as a pre-requisite to their breakdown. Perhaps such departures from the usual shell structure as envisioned in the scenario of Dobaczewski and Nazarewicz [18] predicted to occur near the neutron drip line would lead to deviations from the simple, smooth trajectories of observables observed to date.

All this is to emphasize the importance of understanding these phenomenologies at the microscopic level. There have been virtually no efforts in this direction. Nevertheless, it is clear that, to understand the universality observed and when and how it may break down, presents an important and significant challenge to our microscopic understanding of nuclear structure and its evolution. In some cases, elements of the phenomenology can be understood at either a schematic or a macroscopic model level. For example, Heyde and Sau [23] showed that a smooth trajectory of $E(2_1^+)$ against $N_p N_n$ is reproduced by a simple model of 2-state mixing of normal and mixed symmetry states. Also, the IBA automatically and unavoidably reproduces [24] the linear relation of $E(4_1^+)$ with $E(2_1^+)$, with a slope of 2, for the AHV region for virtually any selection of parameters spanning the majority of the symmetry triangle. An analytic interpretation [25] of this result showed that the only significant pre-requisite is the fact that, in the IBA, the yrast states consume nearly all of the $0\hbar\omega$ E2 strength. That is, that they comprise a set of multi-phonon levels. This is a reassuring result since, as discussed above, the most obvious interpretation of the slope of 2.0 is, indeed, a phonon-based one. Moreover, a recent empirical study of 102 medium and heavy mass nuclei [26] showed that, on average, the $B(E2 : 0_1^+ \rightarrow 2_1^+)$ value is 97% of the total $0\hbar\omega$ $B(E2 : 0_1^+ \rightarrow 2_1^+)$ strength. Nevertheless, these interpretations only point to, but do not comprise, the kind of microscopic understanding that is urgently needed.

5   Conclusions

The brief discussion here has focused on a few recent developments in low energy nuclear structure that raise intriguing challenges for microscopic theories. They concern questions brought to the fore by the on-going resurgence of interest in low spin nuclear structure and they relate to such basic issues as the underlying shell structure of nuclei and the nature and interactions of collective modes.

I am grateful to N. V. Zamfir, W. Nazarewicz, J. Dobaczewski, and A. Aprahamian for illuminating discussions and to Zamfir and my co-authors in the articles cited here for their collaborations.

Work supported by the U. S. Department of Energy under Grant No. DE-FG02-91ER-40609.

References

[1] Goldhaber G., Weneser J. (1955): Phys. Rev. 98, 212; Bohr A., Mottelson
    B.R. (1953): Mat. Fys. Medd. K.Dan.Vidensk. Sebsk. 27, No. 16
[2] Soloviev V.G., Shirikova N. Yu (1981): Z. Phys. A301, 263
[3] Aprahamian A., et al. (1987): Phys. Rev. Lett. 59, 535
[4] Casten R.F., et al. (1992): Phys. Lett. B297, 19
[5] Börner H.G., et al. (1991): Phys. Rev. Lett. 66, 691
[6] Korten W., et al. (1993): Phys. Lett. B317, 19
[7] Wu C.Y., Cline D., Phys. Lett. B, to be published
[8] Casten R.F., von Brentano P. (1994): Phys. Rev. C50, R1280
[9] Burke D.R., Sood P.C. (1995): Phys. Rev. C51, 3525
[10] Casten R.F., von Brentano P. (1995): Phys. Rev. C51, 3528
[11] Arima A., Iachello F. (1975): Phys. Rev. Lett. 35, 1069
[12] von Brentano P. (1996): private communication
[13] Molnar G., et al. (1989): Nucl. Phys. A500, 43
[14] Herrberg R.D., et al. (1995): Nucl. Phys. A592, 211
[15] Frascaria N., et al. (1977): Phys. Rev. Lett. 39, 918
[16] Ritman J., et al. (1993): Phys. Rev. Lett. 70, 533
[17] Blumenfeld Y. (1996): Nucl. Phys. A599, 289c
[18] Dobaczewski J., et al. (1994): Phys. Rev. Lett. 72, 981
[19] Casten R.F., Zamfir N.V., J. Phys. G, to be published
[20] Zamfir N.V., et al. (1995): Phys. Lett. B357, 515
[21] Zamfir N.V., Casten R.F., Brenner D.S. (1994): Phys. Rev. Lett. 72, 3480
[22] Bucurescu D., et al. (1994): Phys. Rev. C49, R1759 and to be published
[23] Heyde K., Sau J. (1986): Phys. Rev. C33, 1050
[24] Zamfir N.V., Casten R.F. (1994): Phys. Lett. B 341, 1
[25] Jolos R.V., et al. (1995): Phys. Rev. C51, R2298
[26] Pietralla N., et al. (1994): Phys. Rev. Lett. 73, 2962

Projected Shell Model
– A Way of Doing Small Scale Shell Model Calculations –

Kenji Hara

Physik-Department, Technische Universität München
D-85747 Garching bei München, Germany

Abstract. The basic idea of the Projected Shell Model (PSM) is presented. Numerical examples are given for some doubly even and odd mass nuclei in the rare-earth region. We demonstrate that the PSM results are easy to interpret in simple physical terms.

1 Introduction

One of the main problems in the (standard) shell model had been how to make a calculation feasible. This problem is to a certain extent eased by the recent development of computer technology which has made it possible to carry out a "Very Large Scale" shell model calculation. Nevertheless, whether such a calculation really helps our physical understanding is open to question. For example, a rotational state may be describable by a coherent superposition of millions of basis states. However, it is hard to analyze a huge number of amplitudes printed over thousands of pages and we thus loose physical insight. This is a serious problem since most of nuclei except for those in the vicinity of magic numbers are deformed and show rotational spectra. The question is not only whether the computer can provide us with a numerical result but also whether we can interpret the result in physical terms. The Projected Shell Model (PSM) aims to give an answer to this question.

Use of the spherical orbitals is not advantageous for deformed nuclei. This is because the spherical representation provides us with no clear hierarchy of states for a deformed system. Classification of the shell model basis can be done most naturally and efficiently if we use (intrinsic) orbitals generated by a deformed potential. We will project out the resulting Slater determinants onto good angular momentum and diagonalize the Hamiltonian within a Hilbert space spanned by such (projected) Slater determinants. The truncation of the basis states can be done in terms of the number of particles and holes and an energy window chosen properly for the physics we want to describe. For example, up to spin ≈ 40, we need at most 50 shell model basis states for the yrast states of normally deformed doubly even rare-earth nuclei and much less for odd mass and doubly odd nuclei. In practice, we use the Nilsson+BCS quasiparticle (qp) representation to take the (static) pairing correlations into account. Three major shells are used for both neutron and proton degrees of freedom. In fact, use of a large single-particle basis is important for the description a rotating body.

The PSM equation for the shell model diagonalization takes the form

$$\sum_{\kappa' K'} \left\{ H^I_{\kappa K \kappa' K'} - E N^I_{\kappa K \kappa' K'} \right\} F^I_{\kappa' K'} = 0 \tag{1}$$

where the Hamiltonian and Norm matrix elements are defined respectively by

$$H^I_{\kappa K \kappa' K'} = \langle \Phi_\kappa | \hat{H} \hat{P}^I_{K K'} | \Phi_{\kappa'} \rangle \quad \text{and} \quad N^I_{\kappa K \kappa' K'} = \langle \Phi_\kappa | \hat{P}^I_{K K'} | \Phi_{\kappa'} \rangle. \tag{2}$$

Here, $\hat{P}^I_{K K'}$ is the angular momrntum projection operator and $|\Phi_\kappa\rangle$ denotes a (multi-) qp state representing a band κ. We define the rotational energy of a band (or the band energy) by the energy expectation value

$$E_{\kappa K}(I) = \frac{\langle \Phi_\kappa | \hat{H} \hat{P}^I_{K K} | \Phi_\kappa \rangle}{\langle \Phi_\kappa | \hat{P}^I_{K K} | \Phi_\kappa \rangle} = \frac{H^I_{\kappa K \kappa K}}{N^I_{\kappa K \kappa K}} \tag{3}$$

which contains no coupling between different bands. A diagram in which the band energies are plotted as functions of spin is called a Band Diagram [1]. It plays an important role when interpreting the numerical result obtained by the PSM equation (1) which takes the coupling between bands into account.

2   Doubly Even Nuclei

In Fig. 1, we show a typical example of a band diagram corresponding to a doubly even nucleus (^{162}Hf) plotted for even spins only.

The band energy of some lowlying 2-qp bands decreases with spin in the low spin region ($I \leq 12$). This behavior is due to the rotational alignment (small K-value) of a decoupled state (particles occupying an intruder sub-shell) and the amount of spin alignment is given by the spin value at which the band energy takes the minimum value [1]. Among such 2-qp bands, the one that crosses first with the g-band is the so-called s-band (a 2-neutron $K = 1$ band). It becomes the yrast band after the crossing. Such a band crossing is responsible for the backbending of the moment of inertia

$$\Theta = \frac{I}{\omega}, \quad \omega = \frac{dE(I)}{dI} \approx \frac{E(I) - E(I - 2)}{2}. \tag{4}$$

Note that the rotational frequency ω represents the slope of the rotational energy in a band diagram. The sharpness of a backbending is determined by the crossing angle between the g- and s-band or the ratio of their slopes, ω_g/ω_s. The larger the crossing angle (or ω_g/ω_s), the sharper the backbending. This is because, after the crossing point, the rotational frequency ω suddenly decreases from that of the g-band (ω_g) to that of the s-band (ω_s) leading to a sudden increase of the moment of inertia (4). Table 1 lists the crossing spin I_c and the ratio ω_g/ω_s for some Er isotopes. Note that the crossing spin increases and the ratio decreases with the mass number (i.e. with the deformation).

Energy (MeV)

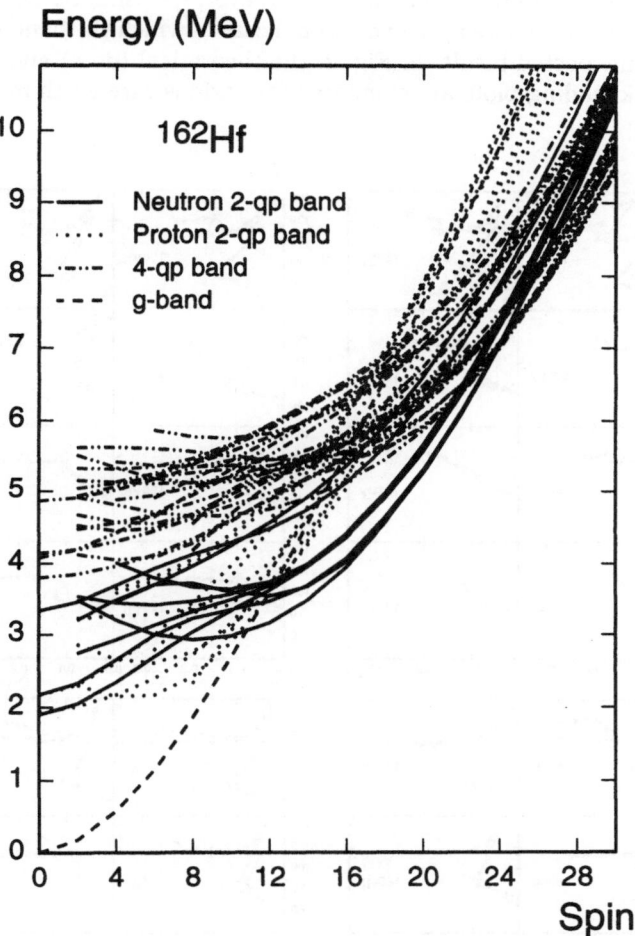

Fig. 1. Band Diagram for ^{162}Hf

Nucleus	^{156}Er	^{158}Er	^{160}Er	^{162}Er	^{164}Er	^{166}Er	^{168}Er	^{170}Er
I_c	12	14	14	14	14	14	14	16
ω_g/ω_s	3.46	2.39	2.33	2.22	2.11	1.71	1.73	1.56

Table 1: The Ratio ω_g/ω_s as a Measure of the g-s Crossing Angle

Strictly speaking, when the Hamiltonian is diagonalized, the g-s coupling weakens the sharpness of the backbending but this is unimportant for a qualitative understanding. As a rule, the essential aspect of the physics can already be seen from the band diagram since, usually, the final shell model diagonalization does not drastically change the main feature. However, we will later show that this is not always the case. There is an exceptional case in which the band coupling may lead to an unexpected result. In Fig. 2, the theoretical (dots) and experimental (circles) backbending plots are compared for various rare-earth nuclei.

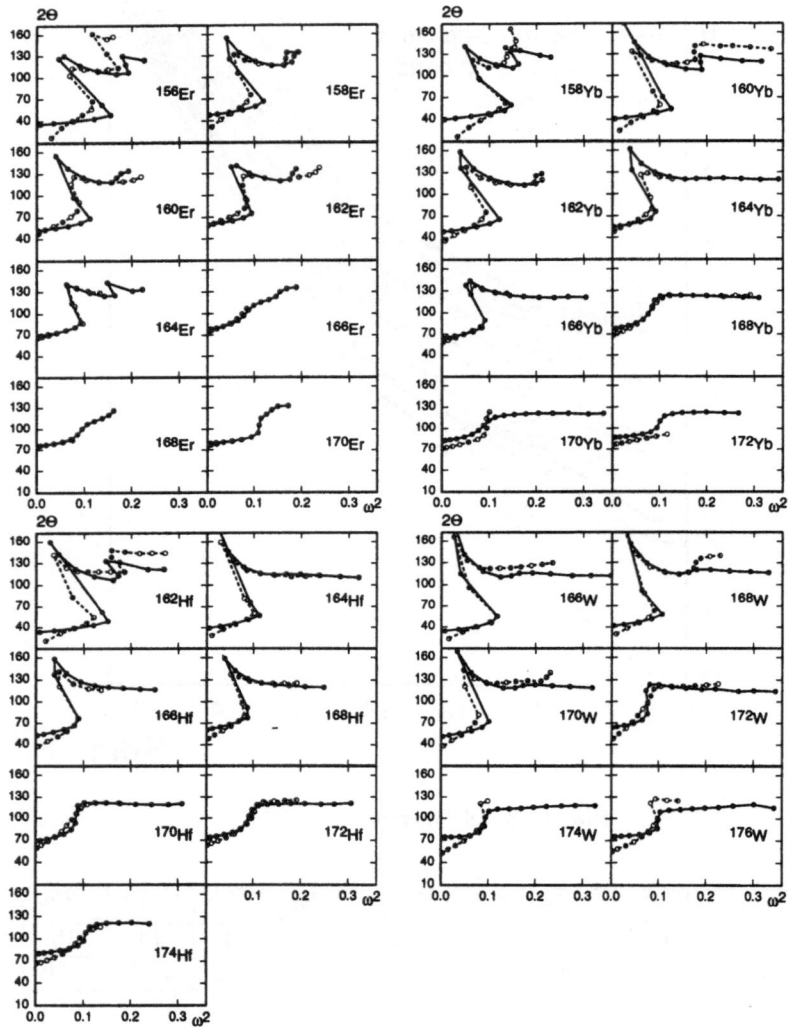

Fig. 2. Backbending Plots for Doubly Even Rare-Earth Nuclei.

Some lighter isotopes show the second backbending. This is due to a crossing between the s-band and an aligned 4-qp (2-neutron⊗2-proton) band. As a matter of fact, such a crossing occurs in all nuclei but the crossing angle is mostly not large enough to cause the second backbending. The crossing spin increases for more deformed isotopes. As a rule, the higher the crossing spin, the smaller the crossing angle. This is the reason why the first backbending turns into an upbending for heavier isotopes. It also means that a backbending higher than the second one will not occur in nature.

3   Odd Mass Nuclei

In Fig. 3, we compare the theoretical (dots) and experimental (circles) transition energies $E(I) - E(I-1)$ for the positive parity yrast states of some odd Er and Yb isotopes, which show the so-called signature dependence ($\Delta I = 1$ staggering).

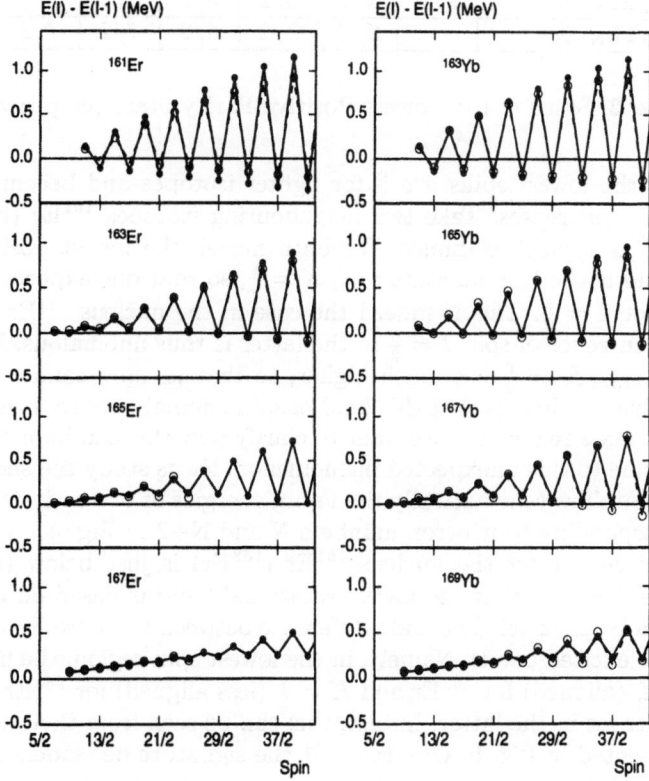

Fig. 3. Transtion Energies $E(I) - E(I-1)$ for odd Er and Yb Isotopes.

Up to spin $\approx \frac{41}{2}$, the relevant bands (configurations) for the shell model diagonalization are 7 decoupled 1-neutron qp states in the intruder sub-shell $i_{13/2}$ with $K = \frac{1}{2}, \cdots, \frac{13}{2}$. It is well-known that the alignment of a decoupled state occurs only in the $K = \frac{1}{2}$ band in the particle-rotor model. Quantum mechanically, the alignment extends beyond $K = \frac{1}{2}$ diminishing rapidly with increasing K-value and one finds that not only the band $K = \frac{1}{2}$ but also the band $K = \frac{3}{2}$ exhibits an appreciable signature dependence. A general discussion of this problem is presented in greater depth in Ref. [1]. Our main interest here is to show that this property of the $i_{13/2} K = \frac{3}{2}$ state plays an essential role for the spectrum and in particular for the determination of the lowest state spin. Table 2 shows the spin systematics for some odd mass nuclei.

N	89	91	93	95	97	99	101
$_{64}Gd_N$	9/2	5/2	5/2	5/2			
$_{66}Dy_N$	9/2	9/2	5/2	5/2	5/2		
$_{68}Er_N$		9/2	9/2	5/2	5/2	7/2	
$_{70}Yb_N$			9/2	9/2	5/2	7/2	7/2

Table 2: Spins of the Lowest Positive Parity State (empirical)

Note that the lowest spins are $\frac{9}{2}$ for lighter isotopes and become $\frac{5}{2}$ as the neutron number increases. Take two neighbouring isotopes ^{161}Er (N=93) and ^{163}Er (N=95) as typical examples. For both nuclei, the lowest positive parity rotational band is the Nilsson state $i_{13/2} K = \frac{5}{2}$, so that one expects the lowest state spin to be $I = \frac{5}{2}$. This is indeed the case in the nucleus ^{163}Er but not in ^{161}Er. The occurrence of spin $I = \frac{9}{2}$ in the latter is thus anomalous. In fact, the Nilsson state $i_{13/2} K = \frac{9}{2}$ lies much higher, so that its spin cannot stem from this band. Table 2 indicates that this kind of spin anomaly occurs systematically in the present mass region and we want to clarify how this can happen. In order to find some clue to this unexpected phenomenon, let us study the shell filling of these nuclei. We show schematically the Fermi energies of two such neighbouring isotopes corresponding to neutron numbers N and N+2 in Fig. 4.

The Fermi energy for the nucleus ^{161}Er (^{163}Er) is just below (above) the $K = \frac{5}{2}$ Nilsson level, so that the lowest rotational band is based on the Nilsson state $K = \frac{5}{2}$ in both nuclei. The main difference between these two isotopes is the location of the excited bands. Namely, in the lowest spin region, the first excited band is $K = \frac{3}{2}$ (aligned) for ^{161}Er and $K = \frac{7}{2}$ (less aligned) for ^{163}Er because of two more neutrons in the latter. In fact, this can be seen from their actual band diagrams presented in Fig. 5. Observe that the signature dependence (favoured or unfavoured state according to the signature rule $I - j =$ even or odd where $j = \frac{13}{2}$ is the intruder sub-shell in question) decreases with increasing K-value.

The mechanism leading to the spin anomaly of ^{161}Er is quite simple. In fact, it is caused simply by the coupling between $K = \frac{5}{2}$ and $K = \frac{3}{2}$ bands. The

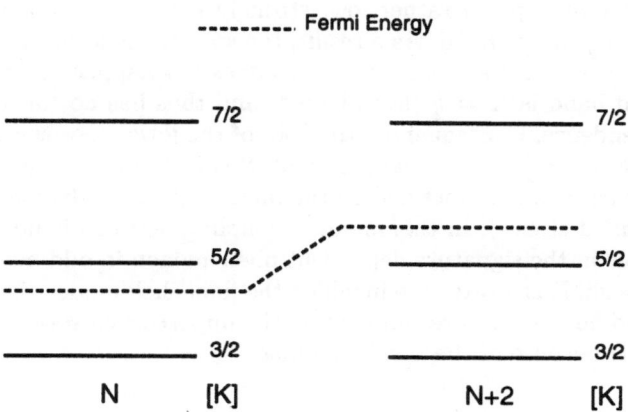

Fig. 4. Fermi Energies for Isotopes N and N+2 (schematic).

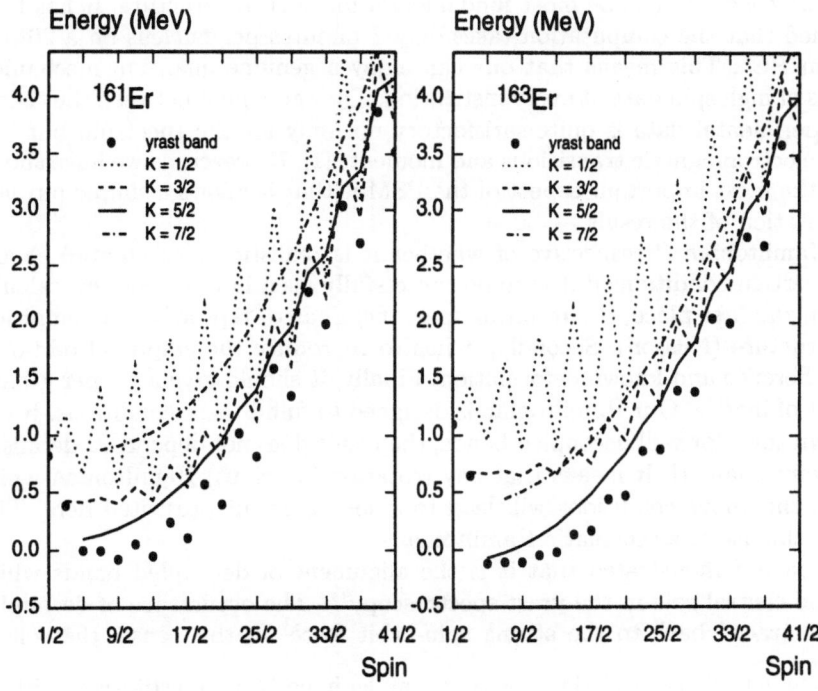

Fig. 5. Band Diagrams for ^{161}Er and ^{163}Er.

$I = \frac{9}{2}$ state of the $K = \frac{3}{2}$ first excited band (aligned), which is a favoured state and therefore appears rather low, strongly couples to and pushes down that of the $K = \frac{5}{2}$ lowest band. As a result, the spin of the lowest state becomes anomalous ($I = \frac{9}{2}$). On the other hand, this does not happen to ^{163}Er because its first excited band is $K = \frac{7}{2}$ (less aligned) and thus has no (or a very weak) signature dependence. Consequently, the spin of the lowest positive parity state remains normal ($I = \frac{5}{2}$). One clearly sees all this in the band diagrams Fig. 5, in which the yrast energies obtained by the final shell model diagonalization are also plotted (filled circles). In this manner, coupling between bands often plays an essential role in the signature dependent phenomenon in odd mass as well as doubly odd nuclei. It is a rare case in which the final shell model diagonalization drastically modifies the features implied by the unperturbed rotational energies (band diagram) because of the band coupling.

4   Conclusion

In the calculations, we have used a Hamiltonian which consists of the harmonic oscillator single-particle Hamiltonian and a sum of schematic (Q·Q + Monopole Pairing + Quadrupole Pairing) forces. These forces represent specific correlations which are considered to be most fundamental for nuclear structure. It has been confirmed that the computation takes only 7 minutes per nucleus on a 90MHz Pentium PC*. This means that one can enjoy a genuine quantum mechanical analysis of high spin data at every institution. The agreement between the theory and experimental data is quite satisfactory not only for the spectrum but also for the electromagnetic transitions and moments [1]. However, as we have shown above, the most important aspect of the PSM is that it allows a simple physical interpretation of the result.

A Hamiltonian (irrespective of whether it is realistic or schematic) should satisfy certain conditions if it is to be successfully used in the structure calculations. In the first place, the resulting mean field has to reproduce the empirical shell structure (Nilsson). Secondly, it has to reproduce the empirical odd-even mass difference and known deformation. Finally, it should give a proper g-band moment of inertia. Our Hamiltonian is designed to fulfill these conditions. It can be shown that, for well decoupled bands, the result does not depend on details of the Hamiltonian [1]. It means that any (rotation invariant) Hamiltonian which satisfies the above conditions will lead to a similar result presented here. This justifies the use of a schematic Hamiltonian.

We have demonstrated that it is the alignment of decoupled bands which plays the central role in the yrast spectroscopy**. The occurrence of decoupled bands is traced back to the strong spin-orbit force. In this sense, the role of

---

* We mention this particularly because it may be important to institutions with no access to sophisticated computing facilities.

** The situation will be different for superdeformed bands as there is no well decoupled bands because of very large deformation.

the spin-orbit force introduced by Mayer and Jensen to account for the magic numbers has been rediscovered in the light of modern spectroscopy. Unlike the aligned bands, those bands which are strongly coupled to the rotating body never come down to the yrast region and are thus unimportant at higher spins. The kinematics of decoupled bands is thoroughly discussed in Ref. [1]. Needless to say, the PSM is not limited to the description of the yrast states. One only needs to select the shell model configuration space appropriately to describe excited bands such as the β- and γ-band as well as the magnetic dipole (1^+) states (low spins and high excitation energies). This will be done in the near future.

We finally mention that application of the PSM is not limited to deformed nuclei. In fact, it is also applicable to spherical nuclei if one uses a spherical single-particle basis. In this case, the angular momentum projection operator replaces the cumbersome CFP calculations and this will accelerate the computation (a single integration instead of multiple summations). Actually, the PSM intrinsic states then reduce to those of the (quasi-particle) M-scheme. Howevr, the diagonalization can be done in a much smaller configuration space compared with that of the M-scheme shell model since one need not include the complete multiplet components to generate good angular momentum states. In fact, if one takes too many PSM intrinsic states, there will be many redundant states when projected. For example, a one-body state $a_{jM}^\dagger |0\rangle$ can be written as $\hat{P}_{MK}^j a_{jK}^\dagger |0\rangle$ with a fixed K (e.g. $K = j$). These two representations are mutually equivalent. The former is the "M-scheme" and the latter PSM which is labeled by a single value of K. This is because the intrinsic states $a_{jK}^\dagger |0\rangle$ having other K-values do generate the same states when projected, which are therefore redundant. In other words, it indicates explicitly that there is only one physically siginficant state. In fact, the actual size of the configuration space for a spin j is 1 (not $2j + 1$ which counts the degree of multiplet). For a one-body system, such a representation is rather artificial and trivial. However, it becomes of significance when used in a many-body system (a two-body system is already a non-trivial example). The occurrence of vanishing eigenvalues for the Norm matrix defined by (2) implies the presence of redundant projected states. One may thus reduce the size of the configuration space until no such eigenvalue appears. This type of the PSM may be referred to as the Projected M-Scheme (PMS). It may open a new way of performing a spherical shell model calculation which goes around the evaluation of the CFP. At present, such an aspect of the PSM is not yet fully exploited and there are many other possible applications that are still to be investigated.

References

[1] K. Hara and Y. Sun, Int. J. Mod. Phys. **E4** (1995) 637; References to earlier works are given in this review article.

Relativistic Mean Field Theory and Applications in Finite Nuclei

P. Ring, A.V. Afanasjev, and J. Meng

Physics Department, Technical University Munich,
D-85478 Garching, Germany

Abstract. Relativistic Mean Field (RMF) theory is applied to modern problems of nuclear structure, such as the description of rotating bands in super-deformed nuclei or the investigation of neutron halos in light exotic nuclei.

1 Introduction

Since the early days of nuclear physics one has known, that the spin plays an important role in our understanding of nuclear spectra, and that, in principle, one should use a relativistic theory for a complete description of the nuclear system even in the low energy domain. Early attempts in this direction of Teller et al [1], however, have been forgotten and it was much later – in the seventies – that Walecka [2], [3] pointed out the power, the simplicity and the elegance of a phenomenological relativistic description of the nuclear system.

In recent years these methods have become very popular. They have been extended so far, that they now can be applied not only to nuclear matter and to the ground state of spherical doubly closed shell nuclei [4], but also to the entire region of the periodic table [5], to exotic nuclei with large neutron excess [6], [7] as well as to deformed [8] and super-deformed nuclei [9] and even to rotating [10] and vibrating [11], [12], [13], [14] nuclei (for a recent review see Ref. [15]).

Relativistic Mean Field (RMF) theory is a phenomenological theory. It is conceptually similar to density dependent Hartree-Fock theory with Skyrme forces, which is nowadays standard for the microscopic description of nuclear properties over the entire periodic table. It shares with this theory, that it contains only a few parameters which are adjusted to data of nuclear matter and of and a few closed shell nuclei. All the other nuclei are described with one parameter set. In most of its applications it also shares with Skyrme theory the disadvantage that pairing is taken into account only in a very primitive way, the so-called *constant gap approximation*, which uses experimental gap parameters to determine BCS occupation probabilities for the calculation of the densities in each step of the iteration. This method works rather well for the ground states of even-even nuclei, where experimental gaps are available. Its predictive power, however, is limited in the region of exotic nuclei far from the stability line, where no experimental gap parameters are known and in the high spin region, where pairing is quenched and where the finite range of the pairing interaction should be taken into account. Therefore, in recent years relativistic mean field (RMF) theory has been extended to Relativistic Hartree Bogoliubov (RHB) theory [16], [17]. However, it has turned out, that in order to get good agreement with experimental data for pairing in finite nuclei one cannot use the meson exchange potentials with the coupling constants adjusted in RMF theory. Therefore it has been proposed in Ref. [17] with great success to use in the pairing channel the same finite range forces of Gaussian as they have been introduces by Gogny and his collaborators [18].

In this talk we discuss a number of recent investigations of problems of low energy nuclear structure within the framework of RMF theory, such as the description of super-deformed rotational bands or the extension of RMF theory to relativistic Hartree-Bogoliubov theory in coordinate space, taking into account properly the coupling to the continuum and using density dependent zero range forces in the pairing channel for the description of halo nuclei.

2   Relativistic Mean Field Theory

In relativistic mean field (RMF) theory the nucleus is described as a system of point-like nucleons, Dirac spinors, coupled to mesons and to the photons. The nucleons interact by the exchange of several mesons, namely a scalar meson σ and three vector particles ω, ρ and the photon. The isoscalar-scalar σ-mesons provide a strong intermediate range attraction between the nucleons. For the three vector particles we have to distinguish the time-like components and the spatial components. For the photons this means the Coulomb field and a possible magnetic field in the case where currents play a role. For the isoscalar-vector ω-meson the time-like component provides a very strong repulsion at short distances for all combination of of particles, pp, nn and pn. For the isovector-vector ρ-meson the time-like components give rise to a short range repulsion for like particles (pp and nn) and a short range attraction for unlike particles (np). They also take care of the symmetry energy. In addition the spatial components of the ω and ρ-mesons lead to an interaction between possible currents, for the ω-meson attractive for all combinations (pp, nn and pn-currents) and for the ρ-meson attractive for pp and nn-currents but repulsive for pncurrents. We have to keep in mind, however, that within mean field theory these currents occur only in cases of time-reversal breaking mean fields as for instance in the case of Coriolis fields at high angular momenta.

The starting point of relativistic mean field theory is the well known local Lagrangian density

$$\mathcal{L} = \bar{\psi}(i\gamma^\mu \partial_\mu - m)\psi - m_\sigma^2 \sigma^2 - \frac{1}{3}g_2\sigma^3 - \frac{1}{4}g_3\sigma^4$$

$$-\frac{1}{4}\Omega_{\mu\nu}\Omega^{\mu\nu} + \frac{1}{2}m_\omega^2\omega_\mu\omega^\mu - \frac{1}{4}\vec{R}_{\mu\nu}\vec{R}^{\mu\nu} + \frac{1}{2}m_\rho^2\vec{\rho}_\mu\vec{\rho}^\mu - \frac{1}{4}F_{\mu\nu}F^{\mu\nu} \quad (1)$$

$$-g_\sigma\bar{\psi}\sigma\psi - g_\omega\bar{\psi}\gamma^\mu\omega_\mu\psi - g_\rho\bar{\psi}\gamma^\mu\vec{\tau}\vec{\rho}_\mu\psi - e\bar{\psi}\gamma^\mu\frac{1+\tau_3}{2}A_\mu\psi,$$

where the non-linear self-coupling of the σ-field, which is important for an adequate description of nuclear surface properties and the deformations of finite nuclei, is taken into account according to Ref. [19]. The field tensors for the vector mesons and the photon field are:

$$\Omega_{\mu\nu} = \partial_\mu \omega_\nu - \partial_\nu \omega_\mu \qquad \vec{R}_{\mu\nu} = \partial_\mu \vec{\rho}_\nu - \partial_\nu \vec{\rho}_\mu \qquad F_{\mu\nu} = \partial_\mu A_\nu - \partial_\nu A_\mu. \qquad (2)$$

In the present state of the art the relativistic mean field theory, the meson and photon fields are treated as classical fields.

The Lagrangian (1) contains as parameters the masses of the mesons m_σ, m_ω and m_ρ, the coupling constants g_σ, g_ω and g_ρ and the non-linear terms g_2 and g_3.

The mesons are considered as effective particles carrying the most important quantum numbers and generating the interaction in the corresponding channels in a Lorentz invariant manner by a local coupling to the nucleons. In this sense, the Lagrangian (1) is an *effective Lagrangian* constructed for the mean field approximation.

3   Relativistic Mean Field Theory in the Rotating Frame

In non-relativistic nuclear physics the cranking model [20] plays an important role in the description of rotating nuclei. It is the symmetry breaking mean field version of a variational theory with fixed angular momentum and can be derived as an approximate variation after angular momentum projection [21]. In the self-consistent version [22] it allows to include alignment effects [23] as well as polarization effects induced by the rotation, such as Coriolis-anti-Pairing or changes of the deformation. But already in the simplified version of the *Rotating Shell Model* with fixed mean fields [24] it is able to describe successfully an extremely large amount of data in the high spin region of deformed nuclei [25]. The cranking idea can be used for a relativistic description too [26], [27]: one simply transforms the coordinate system to a frame rotating with constant angular velocity Ω around a fixed axis is space assuming – as in non-relativistic nuclear physics – that this axis is perpendicular to the symmetry axis of the nucleus in its ground state. Such a transformation in Minkowski space is given in text books [28]:

$$\begin{pmatrix} t \\ x \\ y \\ z \end{pmatrix} \implies \begin{pmatrix} \tilde{t} \\ \tilde{x} \\ \tilde{y} \\ \tilde{z} \end{pmatrix} = \begin{pmatrix} 1\,0 & 0 & 0 \\ 0 & & \\ 0 & e^{it\Omega J} & \\ 0 & & \end{pmatrix} \begin{pmatrix} t \\ x \\ y \\ z \end{pmatrix} \qquad (3)$$

According to the cranking prescription the absolute value of the angular velocity $|\Omega|$ is determined after the self-consistent solution of the equations of motion in the rotating frame by the Inglis condition [20]:

$$\langle \Omega J \rangle_\Omega = |\Omega| \sqrt{I(I+1)} \qquad (4)$$

Using the transformation properties of scalars, vectors, spinors, etc., we obtain in the rotating frame the following quantities

$$\tilde{\sigma}(x) = e^{it\boldsymbol{\Omega L}}\sigma(x) \tag{5}$$

$$\begin{pmatrix} \tilde{\omega}^0(x) \\ \tilde{\boldsymbol{\omega}}(x) \end{pmatrix} = \begin{pmatrix} 1 & 0 \\ -\boldsymbol{\Omega}\times\tilde{r} & 1 \end{pmatrix} \begin{pmatrix} e^{it\boldsymbol{\Omega L}}\omega^0(x) \\ e^{it\boldsymbol{\Omega J}}\boldsymbol{\omega}(x) \end{pmatrix} \tag{6}$$

$$\tilde{\psi}(x) = e^{it\boldsymbol{\Omega J}}\psi(x), \tag{7}$$

where $\boldsymbol{L} = -i(\boldsymbol{r}\times\boldsymbol{p})$ is the orbital and $\boldsymbol{J} = \boldsymbol{L}+\boldsymbol{S}$ is the total angular momentum containing the 4×4-matrices $\boldsymbol{S}$ for the spinor fields with spin $\frac{1}{2}$ and a 3×3-matrices $\boldsymbol{S}$ for vector fields with spin 1. For details see Ref. [26].
Using these quantities we obtain the following Lagrangian in the rotating frame

$$\tilde{\mathcal{L}} = \bar{\tilde{\psi}}\left(\tilde{\gamma}^\mu(\tilde{D}_\mu + g_\omega\tilde{\omega}_\mu) + g_\sigma\tilde{\sigma} + m_N\right)\tilde{\psi}$$

$$+\frac{1}{2}\tilde{\partial}_\mu\tilde{\sigma}\tilde{\partial}^\mu\tilde{\sigma} - U(\tilde{\sigma}) \tag{8}$$

$$-\frac{1}{4}(\tilde{\partial}^\mu\tilde{\omega}^\nu - \tilde{\partial}^\nu\tilde{\omega}^\mu)(\tilde{\partial}_\mu\tilde{\omega}_\nu - \tilde{\partial}_\nu\tilde{\omega}_\mu) + \frac{1}{2}m_\omega^2\tilde{\omega}^\mu\tilde{\omega}_\mu$$

where $\tilde{D}^\mu$ is the covariant derivative with respect to the rotating metric. Neglecting in the following the tilde sign, we derive the classical equations of motion. In the quasi-static limit they are of the following form:

$$\{\boldsymbol{\alpha}(\boldsymbol{p} - g_\omega\boldsymbol{\omega}) + g_\omega\omega_0 + \beta(m_N + g_\sigma\sigma) - \boldsymbol{\Omega J}\}\psi_i = \epsilon_i\psi_i \tag{9}$$

$$\{-\Delta + (\boldsymbol{\Omega L})^2\}\sigma + U'(\sigma) = -g_\sigma\rho_s \tag{10}$$

$$\{-\Delta + (\boldsymbol{\Omega L})^2 + m_\omega^2\}\omega^0 = g_\omega\rho_v \tag{11}$$

$$\{-\Delta + (\boldsymbol{\Omega J})^2 + m_\omega^2\}\boldsymbol{\omega} = g_\omega\boldsymbol{j} \tag{12}$$

These equations are very similar to the RMF equations in the non-rotating frame. There are only three essential differences:

1. The Dirac equation (9) contains a Coriolis term $\boldsymbol{\Omega J}$ in full analogy to non-relativistic cranking theory.
2. The Klein-Gordon equations for the mesons contain terms proportional to the square of the corresponding Coriolis terms. It turns out, however, that they can be neglected completely for all realistic cranking frequencies, because (i) they are quadratic in Ω and (ii) mesons being bosons are to a large extend in the lowest s-states with only small d-admixtures
3. The Coriolis operator in the Dirac equation breaks time-reversal invariance. Currents $\boldsymbol{j}$ are induced, which form the source of magnetic potentials in the Dirac equation (*nuclear magnetism*). In this way the charge current $\boldsymbol{j}_c$ is the source of the normal magnetic potential $\boldsymbol{A}$, the isoscalar baryon current $\boldsymbol{j}_B$ is the source of the spatial components $\boldsymbol{\omega}$ of the ω-mesons and the isovector baryon current $\boldsymbol{j}_3$ is the source of the spatial component ρ_3 of the

ρ-mesons. In contrast to the Maxwellian magnetic field A having a small electro-magnetic coupling, the large coupling constants of the strong interaction causes the fields ω and ρ to be important in all cases, where they are not forbidden by symmetries, such as time reversal. They have a strong influence on the magnetic moments [29] in odd mass nuclei, where time reversal is broken by the odd particle, as well as on the moment of inertia in rotating nuclei, where time reversal is broken by the Coriolis field [9].

4   Applications to Super-deformed Bands in the A = 140 − 150 Mass Region

In order to investigate the applicability of Relativistic Mean Field theory for the description of rotating super-deformed nuclei, we have carried out a systematic investigation of all super-deformed yrast bands in the A = 140 − 150 mass region using the parameter set NL1 [30]. All the experimental data presently available for yrast and, in some cases, also for excited super-deformed rotational bands in this mass region are compared with the results of these calculations. In Fig. 1 we summarize all the results of these calculations for the yrast super-deformed bands in this region. We find large similarities with earlier phenomenological calculations of Nilsson-Strutinsky or Woods-Saxon-Strutinsky type. whose parameters where obtained by fitting of single-particle levels in odd nuclei of deformed regions. Despite this difference, the single-particle ordering in the super-deformed minima of the nuclei in the mass region of interest obtained in the Cranked Relativistic Mean Field (CRMF) approach on one side and in the Cranked Nilsson (CN) and Cranked Woods-Saxon (CWS) approaches on the other side reveals large similarities.

There are, however, differences in the level density close to the Fermi surface. It is lower in the RMF model compared to the Nilsson and the Woods-Saxon potentials. This is connected with the low effective mass in relativistic theories [31]. Because of this low effective mass in the relativistic theories one would not have expected such an good agreement of the moment of inertia $J^{(2)}$ between the relativistic theory on one side and the phenomenological theories and the experiment on the other side. The phenomenological models have an effective mass one by definition and it is also known that the effective masses of realistic nuclear descriptions taking into account higher order corrections, such as the virtual excitation of collective surface vibrations, should lead to values close to one for the effective mass at the Fermi surface of finite nuclei.

Although it is not understood why the low effective mass does not lead to a considerable disagreement between the relativistic results and the experiment, we feel, that this fact could be in some sense connected with a similar effect occurring in the description of isoscalar magnetic moments nuclei differing by one nucleon form spin-saturated doubly magic nuclei. This quantities are well described in the extreme single particle model like the shell model if one uses an effective mass one (Schmidt-values). Relativistic theories with their small effective

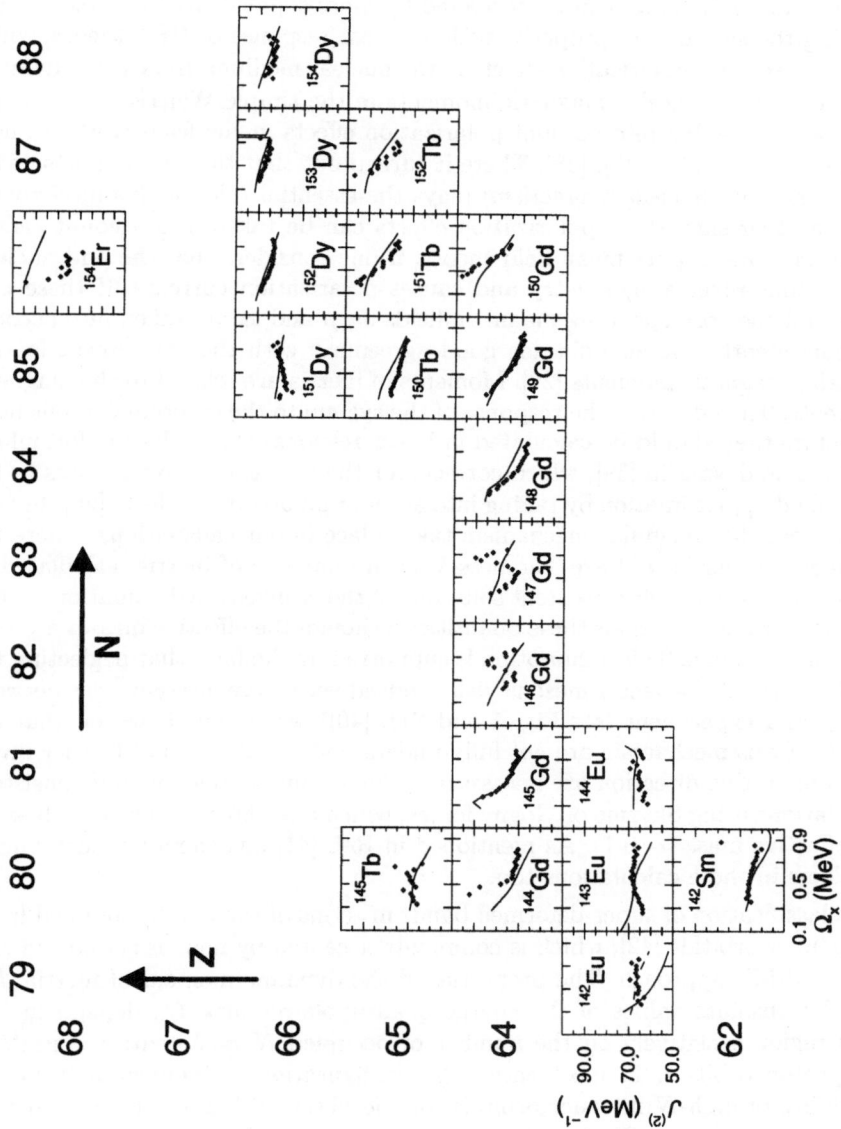

Fig. 1. Systematic overview of the moments of inertia $J^{(2)}$ for the yrast super-deformed bands in the $A \sim 140 - 150$ mass region. Cranked relativistic mean field calculations (full lines) are compared with experimental values (dots)

masses show very large discrepancies from the experimental values [32], [33], [34]. They are connected with the violation of Galilean invariance in the mean field approximation [35] and can be corrected by more sophisticated approximations treating the symmetries properly, such as linear response or RPA theory, which take into account polarization effects in the nuclear medium. In fact, the disturbing problem of isoscalar magnetic moments in the simple Walecka model could be solved by taking into account polarization effects in the framework of linear response theory [36], [37], [38]. There it turned out that the spatial parts of the vector mesons (*nuclear magnetism*) plays the essential role for this mechanism. On the other side, these polarization effects can be taken into account also in the framework of pure mean field theory, if one considers that the odd nucleon breaks time-reversal symmetry and causes polarization currents. If these currents and the corresponding magnetic fields ω, ρ and A are taken into account self-consistently one also obtains good agreement with the experiment for the isoscalar magnetic moments [29]. Moments of inertia are related to the magnetic moments. They describe the response of the system to the external Coriolis field. Therefore they should be calculated in linear response theory by the formula of Thouless and Valatin [39], which corrects for the violation of symmetries in the mean field approximation by taking into account polarization effects [25]. In fact, we suspect that a similar mechanism takes place in our calculations, where the moments of inertia $J^{(2)}$ are Thouless-Valatin moments of inertia, calculated as derivatives of fully self-consistent solutions of the cranked RMF equations. They yield the proper values for these quantities although the effective masses are very different from one. This assumption is supported by the fact, that neglecting the spatial parts of the vector mesons yields indeed very large discrepancies between theory and experiment (see Fig. 1 and Ref. [40]) We admit, however, that the details of this mechanism are not fully understood so far and that further investigations in this direction are necessary. A final remark concerns self-consistent calculations using Skyrme or Gogny forces, which also show rather low values for the effective mass (~ 0.7). As mentioned in Ref. [41] one should expect similar problems in these calculations too.

The classification of super-deformed bands in terms of the number of filled high-N intruder orbitals [42], which is commonly accepted by now, is supported also in the CRMF approach. The properties of the dynamic moment of inertia $J^{(2)}$ and the absolute values of the charge quadrupole moment Q_0 depend in this mass region sensitively on the number of occupied $N = 7$ neutron and $N = 6, 7$ proton orbitals. In most cases, the configuration assignment in terms of the filling of high-N intruder orbitals for the observed bands agrees with that proposed within the CWS and the CN approaches. However, there are some differences for the isotope chain $^{146-148}$Gd which are connected with the fact that at the self-consistent deformation of the configurations of interest the energy difference between the $\nu[642]5/2$ and $\nu[651]1/2$ orbitals is underestimated in the present CRMF calculations. This leads to some difficulties with a quantitative description of observed properties of the dynamic moment of inertia of the bands 1 and 2 in these nuclei. We cannot exclude that this difficulty is connected

Table 1. Charge quadrupole moments Q_0^{CRMF} at $I \sim 30\hbar$ and $I \sim 60\hbar$ of different configurations: CRMF calculations are compared with available experimental values Q_0^{exp} and with results of cranked Woods-Saxon (Q_0^{WS}) and cranked Nilsson (Q_0^{Nils}) calculations. The Q_0^{WS} and Q_0^{Nils} values are given for the configurations which have the same occupation of high-N intruder orbitals as in the CRMF calculations.

Nucleus	Q_0^{CRMF} (eb)		Q_0^{exp} (eb)	Q_0^{WS} (eb)	Q_0^{Nils} (eb)
	$I \sim 30\hbar$	$I \sim 60\hbar$			
^{154}Dy	18.50	18.10			
^{153}Dy	19.67	19.35		19.2(3)	
^{152}Dy	19.05	18.73	19±3	18.9(6)	18.9
^{151}Dy	18.25	17.90		18.0(5)	
^{151}Tb	18.02	17.48		18.0(7)	
^{150}Tb	17.17	16.64		16.8(6)	
^{150}Gd	16.98	16.40	17±3	16.9(8)	16.2
^{149}Gd	16.17	15.50	17±2	16.2(7)	15.7
^{148}Gd	15.35	14.80		15.6(5)	
	15.90	15.25	14.8(3)		16.0
	15.90	15.30	14.6(2)	15.9(7)	16.0
^{146}Gd	14.45	13.95		14.7(1)	
	15.12	14.40	12±2		
	14.85	14.32		14.8(6)	
^{144}Gd	13.83	13.18		13.5(3)	
^{143}Eu	12.90	12.33	13±1	13.3(2)	12.9

with the fact that we neglect pairing correlations in the present calculations. It could also be connected with the small effective mass in relativistic theories and the problems to describe properly the position of specific single particle levels discussed above.

In Table 1 we show quadrupole moments. We find that with an increasing number of occupied high-N intruder orbitals, the charge quadrupole moment Q_0 and the mass hexadecupole moment Q_{40} increases. The absolute values of Q_0 are reasonable close to the experimental data, and they are consistent with the results of CWS and CN calculations.

5   Halo Nuclei and Relativistic Hartree-Bogoliubov Theory

Since the experimental discovery of neutron halo phenomena in ^{11}Li [43], the study of exotic nuclei has become a very challenging topic in nuclear physics. Experimentally a sudden rise in the interaction cross-sections has been observed for light neutron-rich nuclei, specifically while going from ^{9}Li to ^{11}Li, ^{12}Be to ^{14}Be and ^{15}B to ^{17}B. This sudden increase in the interaction cross-sections has been attributed to a relatively large *rms* matter radius as compared to that expected from the conventional mass dependence $1.2A^{1/3}$. This large radius has been interpreted by the fact that filling in more and more neutrons in the nuclear well, the Fermi surface for the neutrons comes close the continuum limit in the nucleus ^{11}Li. Due to the small single-neutron separation energy the tails of the two wave-functions of the last filled orbital ($1p_{1/2}$) reach very far outside of the nuclear well and a neutron halo is formed [43].

Although this simple interpretation is based on the mean field picture, several microscopic theoretical investigations within self-consistent mean field models [44], [45], [46], [47] have failed. In fact it would be a strange accident, if the last occupied neutron level in ^{11}Li, the $1p_{1/2}$ orbit would be so close to the continuum limit, that the tail of the two last uncorrelated neutron wave functions reaches so far out as it is necessary to reproduce the large experimental radius. Therefore, in the non-relativistic scheme, Bertsch et al [44] and Sagawa [46] have introduced an artificial modification to the potential in order to reproduce the small separation energy in their mean field calculations with Skyrme interactions. In this way the authors could qualitatively reproduce the observed trends, though some discrepancies still remain.

Koepf et al [45] where the first to investigate the neutron halo in relativistic mean field (RMF) theory for both spherical symmetrical and axially deformed cases. They found large deformations for the lighter Li isotopes, but a spherical shape for ^{11}Li and as in the non-relativistic investigations the binding energy of the $1p_{1/2}$ was too large as to reproduce a neutron halo. Zhu et al [47] improved the result of these RMF calculations by applying a similar modification to the potential as in Ref. [46] in order to adjust the proper size of the halo.

Bertsch and Esbensen [48] recognized that pairing correlations play an essential role. Performing a quasi-particle continuum RPA calculation based on the core of ^{9}Li and using a density dependent interaction of zero range in the pairing channel they found a large neutron halo ^{11}Li and could reproduce the proper size of the experimental matter radius.

The present investigation is based on a similar idea. Pairing correlations and the scattering of Cooper pairs into the continuum are taken into account in a fully self-consistent way by a continuum Hartree-Bogoliubov in the framework of relativistic mean field theory. A consistent treatment of pairing correlations in the framework of a generalized relativistic mean field approximation has been developed by Kucharek et al [16]. However, applications of this theory to nuclear matter show clearly that a quantitative description of pairing correlations in the nuclear many-body system cannot be achieved in this way with the presently used parameter sets of relativistic mean field theory. The behavior of the meson exchange forces entering this theory is simply not properly adjusted at large momentum transfer. In principle these forces have finite range and kinematical factors guarantee the convergence of the relativistic gap equation. The large masses of the σ- and the ω-mesons, however, do not yield a realistic cut off. Therefore the situation is similar to that of Skyrme forces with zero range: The short range of the relativistic forces produces no problem in the Hartree case, where only momenta up to the Fermi surface are involved, but it causes severe problems in the description of pairing, which allows scattering to very high momentum states.

Therefore it has been proposed to combine the advantages of a relativistic description in the framework of RMF-theory in the Hartree channel with a phenomenological force in the pairing channel In a recent investigation by Gonzalez-Llarena et al [17] a similar concept has been used with great success for the solution of the relativistic Hartree-Bogoliubov equation in an oscillator basis with a finite range force of Gogny's type in the pairing channel. At present the application of finite range forces in the pairing channel of continuum Hartree-Bogoliubov calculations is technically not yet feasible. We therefore use, as the authors of Ref. [48] a density dependent δ-force of the form

$$V(r_1, r_2) = V_0 \delta(r_1 - r_2) \frac{1}{4} [1 - \sigma_1 \sigma_2] \left(1 - \frac{\rho(r)}{\rho_0} \right) \tag{13}$$

Using Greens function techniques it has been shown in Ref. [16] how one can derive a relativistic Hartree-Fock-Bogoliubov theory from such a relativistic Lagrangian: After a full quantization of the system the mesonic degrees of freedom are eliminated and, in full analogy to the non-relativistic case, the higher order Greens functions are factorized in the sense of Gorkov [49]. Finally, neglecting retardation effects one ends up with relativistic Dirac-Hartree-Bogoliubov (RHB) equations

$$\begin{pmatrix} h & \Delta \\ -\Delta^* & -h^* \end{pmatrix} \begin{pmatrix} U \\ V \end{pmatrix}_k = E_k \begin{pmatrix} U \\ V \end{pmatrix}_k, \tag{14}$$

E_k are quasi-particle energies and the coefficients U_k and V_k are four-dimensional Dirac spinors normalized in the following way

$$\int U_k^+ U_{k'} + V_k^+ V_{k'} \, d^3 r = \delta_{kk'} \tag{15}$$

Neglecting the Fock term, as is it mostly done in RMF theory, we obtain for the average field

$$h = \alpha p + g_\omega \omega + \beta(M + g_\sigma \sigma) - \lambda \tag{16}$$

where σ and ω are the meson fields determined self-consistently from the Klein Gordon equations:

$$\{-\Delta + m_\sigma^2\} \sigma = -g_\sigma \rho_s - g_2 \sigma^2 - g_3 \sigma^3, \tag{17}$$

$$\{-\Delta + m_\omega^2\} \omega = g_\omega \rho_B \tag{18}$$

with the scalar density ρ_s and the baryon density ρ_B

$$\rho_s = \sum_k \bar{V}_k V_k, \qquad \rho_B = \sum_k V_k^+ V_k, \qquad (19)$$

where the sum over k runs only over all the particle states in the *no-sea approximation*.

The pairing potential Δ in Eq. (14) is given by

$$\Delta_{ab} = \frac{1}{2} \sum_{cd} V_{abcd}^{pp} \kappa_{cd} \qquad (20)$$

It is obtained from the pairing tensor $\kappa = U^* V^T$ and the one-meson exchange interaction V_{abcd}^{pp} in the pp-channel. More details are given in Ref. [16]. As mentioned above, these forces are not able to reproduce even in a semi-quantitative way proper pairing in the realistic nuclear many-body problem. We therefore replace V_{abcd}^{pp} in Eq. (20) by the density dependent two-body force of zero range given in Eq. (13).

For a zero range force in the pairing channel, the RHB equations (14) are a set of four coupled differential equations for the HFB Dirac spinors $U(r)$ and $V(r)$. They are solved in a self-consistent way by the shooting method and and the Runge-Kutta algorithm. The details will be published elsewhere. The calculation is done in a spherical box of radius $R = 20$ fm with a step size of 0.1 fm and proper boundary conditions. For each spin-parity channel 20 radial wave-functions are taken into account, which corresponds roughly to a cut-off energy of 200 MeV.

We use here the non-linear Lagrangian parameter set NL2 which was designed in Ref: [50] for light nuclei: $M_N = 938$, $m_\sigma = 504.89$, $m_\omega = 780$, $m_\rho = 760$ MeV and $g_\sigma = 10.444$, $g_\omega = 12.945$, $g_\rho = 4.383$, $g_2 = -6.6099$ fm^{-1}, $g_3 = 13.783$. Pairing is neglected for the three protons and the strength V_0 of the pairing force (13) for the neutrons is determined by a calculation in the nucleus ^{7}Li adjusting the corresponding pairing energy $\frac{1}{2} Tr \Delta \kappa$ to that of a RHB-calculation in an oscillator basis using the finite range part of the Gogny force D1S of Ref. [51] in the pairing channel. We this obtain $V_0 = -3000$ [MeV fm^3]. For ρ_0 we use the nuclear matter density 0.152 fm^{-3}.

In Fig. 2 we show the calculated binding energies E_B and the *rms* matter radii for the Li isotopes with mass numbers $A = 6$ to $A = 11$ and compare them with experimental values. In order to correct for the center of mass motion a spurious energy of an harmonic oscillator $E_{cm} = 0.75 \hbar \omega_0$ with $\hbar \omega_0 = 41 A^{-1/3}$ is subtracted. The calculated values show a small under-binding and the odd-even staggering is somewhat exaggerated by our blocking calculations, but in general the agreement is very satisfactory. In full agreement with experiment The matter radius shows a considerable increase when going from the nucleus ^{9}Li to ^{11}Li. In contrast to the earlier mean field calculations of Refs. [44], [46], [47] these results are obtained without any artificial modifications of the potential.

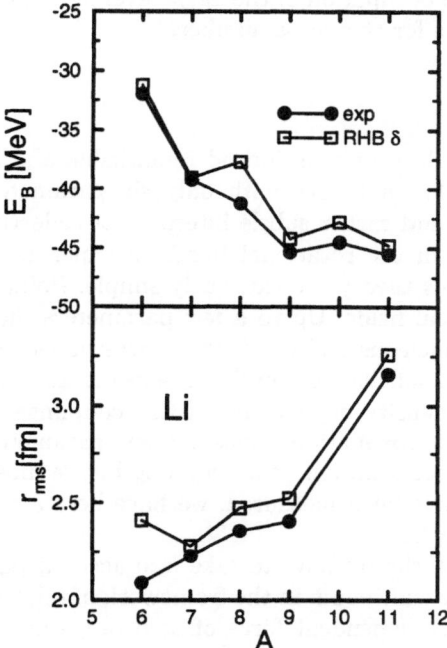

Fig. 2. Binding energies (upper part) and matter radii (lower part) for Li isotopes: RHB calculations (empty squares) are compared with experimental values (full dots).

Investigating the corresponding density distribution for the neutrons in the isotopes ^{9}Li and ^{11}Li, it is clearly seen that the increase of the matter radius is caused by a large neutron halo in the nucleus ^{11}Li. Its density distribution is in very good agreement with the experimental density of this isotope show with its error bars by the shaded area.

In order to understand the microscopic structure of this halo, we have to study the single particle structure and the average fields $S(r)+V(r) = g_\sigma\sigma + g_\omega\omega \pm g_\rho\rho + eA$ for the protons and neutrons together with the energy levels $\epsilon_n = \langle n|h|n\rangle$ in the canonical basis [25]. The Fermi level for the neutrons turns out to be very close to the continuum limit in close vicinity to the $\nu 1p_{1/2}$ and to the $\nu 2s_{1/2}$ level. Pairing correlations cause a partial occupation of both the $\nu 1p_{1/2}$ and the $\nu 2s_{1/2}$ level, i.e. a scattering of Cooper pairs to the continuum.

Going from $A = 9$ to $A = 11$ we observe a continuous increase of the contribution of the $p_{1/2}$ channel to the total *rms* matter radius and in addition a sudden increase of the contribution of the $s_{1/2}$ channel. This means that the halo in ^{11}Li is formed by the sudden occupation of the $2s_{1/2}$ level in this nucleus, which approaches the Fermi surface for this mass number.

6   Conclusions

Summarizing we can conclude that it is indeed remarkable and by no means understood in all details, that a theory with only six parameters is able to reproduce this complicated and rather subtle interplay of collective and single particle degrees of freedom in the rotational bands of super-deformed nuclei. The basic assumptions of this theory are extremely simple: Point-like nucleons are mowing in simple classical fields. Up to a few parameters the structure of the couplings between the nucleons and the fields is determined by the laws of relativity and principle of simplicity. A non-linear self-interaction between the σ-mesons takes care of the density dependence of this couplings. The rotation is treated in the cranking approximation. Since the few parameters have been adjusted in the literature more than ten year ago, long before most of the data on super-deformed bands have been measured, we have here a parameter free description of super-deformation.

For the description of halo nuclei we have to take into account pairing correlations and the coupling to the continuum in the framework of relativistic Hartree Bogoliubov theory. A density dependent force of zero range has been used in the pairing channel, whose strength is adjusted for the isotope ^{6}Li to a similar calculation with Gogny's force D1S in the pairing channel. Good agreement with experimental values is found for the total binding energies and the radii of the isotope chain ^{6}Li to ^{11}Li. In excellent agreement with the experiment we obtain a neutron halo for ^{11}Li without any artificial adjustment of the potential, as it was necessary in earlier calculations. In contrast to these investigations the halo is not formed by two neutrons occupying the $1p_{1/2}$ level very close to the continuum limit, but is is formed by Cooper-pairs scattered mainly in the two two levels $1p_{1/2}$ and $2s_{1/2}$. This is made possible by the fact that the $2s_{1/2}$ comes close down close to the Fermi level in this nucleus and by the density dependent pairing interaction coupling the levels below the Fermi surface to the continuum. In contrast to the very accidental fact that one single particle level is such close to continuum threshold, that the tail of its wave function forms a halo, this is a much more general mechanism, which could possibly be observed also in other halo nuclei. One only needs several single particle levels with small orbital angular momenta and correspondingly small centrifugal barrier close, but not directly at, the to the continuum limit.

The fact that relativistic mean field theory works so remarkably well not only in the field of super-deformations and for the description of halo nuclei, but also in many other areas of nuclear structure physics (see for instance [15]) gives us confidence that this theory is not only a simple phenomenological model but that it contains beyond that basic all the ingredients of a proper theory of the nuclear many-body problem.

Acknowledgments

This work is also supported in part by the Bundesministerium für Bildung und Forschung under the project 06 TM 743 (6).

References

[1] M.H. Johnson and E. Teller; Phys. Rev. **98** (1955) 783
[2] J.D. Walecka; Ann. Phys. (N.Y.) **83** (1974) 491
[3] B.D. Serot and J.D. Walecka; Adv. Nucl. Phys. **16** (1986) 1
[4] J. Boguta; Nucl. Phys. **A372** (1981) 386
[5] Y.K. Gambhir, P. Ring, and A. Thimet; Ann. Phys. (N.Y.) **198** (1990) 132
[6] M.M. Sharma, G.A. Lalazissis, W. Hillebrandt, and P. Ring; Phys. Rev. Lett. **72** (1994) 1431
[7] J. Meng and P. Ring; submitted to Phys. Rev. Lett.
[8] W. Pannert, P. Ring, and J. Boguta; Phys. Rev. Lett. **59** (1987) 2420
[9] J. König and P. Ring; Phys. Rev. Lett. **71** (1993) 3079
[10] W. Koepf and P. Ring; Phys. Lett. **B212** (1988) 397
[11] D. Vretenar, H. Berghammer, and P. Ring; Nucl. Phys. **A581** (1994) 679
[12] D. Vretenar, H. Berghammer, and P. Ring; Nucl. Phys. **A581** (1995) 679
[13] B. Podobnik, D. Vretenar, and P. Ring; Z. Phys. **A354** (1996) 375
[14] P. Ring, D. Vretenar, and B. Podobnik; Nucl. Phys. **A598** (1996) 107
[15] P. Ring; *Relativistic Mean Field Theory in Finite Nuclei*, Progr. Part. Nucl. Phys. **37** (1996) 193
[16] H. Kucharek and P. Ring; Z. Phys. **A339** (1991) 23
[17] T. Gonzalez-Llarena, J.L. Egido, G.A. Lalazissis, and P. Ring; Phys. Lett. **B379** (1996) 13
[18] J. Decharge and D. Gogny; Phys. Rev. **C21** (1980) 1568
[19] J. Boguta and A.R. Bodmer; Nucl. Phys. **A292** (1977) 413
[20] D.R. Inglis; Phys. Rev. **96** (1954) 1059
[21] R. Beck, H.J. Mang, and P. Ring; Z. Phys. **231** (1970) 26
[22] P. Ring, R. Beck, and H.J. Mang; Z. Phys. **231** (1970) 10
[23] B. Banerjee, H.J. Mang, and P. Ring; Nucl. Phys. **A215** (1973) 366
[24] P. Ring and H.J. Mang; Phys. Rev. Lett. **33** (1974) 1174
[25] P. Ring and P. Schuck; *The Nuclear Many-body Problem*, Springer Verlag, Heidelberg (1980)
[26] W. Koepf and P. Ring; Nucl. Phys. **A493** (1989) 61
[27] K. Kaneko, M. Nakano, and M. Matsuzaki; Phys. Lett. **B317** (1993) 261
[28] L.D. Landau and E.M. Lifshitz; *Course of Theoretical Physics*, Pergamon Press, Oxford (1959)
[29] Ulrich Hofmann and P. Ring; Phys. Lett. **214B** (1988) 307
[30] P.G. Reinhard, M. Rufa, J. Maruhn, W. Greiner, and J. Friedrich; Z. Phys. **A323** (1986) 13
[31] M. Jaminon and C. Mahaux; Phys. Rev. **C40** (1989) 354
[32] M. Bawin, C.A. Hughes, and G.L. Strobel; Phys. Rev. **C28** (1983) 456
[33] A. Bouyssy, S. Marcos, and J.F. Mathiot; Nucl. Phys. **A415** (1984) 497
[34] H. Kurasawa and T. Suzuki; Phys. Lett. **154B** (1985) 16
[35] A. Arima, K. Shimizu, W. Bentz, and H. Hyuga; Adv. Nucl. Phys. **18** (1987) 1

[36] J.A. McNeil, R.D. Amado, C.J. Horowitz, M. Oka, J.R. Shepard, and D.A. Sparrow; Phys. Rev. **C34** (1986) 746

[37] S. Ichii, W. Bentz, and A. Arima; Phys. Lett. **192B** (1987) 11

[38] J.R. Shepard, E. Rost, C.Y. Cheung, and J.A. McNeil; Phys. Rev. **C37** (1988) 1130

[39] D.J. Thouless and J.G. Valatin; Nucl. Phys. **31** (1962) 211

[40] W. Koepf and P. Ring; Nucl. Phys. **A511** (1990) 279

[41] B. Desplanques; Z. Phys. **A326** (1987) 147

[42] T. Bengtsson, I. Ragnarsson, and S. Åberg; Phys. Lett. **B208** (1988) 39

[43] I. Tanihata, H. Hamagaki, O. Hashimoto, Y. Shida, N. Yoshikawa, N. Sugimoto, T. Kobayashi, and N. Takahashi; Phys. Rev. Lett. **55** (1985) 2676

[44] G.F. Bertsch, B.A. Brown, and H. Sagawa; Phys. Rev. **C39** (1989) 1154

[45] W. Koepf, Y.K. Gambhir, P. Ring, and M.M. Sharma; Z. Phys. **A340** (1991) 119

[46] H. Sagawa; Phys. Lett. **B286** (1992) 7

[47] Z.Y. Zhu, W.Q. Shen, Y.H. Cai, and Y.G. Ma; Phys. Lett. **B328** (1994) 1

[48] G.F. Bertsch and H. Esbensen; Ann. Phys. **209** (1991) 327

[49] L.P. Gorkov; Sov. Phys. JETP **7** (1958) 505

[50] S.J. Lee, J. Fink, A.B. Balantekin, M.R. Strayer, A.S. Umar, P.G. Reinhard, J.A. Maruhn, and and W. Greiner; Phys. Rev. Lett. **57** (1986) 2916

[51] J.F. Berger, M. Girod, and D. Gogny; Nucl. Phys. **A428** (1984) 32c

Some Thoughts on the Nuclear Shell Model

Michael W. Kirson

Department of Particle Physics, Weizmann Institute of Science, 76100 Rehovot, Israel

When I agreed, a few days ago, to assist in chairing this lunchtime discussion session, it was not really clear to me what would be required of me and, as a result, I have not prepared any material for presentation here. However, while sitting and listening to the many interesting papers presented during the first day and a half of this workshop, I have had some thoughts on the subject which I believe might be worth sharing with the participants, in the hope of stimulating discussion.

It is quite astonishing how vital and active this venerable field remains. Clearly, the explosive growth of computational power has enabled us to carry out calculations of a nature and scope that would have been inconceivable only a few years ago. Effective diagonalisation of shell model hamiltonians in model spaces with dimensions in the millions is coming to be looked on almost as routine. Stochastic methods allow treatment of the full fp shell. But the results of such calculations depend not only on the reliability and feasibility of computing technology but also on the input, specifically the effective interaction matrix elements. One burning question which immediately arises is "How much do the results of large-scale shell model calculations depend on the effective interaction used?"

There is a range of possible answers to this question. At one extreme, the results could be wildly sensitive to even the smallest details of the interaction. This could make it very difficult to trust and understand the output of calculations. At the other extreme, the results could be totally insensitive to even the broadest features of the interaction. This would mean that almost any interaction would produce reasonable results provided the model space was large enough. In this case, we would learn essentially nothing from these calculations, since we would be simply using a digital computer to reproduce the solution of the Schrödinger equation already generated by an analog computer. (The late Amos de Shalit was fond of this characterisation of atomic nuclei as analog computers which solve the Schrodinger equation, with the solution being read by measuring the properties of the nucleus.) Between these extremes would lie various degrees of sensitivity of the results to the input interaction.

During the discussions in the preceding sessions, it has been claimed with some heat by Andres Zucker that large-scale shell model calculations will give essentially the same results with any input interaction having the right monopole properties. This is painfully close to the "learn nothing" extreme described above. However, Andres has clarified that he means that any realistic effective interaction with the right monopole properties will do equally well in reproducing nuclear properties via shell model diagonalisation. This clarification is very reassuring in several ways. First, it does distinguish realistic from arbitrary effective interactions, so that these very computer-intensive calculations do have some chance of revealing something about underlying nuclear forces. Second, it implies that the well-known limitations of effective interactions derived from first principles need not produce unacceptable noise in shell model results.

The longstanding and well-documented failure of existing models of nucleon-nucleon interactions to reproduce the empirical binding energy and equilibrium density of infinite nuclear matter is a serious problem to those who would compute effective interactions for shell model calculations from realistic nuclear potentials. Existing many-body results on nuclear matter are a genuine triumph of fundamental nuclear physics, but the quantitative discrepancy alluded to is of great significance for computation of finite nuclei. The size of a nucleus in its ground state is a tacit input into shell model calculations, determining as it does the scale of the single-particle wave functions. If the scale is chosen to fit the observed nuclear size, it will be incompatible with the equilibrium density associated with the underlying nucleon-nucleon interaction. This will make the calculation inherently unstable to breathing mode excitations, as Larry Zamick has been pointing out for many years. This is precisely Zucker's monopole problem - the monopole properties of the interaction have to be carefully tailored to match the size scale of the nucleus under discussion. Perhaps we have been insufficiently aware of this basic truth that has been learned from shell model calculations with realistic effective interactions.

There would seem to be a further lesson to be drawn from these thoughts. It is important to try to compute effective interactions for shell model calculations from first principles. Some sort of reasonable two-body nuclear force should serve as input, with "standard" reaction-matrices and core polarisation serving then to define the effective interaction. This interaction must then be tweaked in its monopole part to have the right equilibrium properties. It can then be used in large-scale shell model computations. Schematic interactions (pairing plus quadrupole, sum of Gaussians, Skyrme, etc., etc.) can be acceptable only to the extent that they mimic sufficiently well the properties of the realistic effective interaction. It would be most instructive to analyse the extent to which such features as the tensor-force component of the effective interaction determine the details of nuclear spectroscopy. It must also be remembered at all times that the effective interaction appropriate to a particular shell model calculation is inherently dependent on the model space used. Changing the model space without a corresponding change in the effective interaction is an inconsistent procedure which puts the results of calculation very much in question. It should never be done without a careful check on the sensitivity of the effective interaction to the change in the model space.

A further question which arises from the success of large-scale computations is "What can we really learn about the structure of the nucleus from the computed eigenvectors?" It is evidently not useful to stare at the several million expansion coefficients which define a particular eigenvector in the shell model basis used. Of course, matrix elements of physically important operators between eigenvectors convey much useful information, which can be compared with spectroscopic data. But more is required for an instructive understanding of the structure of individual states and the relationships between different states, preferably independent of the details of the shell model basis used. More thought should be given to how such insights could be obtained. Many years ago, Zucker showed that (then-large) wavefunctions of a nucleus such as ^{26}Mg looked very much like ^{24}Mg wavefunctions with two extra sd-shell neutrons. This may indicate one way of analysing large-scale eigenvectors. Better ideas can probably be found.

Most nuclei live in model spaces which cannot conceivably be handled by any rational extension of existing large-space methods. Igal Talmi's enlightening discussion of rare earth nuclei (with 2^+ spaces of dimension 10^{18} or more) shows the hopelessness of such a task. Realistic truncations of the model space must be found if anything like shell model methods is to be applied in these fascinating nuclei. Perhaps existing shell model codes can be implemented quite close to doubly-closed shells, with suitable effective interactions, and the resulting eigenfunctions studied for hints of convincing truncations. The interacting-boson model (itself a kind of bosonic shell model) has suggested a possible truncation, which is implemented in the fermion dynamic symmetry model. There should be a way to use today's impressive shell model technology to test these and other ideas and to probe the nature of the effective interactions to be used within the twice truncated space.

Quite a few years ago, Michael Fisher, when still at Cornell, commented to me that nuclear physics had not yet found the correct approach to nuclear properties. He compared the situation to the early years of studying phase transitions, when much effort was devoted to attempts to understand precise transition temperatures and specific properties of particular materials. The breakthrough came with the realisation that the proper subject of study was the critical exponents, with the concomitant development of renormalization group methods, universality classes and so on. He felt we had not yet discovered the right questions to ask. This may well still be true.

The technical triumphs represented by today's shell model codes need to be translated into physical insight. We need to give serious thought to what we can really learn from these truly impressive computational achievements.

Conventional Shell Model: Some Issues

M. Vallières[1], X. W. Pan[1], D. H. Feng[1], A. Novoselsky[2]

[1] Department of Physics, Drexel University, Philadelphia, PA 19104, USA
[2] Department of Physics, Hebrew University, Jerusalem, Israel

Abstract. We discuss some important issues in Shell-Model calculations related to the effective interactions used in different regions of the periodic table; in particular the quality of different interactions is discussed, as well as the mass dependence of the interactions. Mention will be made of the recently developed Drexel University Shell-Model (DUSM).

1 Introduction

There exists ample experimental evidence for the validity of the Shell-Model approach in describing nuclei. Fig. 1 gives an example of the experimental binding energies, the two-neutron separation energies and the first excitation energies for many nuclei. It is clear that the manifestations of shell closure is evident in the spectroscopic variables; the binding energies do not show the effect in any significant way. The bulk properties generally reflect a smooth A dependence, best handled by the configuration space Hartree-Fock approach, while the spectroscopical properties reflect the single particle motion.

Mythologically, the Shell-Model uses a very intuitive approach to study the nuclear many-body dynamics in terms of valence particles. It assumes that the nucleons belonging to a closed core do not participate in the establishment of the spectrum of a nucleus, and that the valence nucleons are not excited beyond a certain Hilbert space defined by the physics, and, unfortunately, at times by the available computer power. The many body solution of the Schrodinger equation is then expressed in terms of Slater determinants corresponding to all possible occupations of the model space by the valence nucleons. The calculation is normally done in second quantized notation, and often leads to enormous matrices to diagonalized.

In the 60's and 70's, Shell-Model calculations were performed by selecting moderate configurations of valence particles and using microscopically-oriented interactions. The results were certainly sensitive to the nuclei in question. A major concern for Shell-Model calculations has been and continues to be to physically explain the structure of light nuclei and/or nuclei near closed shell under the framework of the conventional Shell-Model (Kuo and Brown 1966, Kuo 1967, Goldstein and Talmi 1956). Nowadays, with the dramatic development of efficient algorithms to handle large Shell-Model basis, and the availability of supercomputing power, may groups tend to pursue large basis Shell-Model calculations, while employing some empirical interactions (Wildenthal et al. 1988)

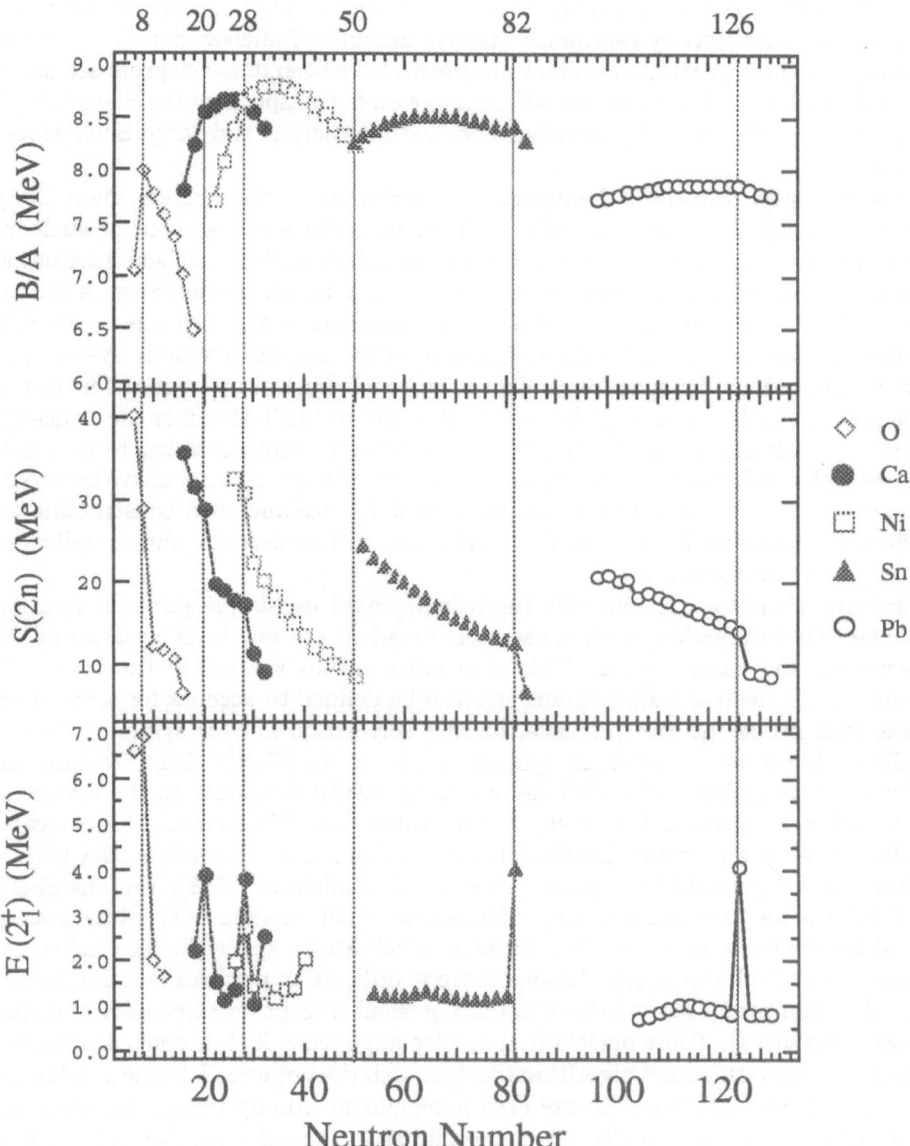

Fig. 1. Experimental data of binding energy, two-neutron separation energy and the first excitation energy. The data are from (Audi and Wapstra 1993, and NNDC).

or semi-microscopic ones (Zuker 1996). The major concern here turns out to be "How to quantitatively reproduce massive amount of nuclear data". As is evidenced by these proceedings, great progresses have been made or promised along this direction. In this paper, we will straddle on both aspects of the Shell-Model, i.e., what is the interplay between effective interaction and large Shell-Model basis.

The first step towards a fundamental description of the nuclear many-body systems using the nucleon degrees of freedom is the establishment of nucleon-nucleon effective interactions, presumably based on nucleon-nucleon interaction derived from underlying dynamics; this can take various forms. From a fundamental point of view, one uses the derived realistic NN potentials to directly solve Faddeev equations for the eigenvalues of the nuclear few-body systems, or to determine a model-space dependent microscopic nuclear interaction for the nuclear many-body systems (to be used in the nuclear Shell-Model or the Bruckner Hartree-Fock approach). For nuclear many-body systems, starting from a fundamental stand point, one can take into account the strongly repulsive bare NN interaction by first selecting an effective model space and then constructing an effective Hamiltonian (nuclear G reaction matrix) to describe the complicated many-body bound states.

Effective Hamiltonians can also be defined based on simple parameterization of the NN interaction, with parameters fitted to the few body systems or directly for the large systems. This alternative avoids the use of the bare NN potentials. Effective Hamiltonians can also be defined to account for some characteristic properties (for instance, pairing correlation and/or symmetry issues (Elliott 1958)) of a many-body system. At times, the effective Hamiltonians are the result of global fitting (Wildenthal et al. 1988); this leads to the empirical Shell-Model approach. Let us say at the outset that this approach has successfully achieved an unified description for most of s-d shell stable nuclei with a fixed mass-dependent interaction (Brown and Wildenthal 1988). Yet, the Shell-Model has yet to produce a deep microscopic understanding of this interaction. Besides, we know that the huge infeasible Shell-Model spaces in the traditional large-scale Shell-Model calculations make it difficult to perform systematic calculations for nuclei beyond those in the f-p shell (except for some heavier nuclei near close shells). These nuclei often require huge scale Shell-Model calculations involving the best recent algorithms to deal with the large model spaces, while so many two-body matrix elements (195 independent 2-body matrix elements for f-p shell) cannot be globally fitted as easily as in the s-d shell. Beyond the f-p shell, some dramatic truncation of the Shell-Model is required to even render the effective Shell-Model possible. Therefore, the microscopic Shell-Model study with NN interactions is not only tackling a fundamental nuclear many-body problem, but it may provide helpful information for the empirical approach.

In this report, we will analyze the physics implied by different NN interactions. We will carry out systematic Shell-Model calculations for modest valence-particles system, where exact calculations can eliminate some uncertainties, starting from various NN interactions (i.e, using the NN interactions as input to obtain Bruckner's G-matrix as microscopic effective two-body interactions). We argue that different NN forces can give substantial discrepancies in nuclear low-lying spectroscopy. The aim of our calculations is to provide us with some lessons of and insights about the microscopic study of the nuclear dynamics.

This report is organized as follows: In sect. 2, we will first mention the recently developed *Drexel University Shell-Model* (DUSM) algorithm and code. In sect 3, we will briefly review some issues regarding effective Shell-Model interactions and the physics they encompass. Sect.4 will focus on conventional Shell-Model calculations for s-d shell nuclei and f-p shell nuclei using various interactions. Finally, comments and perspectives will be given in sect. 5.

2 Drexel University Shell-Model Code (DUSM)

Nuclei beyond the s-d shell require new techniques to handle the very large model spaces involved. Using massive amount of CPU on the fastest computers while keeping the traditional algorithms is simply not sufficient to model large nuclei. For this reason, there is renewed interest in recent years to develop better, faster and more robust Shell-Model algorithms. One of the latest such algorithms is the Shell-Model Monte-Carlo (SMMC) approach (Koonin 1996). The Shell-Model Monte Carlo code seeks a solution of the many body problem via a Monte-Carlo variational approach. It works well in reproducing the ground state properties of nuclei (Koonin 1996) and the aggregate transition strengths to excited states at a specific temperature. Recently, this approach has been extended by the quantum Monte Carlo diagonalization method (QMCD), which can produce wave functions of both the ground state and low-lying excited states (Otsuka *et al* 1996).

The Drexel University Shell-Model (DUSM) (Valliéres and Novoselsky 1993) is another recently completed code to perform Shell-Model calculations which embodies an entirely new algorithm based on permutation group concepts. This is the code we used to perform the calculations described in this report despites their simplicity. This code promises great performance This process is the diagonalization of a given interaction under convention cases where a full exact Shell-Model solutions are sought for even-even, even-odd and odd-odd nuclei. It outperforms standard Shell-Model code Brown et al. 1988 by large (4 or 5) factors in CPU usage with much reduced disk space and I/O requirements for J-T (spin-isospin) coupled spaces.

The DUSM algorithm starts from the simple observation that Shell-Model calculations are often "multi-shell" in only a single subspace. For instance, a "$J - T$" calculation is multi-shell in the J subspace but single shell in the T subspace ($t = \frac{1}{2}$ for all nucleons). Or a Shell-Model calculation for an electronic system is multi-shell in L and single-shell in S. The wavefunctions in each subspace have arbitrary permutational symmetry; these symmetry patterns are simply restricted to be conjugate, so as to eventually couple to total antisymmetric wavefunctions. The DUSM coupling scheme is shown in the grayed area of Fig.

2. The DUSM algorithm then proceeds to calculate wavefunctions, coupling coefficients and matrix elements of elementary operators in each shell for each
subspace and then in the total space according to the following steps:

1. Compute Coefficients of Fractioal Parentage (CFP) for each shell in each
   subspace by diagonalysing the Casimir operator of $SU(2j+1)$ in product
   basis.
2. Compute the matrix elements in all single states of the elementary operators
   (all combinations of second quantized creation and annihilation operators)
3. Compute the coupling coefficients for the coupling among shells (Outer Product Isoscalar Factors - OISF)
4. Compute the coupling coefficients for the coupling between subspaces (Inner
   Product Isoscalar Factors - IISF)
5. Compute the matrix elements of the Hamiltonian via a "Sum over Path" in
   permutation diagrams concepts
6. Diagonalyze the hamiltonian matrix
7. Loop over all total quantum number (J and T for instance) to obtain the
   global solution

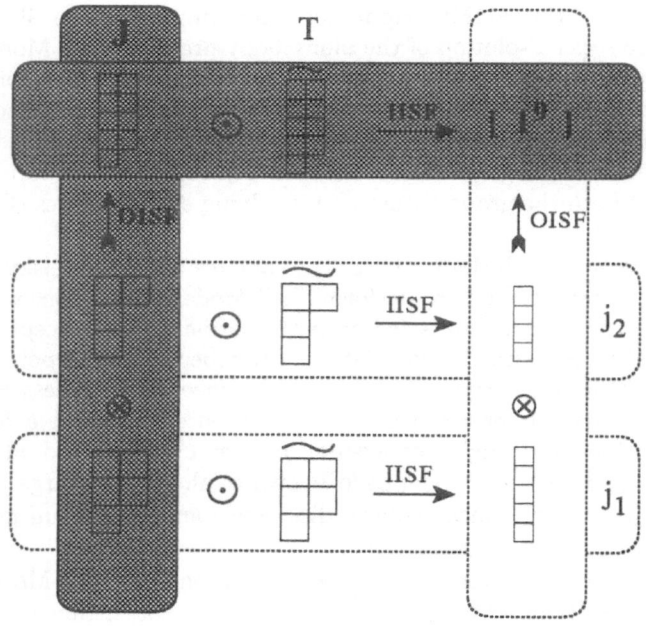

Fig. 2. Two coupling schemes for multishell calculations.

This algorithm presents many advantages over traditional schemes: in particular it avoids computing and storing total space CFPs. This implies much reduced I/O during execution and much less disk storage. Single shell calculations are known to be possible up to very large J and number of particle; DUSM inherits this advantage. The DUSM algorithm uses group theory concepts at all levels to obtain the coupling coefficients, diagonalizing the matrix representation of casimir operators of the appropriate groups at all steps, implying very stable numerical schemes. This is an advantage over using Racah formulas to derive the CFP (Ji and Vallieres 1987) in that no explicit orthogonalization is ever needed. The algorithm can furthermore be implemented (with pointers) without any search since the approach specifies fully the range over all intermediate sums. The approach is fully coupled; it provides full spectra and transitions among the states if required.

This code has recently been completed; we use it in this report to compute some simple cases appropriate to understand the physics of the NN interaction. This does not illustrate by any means the capabilities of DUSM.

3   Effective Nuclear Shell-Model Interactions

3.1   Origin of Effective Interactions

The description of the nucleon-nucleon interaction remains one of the most fundamental themes in low-energy hadron physics. The resulting force has important consequences for nuclear physics. With the advent of quantum chromodynamics (QCD), one would naturally hope to derive the NN force which is based on the quark-gluon dynamics. Typically, under the assumption of spontaneous breaking of dynamical symmetry and with 't Hooft's large N_c expansion approximation, and by formally integrating out the quark and gluon fields, one can obtain an effective chiral lagrangian of low-energy hadron dynamics. "Derivations" from the QCD lagrangian to a hadron dynamics, lagrangian, which contains QCD string structured mesons and baryons, have been tried (for instance, Karchev and Slavnov 1985), but such kind of efforts are not close to yielding practical interactions.

Due to the non-perturbative nature of low-energy QCD, the description of the NN force is carried out in practice directly via the nucleon-meson process, i.e., using the meson degree of freedom to describe the interaction between two free nucleons. Using this meson based description, the nucleon-nucleon interactions follow from three different approaches: second order perturbation theory (e.g., Yukawa potential) (Yukawa 1935), dispersion theory (e.g. Paris potential) (Vinh Mau et al. 1973), and the field-theoretical meson-exchange model for the NN interaction (e.g., Bonn potential (Machleidt et al. 1987) and Nijmegen potential (Stoks and de Swart 1993)).

There are of course phenomenological approaches to the nuclear potential as well; in particular we quote the hard-core Hamada-Johnston (HJ) potential (Hamada and Johnston 1962), the Reid soft-core (RSC) potential (Reid 1968), and most recent Argonne V_{18} Potential (Wiringa 1996). It is worth mentioning that chiral

symmetry perturbation theory has recently been used to explore the NN interactions and $3N$ interactions (Weinberg 1990, Kolck et al. 1994). A review on the
construction and comparison of various NN interactions can be found in these
proceedings (Wiringa 1996).

At the mesonic level, there are many available modern NN potentials which
are the starting points for a microscopic study of the nuclear dynamics. The
NN potentials are also the starting points for building effective nucleon-nucleon
interactions of the nuclear many-body system. However, the true playing ground
for the NN potentials is in the study of nuclear few-body system, where methods
used to obtain the nuclear few-body solutions (i.e., in configuration-space or
momentum-space) provide very consistent and nearly exact solutions for the
bound-state problem. The results of the few-body problems are more or less
free from approximations and let one concentrate on the study of the nucleon
dynamics. For instance, one common debate is on the fact that the calculated
binding energies based on a large class of NN potentials are about 1 MeV
underbound (the experimental binding energy is 8.536 MeV for ^{3}H). Whether
the additional binding comes from effective three nucleon interactions, or from
relativistic corrections, which is not included in the Schrödinger approach, is still
debatable.

The transition from a nuclear few-body system to a nuclear many-body system
requires new methodologies to obtain a solution since the traditional few-body
techniques no longer apply due to obvious computational difficulties. In particular, the nucleon-nucleon interactions are normally not used directly in the nuclear
many-body problem. Instead, effective interactions either "derived" from or motivated by the fundamental interactions are normally used in the many-body
context (Kirson 1985). Hence such interactions must necessarily be intimately
linked to the Hilbert space of the calculation. Perhaps the most important ingredient of the many-body systems is to introduce the concept of the *mean field*. It
is often sufficiently accurate to obtain a valid solution of the many-body problem. Based on the *mean field* assumption, the many body Schrödinger equation
for the nucleus can be rewritten as

$$H\psi_\lambda(1, 2, \cdots, A) = \left(\sum_{i=1}^{A} t_i + \frac{1}{2} \sum_{i \neq j=1}^{A} v_{ij} \right) \psi_\lambda(1, 2, \cdots, A) \tag{1}$$

$$= (H_0 + H')\psi_\lambda(1, 2, \cdots, A) = E_\lambda \psi_\lambda(1, 2, \cdots, A), \tag{2}$$

with

$$H_0 \phi_\mu(1, 2, \cdots, A) = \sum_{i=1}^{A} (t_i + u_i) \, \phi_\mu(1, 2, \cdots, A) = E_\mu \phi_\mu(1, 2, \cdots, A) \tag{3}$$

$$H' = \left(\frac{1}{2} \sum_{i \neq j=1}^{A} v_{ij} - \sum_{i=1}^{A} u_i \right), \tag{4}$$

Table 1. Some major Shell-Model interactions

Interaction	Label	shell	references
Kuo-Brown	KB	sd	Kuo and Brown 1966
Modified Kuo-Brown	RK	sd	Kuo 1967
folded G-matrix	SKSP4	sd	Shurpin et al. 1983
Wildenthal	W	sd	Wildenthal et al. 1988
"best-fit" interaction	SDPOTB	sd	Wildenthal et al. 1988
Modified surface-delta	MSDI	sd etc.	Arvien and Moszkowski 1966
Kuo-Brown	KBFP	fp	Kuo and Brown 1968
Richter	FPD6 & FPM13	fp	Richter et al. 1991
Modified KB	KB3	fp	Poves and Zuker 1981
cross-shell	WBT & WBP	spsdfp	Warburton and Brown 1992
	WBMB	sdfp	Warburton et al. 1990

where t_i is the kinetic energy, and u_i is the mean field that the i-th nucleon feels in the many-body system, i.e., u_i represents an average potential contributed by all nucleons in a many-nucleon system. H' is the residual interaction.

Basically, there are two ways to deal with the above many-body equations: One is to focus on determining an optimal single-particle mean field H_0, namely, the Hartree-Fock method. The other is to concentrate on deriving a realistic two-body residual interaction H' in order to reproduce the spectroscopy of the many-body system. In the later case, the technique assumes the solution to be a configuration mixture of Slater determinants. The solution divides neatly into two parts: the calculation of few body terms (one- and two- for most interactions) in a harmonic oscillator basis, followed by the many-body solution in second quantized form of eq.(3). This is the conventional Shell-Model approach. Prior to performing the Shell-Model calculation, one needs to derive the microscopic two-body interactions for a given model space.

Both the Hartree-Fock and the Shell-Model approaches can also be used in conjunction with phenomenological interactions. In the first approach, the use of the Skyrme interaction leads to manageable calculations. It was shown that this interaction is related to the fundamental interaction (Negele 1970). On the other hand, for the Shell-Model, an approach based on a fit of the matrix elements in second quantized form (Wildenthal et al. 1988) attempts to differentiate the quality (or lack of) the fit to experimental data as coming from either the many body aspects of the calculations or from the quality of the two body interactions. These approaches are fundamental in many aspects.

A list of some widely-used shell model interactions is given in Table 1. In the 60's, Kuo and Brown presented a classic example to pertubatively derive the microscopic effective interaction (i.e., Shell-Model reaction matrix elements) (Kuo and Brown 1966). Later, a more exact and systematic way to derive the effective two-body interaction for a given model space was developed, namely the Folded-diagram method. This method is reviewed in (Kuo 1996). For instance, using

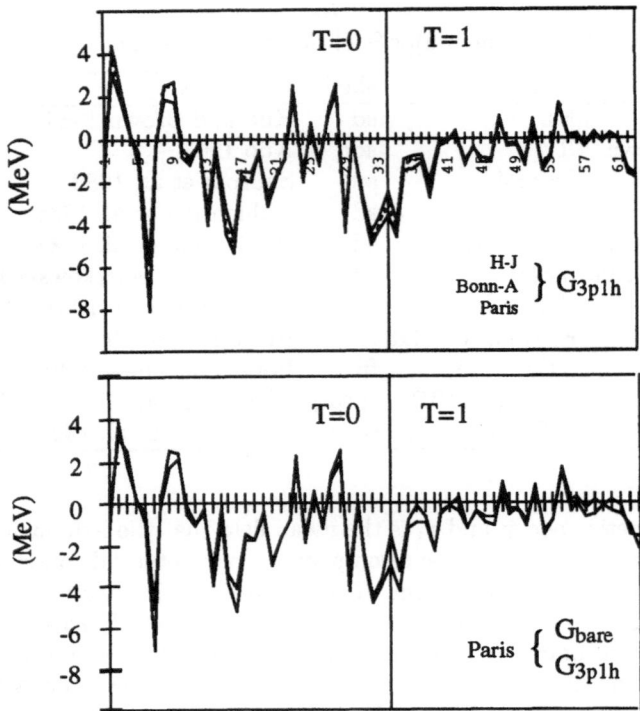

Fig. 3. Comparison of 63 two-body matrix elements in the *s-d* shell.

the folded-diagram method, one can determine the 63 independent two-body matrix elements in the *s-d* shell, which together with the three single-particle energies are the essential input in the *s-d* shell calculations. In Fig. 3, the upper panel gives a comparison of those 63 G-matrix elements, which include bare-G and particle-hole polarization (i.e., G_{3p1h}) and derived from Hamada-Johnston, Paris and Bonn-A potentials. It shows quite vividly that the different NN potentials provide very consistent 2-body matrix elements. Only for some matrix elements, especially in the $T = 0$ case, are there some discrepancies. However, these small discrepancies can produce significant differences in the binding energy calculations as we will see in the next section. The lower panel in Fig. 3 shows the discrepancies of the G-matrix with the G_{3p1h} and without it (i.e., G_{bare}). It has been confirmed that the G-matrix with the G_{3p1h} can significantly improve the calculations both in binding energy and excitation energy (Kuo and Brown 1966). By comparing these two panels one can draw the conclusion that the overall differences are about same magnitude. That is to say, the differences due to different NN potentials may not be negligible in application to Shell-Model calculations.

On the other hand, semi-empirical "best-fit" interactions for s-d shell are constructed as the result of the linear combination fitting to 447 s-d shell binding-energy data (Wildenthal et al. 1988). The Wildenthal interaction is very successful in reproducing stable nuclei with a fixed mass dependence. Fig. 4 presents a comparison of some two-body matrix elements from G-matrix, Wildenthal interaction and modified surface delta interaction. One can see a clear consistency between the microscopic interaction and empirical ones in diagonal parts of two-body matrix elements, while some substantial differences for off-diagonal matrix elements are visible.

3.2   Comparisons to Experimental Data

We now describe calculations done in various model spaces via DUSM using different effective interactions. The calculations for ^{18}O and ^{18}F using bare G-matrix, second-order pertubative G-matrix and folded G-matrix, which all start with Bonn-A, are shown in Fig. 5. The calculated Shell-Model binding energy (B_{SM}) are also given at the bottom of the figure. By Shell-Model binding energy we mean the energy eigenvalues after the diagonalization under conventional Shell-Model basis. Physically, it is the binding energy related to the shell model core. i.e., $B_{SM} = B_{total} - B_{core} - E_C$, where B_{total} is the total binding energy and E_C is the Coulomb energy. Therefore, an important factor in the comparison of the experimental binding energies with results from large scale Shell-Model calculations lies in the evaluation of the Coulomb energy. In (Pan et al. 1996), we use the finite-range droplet model (FRDM) to estimate the Coulomb energies for ^{18}O and ^{18}F, and provide the estimated experimental Shell-Model binding energies (-12.0 MeV for ^{18}O and -12.8 MeV for ^{18}F). From Fig. 5, one can see that the calculation with bare G-matrix is not in agreement with experimental spectra, and that the level scheme is significantly improved with the inclusion of G_{3p1h}, namely, the particle-hole polarization. The results (both the level schemes and binding energies) are further improved by folded G-matrix.

In Fig. 6, we perform Shell-Model calculations for the same nuclei but using different potentials. The results from the Wildenthal interaction and the Kuo-Brown interaction are compared with the ones from the folded G-matrix of the Paris, Bonn-A, Reid and V_{18} NN potentials.

The above Shell-Model results for the binding energies are at best confusing: when one starts from a NN soft-core potential, based on effective meson theory with coupling constants being determined by analyzing a few thousands NN scattering data values and the deuteron properties, the binding energy resulting from the microscopic Shell-Model calculations are overbound. On the other hand, starting from a NN hard-core potential, which is given perturbatively (i.e., the early KB effective interactions), the binding energy is reproduced very well. Not only are the binding energies somewhat better, but the resulting spectra also demonstrate that KB's presents the best agreement compared with results from other modern potentials. We illustrate this point in Fig.4 and Fig.5 where

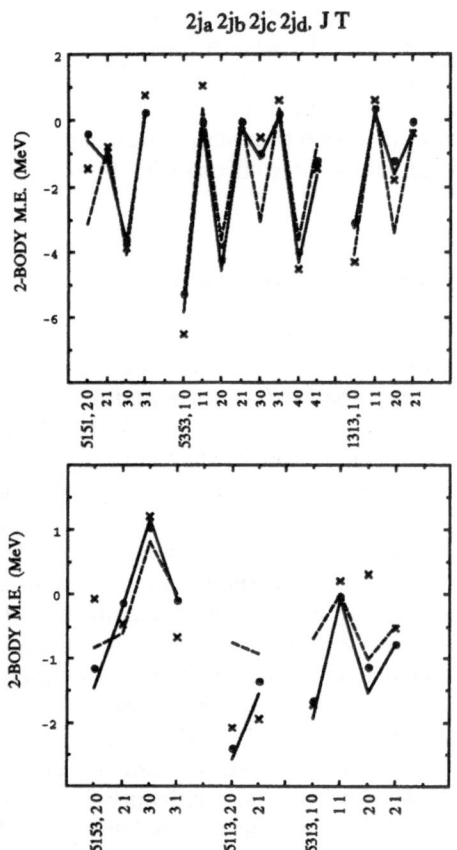

Fig. 4. Some Two-body matrix elements in the *s-d* shell. × represent the Wildenthal interaction (Wildenthal et al. 1988), • the Kuo-Brown interaction (Kuo and Brown 1966), the solid lines show the modified Kuo-Brown interaction (Kuo 1967), and the dash lines show the modified surface-delta interaction listed in (Halbert *et al.* 1971).

we show the theoretical and experimental spectra of ^{18}O and ^{18}F. It is worth mentioning that KB not only provides the best microscopic shell model results for nuclei in *s-d* shell; recently, systematic Shell-Model calculations for *f-p* nuclei have been carried out by using the Kuo-Brown interaction for the *f-p* shell (Nakada et al. 1994) showing a similar trend.

We may point out that a main difference between the early KB effective interactions and the more recent ones (for example, Kuo 1994) is about the treatment

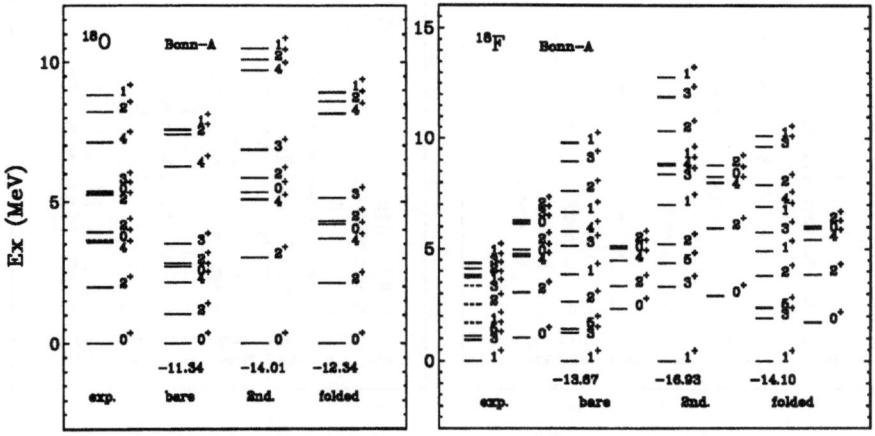

Fig. 5. Spectra of ^{18}O and ^{18}F.

of the folded diagrams. For the former the folded diagrams were ignored, with
the effective interaction given merely by the bare-G and the second-order core
polarization diagrams. These diagrams are usually referred to as G and G_{3p1h}
in the literature (Shurpin et al. 1983). For the later, certain types of folded diagrams are included to all orders using a $\hat{Q}$-box formulation (Shurpin et al. 1983).
What we have found in this work may indicate the need of a further investigation
of the folded diagrams.

3.3 Mass Dependence

Various mass dependence of the Shell-Model effective interactions have been
proposed: first, one may neglect any dependence, i.e., one may use G-matrix
elements with fixed $\hbar\omega$ value for a given major shell. The same G-matrix is
then used for nuclei with different masses in a given region, for instance, the
f-p shell (Nakada et al. 1994). Shell-Model calculations in the s-d shell on the
other hand show that such a simple G-matrix is not applicable in this region. A
second possibility is to use a G-matrix with adjusted $\hbar\omega$ based on the nuclear
mass. Another possibility is to use an effective interaction determined from a
G-matrix approach or an empirical fit with no mass dependence and a monopole
term to reflect mass dependence. This is the approach used by Zucker (Poves
and Zuker 1981) for the f-p shell; it is discussed elsewhere in these proceedings.
The empirical Shell-Model uses yet another mass dependence which amounts
to a global scaling of all matrix elements with a simple mass dependence. For
instance, the Wildenthal interaction uses the Two Body Matrix Elements for
A=18 nuclei, and extends it to different mass nuclei in the s-d shell by using the
simple mass-dependence $TBME(A) = TBME(18)(A/18)^{-0.3}$.

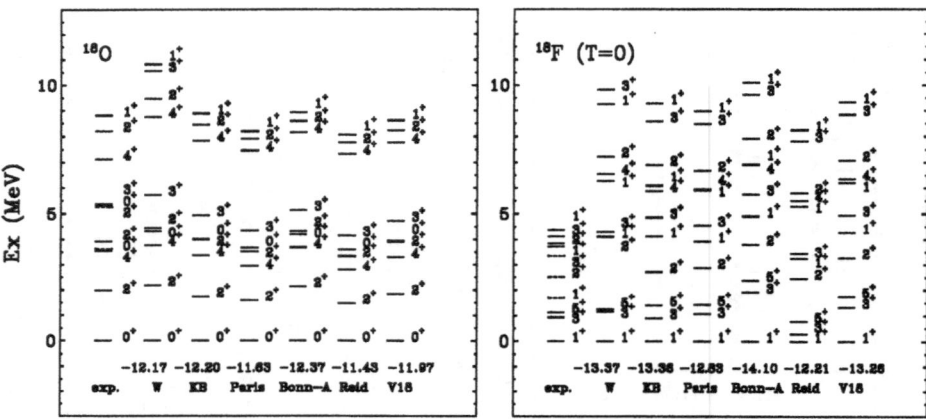

Fig. 6. Spectra of ^{18}O and ^{18}F. The theoretical levels are calculated with different interactions

Fig.7 compares shell Model calculations using mass-independent Two Body Matrix Element (TBME) (i.e., Wildenthal interaction at mass 18) and mass-dependent TBME (i.e., $TBME(A) = TBME(18)(A/18)^{-0.3}$) with the experimental spectra. The mass 18 interaction fits the spectra of mass 18 nuclei relatively well (see Fig. 6), yet the discrepancies become large with increasing mass. The simple mass dependence certainly improves the agreement with experiment as seen in Fig. 7. A good global fit to the s-d shell nuclei only became possible when this mass dependence was introduced in the matrix elements (Wildenthal). Fig.8 shows the spectra for ^{28}Si using G-matrix with $\hbar\omega$ value $G(18)(\hbar\omega=14)$, adjusted for mass 18, and G-matrix with $\hbar\omega$ $G(28)(\hbar\omega=12)$, adjusted for mass 28, as compared to the experimental spectrum. It is evident that the results using the interactions with either $\hbar\omega$ values are in poor agreement with experimental results. The changes in the spectrum of ^{28}Si induced by the the rescaling of the $\hbar\omega$ due to the mass effect does not significantly improve the quality of the spectrum. This mass dependence is the most logical one, since it scales the physical size of the basis according to mass. Yet, it is not producing the desired effect on the spectra.

A mass dependence of the type used by Wildenthal, and applied to these G-matrix elements, will improve the agreement with the data much better than what seen in Fig. 8. The explanation for this is far from being clear. In fact the $\hbar\omega$ mass scaling does not produce a uniform scaling of the matrix elements as seen in Fig. 9. The Wildenthal interaction, with its simple mass dependence, is not explained in terms of a G-matrix approach, and the present folded-diagram approach has difficulties in producing the right mass dependence for Shell-Model calculations.

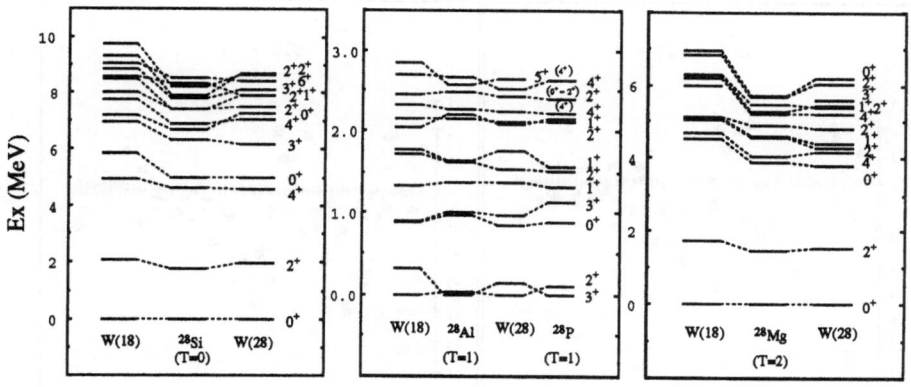

Fig. 7. Spectra of $A = 28$ nuclei $(T = 0 - 2)$. Both mass dependent W(A) and independent interactions W(18) are presented for comparison.

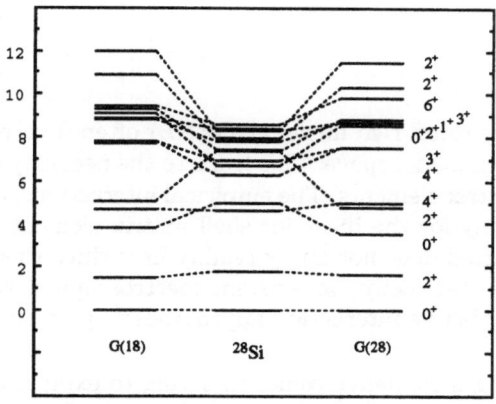

Fig. 8. Spectra of ^{28}Si. Both G-matrix G(18) and G(28) are presented for comparison.

4   Beyond the s-d Shell

4.1   Nuclei Far from Stability

The spectroscopic description of nuclei far from stability presents new challenges for the Shell-Model. First, the not-so-large proton rich or neutron rich nuclei soon

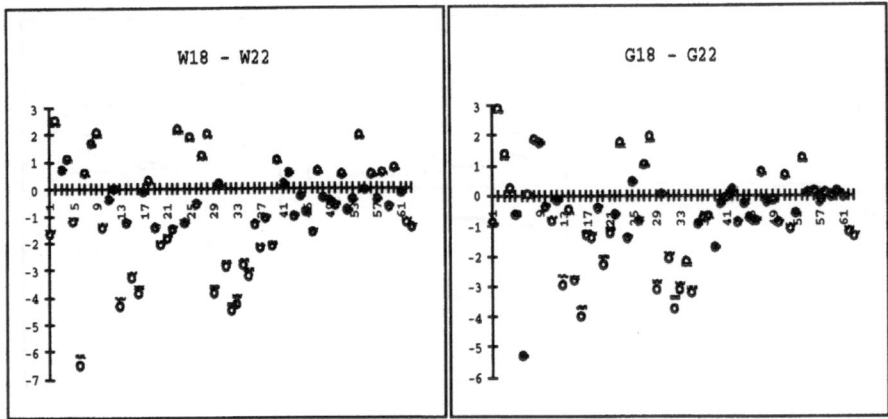

Fig. 9. comparison between the TBMEs of Wildenthal interaction and G-matrix for *s-d*-shell.

to be measured (the proton rich nuclei at ORNL) often involve a mix of p and s-d shell single nucleon model space. This leads to the necessity of having in- and cross-major shells matrix elements. The empirical interactions, of the Wildenthal type, provide help only for the in-major shell matrix elements. Using the MEs as quantities to be fitted does not apply readily here since there are many such matrix elements, and practically non existent spectroscopic data yet. This leaves a large task for the effective interaction approaches to perform.

Even a more daunting a challenge comes in trying to explain the spectroscopic properties of nuclei very far the region of stability in that the single nucleon orbitals become barely bounded. This raises fundamental issues in effective interaction derivation and even in the applicability of the Shell-Model approach itself. In deriving the effective interaction from fundamental NN interaction, one often assumes a harmonic oscillator basis for simplicity. This approximation definitely breaks down for the barely bound orbitals. Kuo (Kuo 1996) in these proceedings, introduces a two frequency (two sets of harmonic oscillator basis states with different frequencies) to estimate this effect. The proximity of the continuum, or the inclusion of the continuum in the q-space, remains a problem.

Note that the effect just mentioned is also plaguing other approaches. In this workshop, Dobaczewski addressed the issue in the context of the Hartree-Fock approach. In this case it can be argued that a self-consistent quasi-particle treatment of the pairing interaction helps solve the problem.

4.2   f-p Shell Nuclei

The application of the Shell-Model techniques to the f-p shell nuclei is one of the most active field of research in connection to modern Shell-Model activities. There are a few issues we would like to touch upon in this context. First, there is the fact that these calculations are very demanding. Codes have been recently developed to deal with this problem. We talked above about the DUSM approach which works in a coupled scheme. Andréas P. Zuker (Zuker 1996) presented a summary of his work in this area which was accomplish with the help of the code ANTOINE, and m-scheme code capable of dealing with matrices of dimensions 10^6 to 10^7. Koonin et al. (Koonin 1996) have developed a Monte-Carlo Approach which allows to study ground state properties of large nuclei, as well as transitions to states located at some excited energies (at some temperature). We have also heard about the Quantum Monte-Carlo Diagonalization Method that Otsuka et Al. (Otsuka *et al* 1996) have recently developed which allows to compute low lying excited states as well as the ground state of nuclei. The significant point about this list of algorithms and codes is the vitality of the field they exemplify; there is a renewed interest in recent years to developed new approaches to tackle this very important problem. Finished are the days of having to choose between one or two of the pioneering codes developed many years ago. A second point to note is the continued nagging problem of the interaction to choose. Of course, the issue of the mass dependence of the interaction we raised previously is less important in the f-p shell since the number of nucleon is larger to start with. A global fit of the interaction, as Wildenthal (Wildenthal et al. 1988) did in the s-d shell is on the other hand out of question due to the very large number of matrix elements, and the difficulty of the Shell-Model calculations. Miraculously, the effective interaction seems to come to the rescue, with no adjusting as was explained by Otsuka (Otsuka *et al* 1996), or with a "monopole adjustment" as explained by Zuker (Zuker 1996). The jury is out as to what ought to be done in this case. Note also that the effective interaction as used by Otsuka in the f-p shell was not as good when applied to the s-d shell. A significant aspect of the f-p shell nuclei calculations via Shell-Model is the appearance of the onset of collectivity in nuclei large enough to exhibit collective behavior. In particular, Zuker (Zuker 1996) presented Shell-Model descriptions of rotational bands in ^{47}V and ^{49}Cr and back bending effects in ^{50}Cr that rival those calculated via Hartree-Fock-Bogoliubov. These calculations point to the fact that Shell-Model is finally reaching the point of computing collective aspects of nuclear structure.

5   Comments

The fundamental question in Shell-Model studies remains that of the interaction to use. Effective interactions derived from the NN interactions have had a long history in nuclear physics, and seem to have success in describing nuclear

308   M. Vallières, X. W. Pan, D.H. Feng, A. Novoselsky

spectroscopy, specially in larger nuclei. Yet there remain numerous unanswered questions.

In the area of Shell-Model techniques, there is renewed activity in improving algorithms to go further in nuclear size. Much success has been met in this direction has is evidenced by many of the talks in this workshop.

Many challenges remain, specially in the regions of neutron- and proton-rich nuclei, and in the f-p shell, remain many.

References

Ajzenberg-Selove, F. (1987): Nucl. Phys. **A475**, 1

Arvien R., Moszkowski S.A. (1966): Phys. Rev. **145**, 830

Audi, G. and Wapstra, A.H. (1993): Nucl. Phys. **A565**, 1, The data on excitation energy are taken from National Nuclear Data Center, Brookhaven National Laboratory

Brown,B.A. ,Richter,W.A., Julies, R.E. and Wildenthal, B.H. (1988): Ann. Phys. **182**, 191

Brown B.A., Wildenthal B.H. (1988): Ann. Rev. Nucl. Part. Sci. **38**, 29

Brown B.A., Etchegoyen A., Rae W.D.M., Ormand W.E., Winfield J.S., Zhao L.(1988): *OXBASH code*, MSUNSCL Report **524**

Cottingham, W.N., Lacombe, M., Loiseau, B., Richard, J.M. and Vinh Mau, R. (1973): Phys. Rev. **D8**, 8

Elliott J.P. (1958): Proc. Roy. Soc. **A245**, 128, 562

Goldstein and Talmi (1956): Phys. Rew. **102**, 589

Halbert, E.C., McGrory, J.B., Wildenthal, B.H., Pandya, S.P. (1971): Adv. Nucl. Phys. **4**, 315

Hamada, T. and Johnston, I.D. (1962): Nucl. Phys. 34, 382

Ji X., and Vallières M. (1987): Phys. Rev. **C35**, 1583.

Kahana S., Lee H.C., Scott C.K. (1968): Phys. Rev. **180** 956

Karchev, N.I. and Slavnov, A.A. (1985): Theor. Math. Phys. **65**, 192

Kirson M.K. (1985): For example, in *Nuclear Shell-Models*, ed. by M. Vallières and B.H. Wildenthal, (World Scientific 1985).

Koonin, S.E. (1996): these proceedings and the references therein

Kuo, T.T.S. and Brown,G.E. (1966): Nucl. Phys. **85**, 40

Kuo, T.T.S. (1967): Nucl. Phys. **A103**, 71

Kuo, T.T.S. and Brown,G.E. (1968): Nucl. Phys. **A114**, 241

Kuo T.T.S. (1994): Nucl. Phys. **A570**, 173c

Kuo, T.S.S., (1996): these proceedings and the references therein

Machleidt, R., Holinde, K. and Elster, Ch. (1987): Phys. Rep. 149 ,1

Nakada H., Otsuka T., Sebe T. (1994): Nucl. Phys. **A571**, 467

Negele J.W. (1970): Phys. Rev. **C1**, 1260.

Ordòñez, C., Ray, L., Klock, U. van (1994): Phys. Rev. Lett.**72**,1982

Otsuka, T., Honma, M and Mizusaki, T. (1996): these proceedings

Pan X.W., Kuo T.S.S., Vallières M., Feng D.H., Novoselsky A. (1996): Phys. Rep. **264** , 311

Poves A. Zuker A. (1981): Phys. Rep. **70**, 235

Reid, R.V. Jr. (1968): Ann. Phys. (N.Y.) 50, 411

Richter W.A., Van der Merwe M.G., Julies R.E. and Brown B.A. (1991): Nucl. Phys. **A523**, 325

Shurpin J., Strottman D., Kuo T.T.S. (1983): Nucl. Phys. **A408**, 310

Stoks, V. and de Swart, J.J. (1993): Phys. Rev. C47, 761.

Valliéres M., Novoselsky A. (1993): Nucl. Phys.**A570**, 345c

Weinberg, S. (1990): Phys. Lett.**B251**,288

Warburton E.K., Becker J.A., Brown B.A. (1990): Phys. Rev. **C41**, 1147

Warburton E.K., Brown B.A. (1992): Phys. Rev. **C46**, 923

Wildenthal, B. H.: private communication.

Wiringa, R.B. (1996): These proceedings

Yukawa H. (1935): Proc. Math. Soc. Japan **17**, 48.

Zuker, A.P. (1996): these proceedings and the references therein

Lecture Notes in Physics

For information about Vols. 1–449
please contact your bookseller or Springer-Verlag

Vol. 450: M. F. Shlesinger, G. M. Zaslavsky, U. Frisch (Eds.), Lévy Flights and Related Topics in Physics. Proceedigs, 1994. XIV, 347 pages. 1995.

Vol. 451: P. Krée, W. Wedig (Eds.), Probabilistic Methods in Applied Physics. IX, 393 pages. 1995.

Vol. 452: A. M. Bernstein, B. R. Holstein (Eds.), Chiral Dynamics: Theory and Experiment. Proceedings, 1994. VIII, 351 pages. 1995.

Vol. 453: S. M. Deshpande, S. S. Desai, R. Narasimha (Eds.), Fourteenth International Conference on Numerical Methods in Fluid Dynamics. Proceedings, 1994. XIII, 589 pages. 1995.

Vol. 454: J. Greiner, H. W. Duerbeck, R. E. Gershberg (Eds.), Flares and Flashes, Germany 1994. XXII, 477 pages. 1995.

Vol. 455: F. Occhionero (Ed.), Birth of the Universe and Fundamental Physics. Proceedings, 1994. XV, 387 pages. 1995.

Vol. 456: H. B. Geyer (Ed.), Field Theory, Topology and Condensed Matter Physics. Proceedings, 1994. XII, 206 pages. 1995.

Vol. 457: P. Garbaczewski, M. Wolf, A. Weron (Eds.), Chaos – The Interplay Between Stochastic and Deterministic Behaviour. Proceedings, 1995. XII, 573 pages. 1995.

Vol. 458: I. W. Roxburgh, J.-L. Masnou (Eds.), Physical Processes in Astrophysics. Proceedings, 1993. XII, 249 pages. 1995.

Vol. 459: G. Winnewisser, G. C. Pelz (Eds.), The Physics and Chemistry of Interstellar Molecular Clouds. Proceedings, 1993. XV, 393 pages. 1995.

Vol. 460: S. Cotsakis, G. W. Gibbons (Eds.), Global Structure and Evolution in General Relativity. Proceedings, 1994. IX, 173 pages. 1996.

Vol. 461: R. López-Pen˜a, R. Capovilla, R. Garci´a-Pelayo, H. Waelbroeck, F. Zertuche (Eds.), Complex Systems and Binary Networks. Lectures, México 1995. X, 223 pages. 1995.

Vol. 462: M. Meneguzzi, A. Pouquet, P.-L. Sulem (Eds.), Small-Scale Structures in Three-Dimensional Hydrodynamic and Magnetohydrodynamic Turbulence. Proceedings, 1995. IX, 421 pages. 1995.

Vol. 463: H. Hippelein, K. Meisenheimer, H.-J. Röser (Eds.), Galaxies in the Young Universe. Proceedings, 1994. XV, 314 pages. 1995.

Vol. 464: L. Ratke, H. U. Walter, B. Feuerbach (Eds.), Materials and Fluids Under Low Gravity. Proceedings, 1994. XVIII, 424 pages, 1996.

Vol. 465: S. Beckwith, J. Staude, A. Quetz, A. Natta (Eds.), Disks and Outflows Around Young Stars. Proceedings, 1994. XII, 361 pages, 1996.

Vol. 466: H. Ebert, G. Schütz (Eds.), Spin – Orbit-Influenced Spectroscopies of Magnetic Solids. Proceedings, 1995. VII, 287 pages, 1996.

Vol. 467: A. Steinchen (Ed.), Dynamics of Multiphase Flows Across Interfaces. 1994/1995. XII, 267 pages. 1996.

Vol. 468: C. Chiuderi, G. Einaudi (Eds.), Plasma Astrophysics. 1994. VII, 326 pages. 1996.

Vol. 469: H. Grosse, L. Pittner (Eds.), Low-Dimensional Models in Statistical Physics and Quantum Field Theory. Proceedings, 1995. XVII, 339 pages. 1996.

Vol. 470: E. Martı´nez-González, J. L. Sanz (Eds.), The Universe at High-z, Large-Scale Structure and the Cosmic Microwave Background. Proceedings, 1995. VIII, 254 pages. 1996.

Vol. 471: W. Kundt (Ed.), Jets from Stars and Galactic Nuclei. Proceedings, 1995. X, 290 pages. 1996.

Vol. 472: J. Greiner (Ed.), Supersoft X-Ray Sources. Proceedings, 1996. XIII, 350 pages. 1996.

Vol. 473: P. Weingartner, G. Schurz (Eds.), Law and Prediction in the Light of Chaos Research. X, 291 pages. 1996.

Vol. 474: Aa. Sandqvist, P. O. Lindblad (Eds.), Barred Galaxies and Circumnuclear Activity. Proceedings of the Nobel Symposium 98, 1995. XI, 306 pages. 1996.

Vol. 475: J. Klamut, B. W. Veal, B. M. Dabrowski, P. W. Klamut, M. Kazimierski (Eds.), Recent Developments in High Temperature Superconductivity. Proceedings, 1995. XIII, 362 pages. 1996.

Vol. 476: J. Parisi, S. C. Müller, W. Zimmermann (Eds.), Nonlinear Physics of Complex Systems. Current Status and Future Trends. XIII, 388 pages. 1996.

Vol. 477: Z. Petru, J. Przystawa, K. Rapcewicz (Eds.), From Quantum Mechanics to Technology. Proceedings, 1996. IX, 379 pages. 1996.

Vol. 479: H. Latal, W. Schweiger (Eds.), Perturbative and Nonperturbative Aspects of Quantum Field Theory. Proceedings, 1996. X, 430 pages. 1997.

Vol. 480: H. Flyvbjerg, J. Hertz, M. H. Jensen, O. G. Mouritsen, K. Sneppen (Eds.), Physics of Biological Systems. From Molecules to Species. X, 364 pages. 1997.

Vol. 482: X.-W. Pan, D. H. Feng, M. Vallières (Eds.) Contemporary Nuclear Shell Models. Proceedings, 1996. XII, 309 pages. 1997.

New Series m: Monographs

Vol. m 1: H. Hora, Plasmas at High Temperature and Density. VIII, 442 pages. 1991.

Vol. m 2: P. Busch, P. J. Lahti, P. Mittelstaedt, The Quantum Theory of Measurement. XIII, 165 pages. 1991. Second Revised Edition: XIII, 181 pages. 1996.

Vol. m 3: A. Heck, J. M. Perdang (Eds.), Applying Fractals in Astronomy. IX, 210 pages. 1991.

Vol. m 4: R. K. Zeytounian, Mécanique des fluides fondamentale. XV, 615 pages, 1991.

Vol. m 5: R. K. Zeytounian, Meteorological Fluid Dynamics. XI, 346 pages. 1991.

Vol. m 6: N. M. J. Woodhouse, Special Relativity. VIII, 86 pages. 1992.

Vol. m 7: G. Morandi, The Role of Topology in Classical and Quantum Physics. XIII, 239 pages. 1992.

Vol. m 8: D. Funaro, Polynomial Approximation of Differential Equations. X, 305 pages. 1992.

Vol. m 9: M. Namiki, Stochastic Quantization. X, 217 pages. 1992.

Vol. m 10: J. Hoppe, Lectures on Integrable Systems. VII, 111 pages. 1992.

Vol. m 11: A. D. Yaghjian, Relativistic Dynamics of a Charged Sphere. XII, 115 pages. 1992.

Vol. m 12: G. Esposito, Quantum Gravity, Quantum Cosmology and Lorentzian Geometries. Second Corrected and Enlarged Edition. XVIII, 349 pages. 1994.

Vol. m 13: M. Klein, A. Knauf, Classical Planar Scattering by Coulombic Potentials. V, 142 pages. 1992.

Vol. m 14: A. Lerda, Anyons. XI, 138 pages. 1992.

Vol. m 15: N. Peters, B. Rogg (Eds.), Reduced Kinetic Mechanisms for Applications in Combustion Systems. X, 360 pages. 1993.

Vol. m 16: P. Christe, M. Henkel, Introduction to Conformal Invariance and Its Applications to Critical Phenomena. XV, 260 pages. 1993.

Vol. m 17: M. Schoen, Computer Simulation of Condensed Phases in Complex Geometries. X, 136 pages. 1993.

Vol. m 18: H. Carmichael, An Open Systems Approach to Quantum Optics. X, 179 pages. 1993.

Vol. m 19: S. D. Bogan, M. K. Hinders, Interface Effects in Elastic Wave Scattering. XII, 182 pages. 1994.

Vol. m 20: E. Abdalla, M. C. B. Abdalla, D. Dalmazi, A. Zadra, 2D-Gravity in Non-Critical Strings. IX, 319 pages. 1994.

Vol. m 21: G. P. Berman, E. N. Bulgakov, D. D. Holm, Crossover-Time in Quantum Boson and Spin Systems. XI, 268 pages. 1994.

Vol. m 22: M.-O. Hongler, Chaotic and Stochastic Behaviour in Automatic Production Lines. V, 85 pages. 1994.

Vol. m 23: V. S. Viswanath, G. Müller, The Recursion Method. X, 259 pages. 1994.

Vol. m 24: A. Ern, V. Giovangigli, Multicomponent Transport Algorithms. XIV, 427 pages. 1994.

Vol. m 25: A. V. Bogdanov, G. V. Dubrovskiy, M. P. Krutikov, D. V. Kulginov, V. M. Strelchenya, Interaction of Gases with Surfaces. XIV, 132 pages. 1995.

Vol. m 26: M. Dineykhan, G. V. Efimov, G. Ganbold, S. N. Nedelko, Oscillator Representation in Quantum Physics. IX, 279 pages. 1995.

Vol. m 27: J. T. Ottesen, Infinite Dimensional Groups and Algebras in Quantum Physics. IX, 218 pages. 1995.

Vol. m 28: O. Piguet, S. P. Sorella, Algebraic Renormalization. IX, 134 pages. 1995.

Vol. m 29: C. Bendjaballah, Introduction to Photon Communication. VII, 193 pages. 1995.

Vol. m 30: A. J. Greer, W. J. Kossler, Low Magnetic Fields in Anisotropic Superconductors. VII, 161 pages. 1995.

Vol. m 31: P. Busch, M. Grabowski, P. J. Lahti, Operational Quantum Physics. XI, 230 pages. 1995.

Vol. m 32: L. de Broglie, Diverses questions de mécanique et de thermodynamique classiques et relativistes. XII, 198 pages. 1995.

Vol. m 33: R. Alkofer, H. Reinhardt, Chiral Quark Dynamics. VIII, 115 pages. 1995.

Vol. m 34: R. Jost, Das Märchen vom Elfenbeinernen Turm. VIII, 286 pages. 1995.

Vol. m 35: E. Elizalde, Ten Physical Applications of Spectral Zeta Functions. XIV, 228 pages. 1995.

Vol. m 36: G. Dunne, Self-Dual Chern-Simons Theories. X, 217 pages. 1995.

Vol. m 37: S. Childress, A.D. Gilbert, Stretch, Twist, Fold: The Fast Dynamo. XI, 410 pages. 1995.

Vol. m 38: J. González, M. A. Martín-Delgado, G. Sierra, A. H. Vozmediano, Quantum Electron Liquids and High-T_c Superconductivity. X, 299 pages. 1995.

Vol. m 39: L. Pittner, Algebraic Foundations of Non-Commutative Differential Geometry and Quantum Groups. XII, 469 pages. 1996.

Vol. m 40: H.-J. Borchers, Translation Group and Particle Representations in Quantum Field Theory. VII, 131 pages. 1996.

Vol. m 41: B. K. Chakrabarti, A. Dutta, P. Sen, Quantum Ising Phases and Transitions in Transverse Ising Models. X, 204 pages. 1996.

Vol. m 42: P. Bouwknegt, J. McCarthy, K. Pilch, The W_3 Algebra. Modules, Semi-infinite Cohomology and BV Algebras. XI, 204 pages. 1996.

Vol. m 43:

Vol. m 44: A. Bach, Indistinguishable Classical Particles. VIII, 157 pages. 1997.

Vol. m 45: M. Ferrari, V. T. Granik, A. Imam, J. C. Nadeau (Eds.), Advances in Doublet Mechanics. XVI, 214 pages. 1997.